Sustainability of Fossil Fuels

Sustainability of Fossil Fuels

Special Issue Editor

Pavel A. Strizhak

MDPI • Basel • Beijing • Wuhan • Barcelona • Belgrade

Special Issue Editor
Pavel A. Strizhak
National Research Tomsk
Polytechnic University
Russia

Editorial Office
MDPI
St. Alban-Anlage 66
4052 Basel, Switzerland

This is a reprint of articles from the Special Issue published online in the open access journal *Energies* (ISSN 1996-1073) from 2018 to 2019 (available at: https://www.mdpi.com/journal/energies/special_issues/Fossil_Fuels)

For citation purposes, cite each article independently as indicated on the article page online and as indicated below:

LastName, A.A.; LastName, B.B.; LastName, C.C. Article Title. *Journal Name* **Year**, *Article Number*, *Page Range*.

ISBN 978-3-03921-219-4 (Pbk)
ISBN 978-3-03921-220-0 (PDF)

Contents

About the Special Issue Editor

Pavel A. Strizhak is Doctor of Science, Full Professor. Scientific activities: Investigation of complexes of interrelated physical and chemical processes during liquid fuel ignition by local energy sources; Investigation of heat and mass transfer with physical and chemical transformations under the influence of the dispersed combustion deterrents on high-temperature flames; Automatization of heat and power and production processes.

Preface to "Sustainability of Fossil Fuels"

In this Special Issue, we will try to provide readers with the results of fundamental and applied research and reviews in the field of energy production from the combustion of fossil fuels (coal, peat, oil), waste-derived fuels, and biomass.

Pavel A. Strizhak
Special Issue Editor

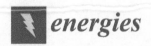

Article

Impact of Forest Fuels on Gas Emissions in Coal Slurry Fuel Combustion

Galina Nyashina and Pavel Strizhak *

Department of Power Engineering, National Research Tomsk Polytechnic University,
Tomsk 634050, Russia; gsn1@tpu.ru
* Correspondence: pavelspa@tpu.ru; Tel.: +7-3822-701-777 (ext. 1910)

Received: 15 August 2018; Accepted: 17 September 2018; Published: 19 September 2018

Abstract: Anthropogenic emissions from coal combustion pose a serious threat to human wellbeing. One prospective way to solve this problem is by using slurry fuels instead of coal. The problem is especially pressing in China and Russia, so these countries need reliable experimental data on the SO_x and NO_x emissions reduction range more than others do. The experiments in this research are based on the components that are typical of Russia. Experimental research was conducted on the way typical forest fuels (ground pine needles, leaves and their mixtures, bark, sawdust, and charcoal) affect the gas emissions from the combustion of slurry fuels based on the wastes. It was established that using forest fuels as additives to coal-water slurries reduces SO_x and NO_x emissions by 5–91% as compared to coal or to slurries based on used turbine oil. It was revealed that even small concentrations of such additives (7–15%) could result in a several-fold reduction in SO_x and NO_x. The higher the temperature, the more prominent the role of forest biomass. The calculated complex criterion illustrates that forest fuels increase the performance indicator of fuel suspensions by 1.2–10 times.

Keywords: coal; slurry fuel; combustion; forest fuels; biomass; anthropogenic emission concentration

1. Introduction

1.1. Environmental Issues of Coal-Fired Power Industry

Energy issues are critical in many economic, social, and environmental spheres. It is the efficiency of the energy complex that, to a great extent, governs the economic potential of countries and welfare of people [1,2]. Today, fossil fuels such as oil, coal, gas, oil shale, peat, uranium, etc. are the main energy sources. The present and future of the power industry rely strongly on its resourcing. According to studies [1–3], the share of coal in the global fuel and energy balance makes up 25–35%. Its main consumers are metallurgy and power engineering. 40–45% of the world's electricity is generated using coal [1–3]. One of the main concerns associated with using coal is the harm that its production, processing, and combustion do to the environment. The most important environmental issues (climate change, acid rains and the overall pollution) are directly or indirectly linked to this energy resource [4,5]. Major environmental problems are caused by solid wastes of coal-fired thermal power stations, such as ash and slag. Emissions from coal-fired thermal power stations are largely responsible for benzopyrene, a strong carcinogenic substance causing oncological diseases [5]. These abrasive materials can destroy lung tissue and cause a disease called silicosis. A negative impact of coal-fired power plants on the humankind and environment leads to illnesses, human migration, extinction and migration of animals, and reduction of eco-friendly woodlands [5,6]. This has caused power-generating enterprises in many countries to improve the devices monitoring air pollution [4–6]. Programs to increase energy efficiency and reduce emission with no negative effect on the rapid economic growth come into the picture [7]. In this research field, the authors [8–10] suggest focusing on solving the main fundamental and practical tasks, primarily with the environment in mind, to determine effective conditions of using high-potential coal-water slurries instead of the traditional energy sources (fuel oil, gas, and coal).

Coal-water slurry (CWS) is used to mean a mixture of a ground coal component with water. Coal sludge, filter cakes, low-grade coal, solid residuals from processing traditional energy resources (coal and oil) as well as resins can serve as a fuel base [8,10]. In the studies [8,10] authors were the first to suggest adding 10–15% of a liquid fuel component to a slurry to provide the necessary level of generating capacity and increase energy efficiency when using coal-water slurry and coal-water slurry containing petrochemicals (CWSP). The ranges of varying relative mass fractions of CWS and CWSP components make up 60–30% for coal dust, 30–50% for a dry filter cake, 0–15% for a combustible liquid, 30–50% for water, and 0.5–1% for plasticizer [10].

The production of composite liquid fuels (CLFs) from wastes is of great social, economic, and international importance [8–10]. First, it will reduce the vast volumes of accumulated wastes. Second, it will extend the resource potential of power-generating facilities. Third, fire and explosion safety gets higher, since water slurries are used instead of easily flammable and fire-hazardous fuels (coal dust, gas, or fuel oil). Fourth, mixing waste with water considerably reduces the environmental load on the surrounding nature. However, the use of liquid flammable wastes and low-reactive components as part of CWSP fuels inevitably leads to the growth of anthropogenic emission concentration. Involving plant additives is deemed to be the main solution to this problem [11–14].

The general trend of recent studies [11–16] on the current topic is to use large volumes of plant additives to produce energy and minimize the negative influence on the environment. Mixing plant waste with coal fuels leads to energy source diversification, as the range of raw materials is vast and growing [12,15,16]. The analysis of the global energy situation reveals the following benefits of co-firing coal fuel with plant biomass to produce energy [11–19]: a low-cost and low-risk renewable energy source is used; otherwise unused waste gets involved in the energy generation; emission concentration decreases; job opportunities in local neighborhoods are created, and external factors connected with fossil fuel combustion are reduced. Generally, most scientific groups exploit a traditional solid coal fuel mixing it with biomass (straw, sawdust, rice hulls and vegetable oil waste). Most of the well-known studies [11–19] investigate the processes of preparation, ignition, combustion, emissions, and ash production at the co-firing of coal and biomass with varying mass fractions of each of the mixture component. There are much fewer studies dealing with the development of CLFs containing various plant additives [19,20]. Until now, no data has been published on the complex analysis results of all main performance indicators of burning slurry fuels containing wood additives.

Over 25% of the world's woodland belongs to Russia, and it is deeply engaged in wood material processing. Two fifth of the country's territory is taken up by forests, 80% of which is coniferous. Russia's forest volume is mainly concentrated in Siberia and the North of the European part of the country [21]. Wood is a renewable resource, but due to large volumes of its production the question of effective and complex use of forest resources is getting increasingly important. The group of forest fuels comprises, besides trees themselves, their plant waste (stumps, branches, twigs, and tops) and woody debris (pine needles, leaves, brushwood, and bark), as well as industrial wastes (offcuts, chips, shavings, sawdust, pallets, etc.) [11,22,23]. The low demand for wood waste emerging from logging and wood processing is explained by insufficient development of recycling enterprises and is detrimental to the economy and the environment.

At present, a big mass of forest fuels in the world is used as fuel in hot-water and steam boilers. These take part in the technological production cycle and satisfy the domestic needs of wood processing enterprises, which reduces their thermal energy expenses [24]. Lumps of wood rather than sawdust, shavings, small pieces, or bark are used as fuel. Besides, a high content of chlorine and alkaline metals in wood waste accounts for high-temperature corrosion of heating surfaces of boiler units. However, the given limitations do not refute the fact that 5–15% addition of wood waste to a fuel can have a great effect in terms of anthropogenic emission minimization [12,13,16].

The analysis shows [21] that many countries (such as China, India, Japan, the USA, Australia, and Russia) using large volumes of coal fuel, have internal resources to forest biomass into the fuel sector [21]. Using plant additives together with coal slurries at thermal power stations or boiler plants can be considered a promising solution to ecological problems connected with anthropogenic emissions and waste recovery (both plant and coal wastes) [11–19].

1.2. Forest Fuels

Forest fuels can be regarded as a hydrocarbon fuel with known coefficients of emission, elemental composition, chemical formula, and heat effects. The organic part of vegetation consists primarily of hydrocarbons and a smaller amount of proteins, fats, waxes, and resins which constitute plant cells or fill intercellular space of plant tissue. Cellulose $(C_6H_{10}O_5)_x$ is the main crystalline component of cell walls. Hemicellulose is a matrix ramified polysaccharide with an amorphous structure. Lignins are complex phenolic polymers filling the spaces between the previous components. The content of cellulose and hemicellulose in the organic part of metaphytes reaches 60%, while that of lignin is 20–30% (depending on the species and age of timber). Plant tissue is saturated with water containing dissolved mineral salts that produce ash during combustion [25,26].

Timber is the most common type of forest biomass [27,28]. In terms of power generation, timber is considered the most appropriate type of plant biomass due to its rather high density, high calorific value, and low content of nitrogen, sulfur, and ash [28,29]. However, the amount of timber is limited and, besides being used in the power industry, it is also the main resource for pulp and paper industry and construction engineering [28]. Thus, waste from timber and construction industries looks most attractive in terms of burning, since a great amount of it has been accumulated and lies idle [30]. The main and most common timber industry waste includes sawdust, shavings, board offcuts, slabs, firewood, and bark [27]. Both needles and leaves also can be considered as fractions of logging residues in forestry (if branches, twigs, tops). They are not usually used as forest fuels directly but can be a portion of chipped logging residues that are used as fuels. This waste is a nuisance to many logging companies, as its recovery entails additional expenses, eventually increasing the product prime cost. Therefore, it is sensible to consider using this waste as an answer to many environmental, economic, and social issues [12,16].

1.3. Aim of the Research

The aim of this research is to determine the influence of forest fuels on the emissions from slurry fuels combustion. The main objective was to evaluate the prospects of using the given additives, as well as to determine their rational concentrations in slurry fuels.

2. Experimental Approach

2.1. Materials

Forest fuel from pine and birch trees from forests in Siberian Federal District (Russia) was used as additives in the present study. The origin of the used needles and sawdust are coniferous trees, mostly pine of the Pinus sylvestris species (Scots pine). The species of birches typical for Siberia is Betula pendula, commonly known as silver birch, which is native to Europe and parts of Asia.

The samples of pine sawdust, needles and birch leaves were taken from a timber processing factory in Tomsk, Russia. In the study, charcoal of grade A (Chernogolovka, Moscow region) was used. For its production, hardwood was used, in particular birch. The oak bark came from debarking operations of Quercus robur logs in the same industrial facility.

All samples of forest biomass were air dried, then milled by Rotary Mill Pulverisette 14 (rotor speed 6000–20,000 rpm). After milling, the samples were sieved. The average particle size was about 100 μm. The milled samples, spread in a thin layer, were exposed to air for several days to equilibrate with atmospheric moisture. Additional drying of the samples was not carried out.

In this study, authors used the flotation waste (filter cake) of coking coal as a main fuel component of CWSP. This waste is typical for coal processing plants in many regions; it has a fairly low ash content and better ignition and combustion characteristics in comparison with the flotation waste of some other coal ranks [10]. Waste turbine oil was used as liquid combustible component of CWSP.

Tables 1–3 present the properties of forest biomass used as additives (birch leaves, pine needles, pine sawdust, oak bark, and charcoal). Tables 4 and 5 present the properties of the main CWSP components (filter cake and used turbine oil).

Table 1. Properties of forest fuels (on the base of data [31–35]).

Additive	Ultimate Analysis					Proximate Analysis		
	C^{daf}, %	H^{daf}, %	N^{daf}, %	O^{daf}, %	$S_t{}^d$, %	A^d, %	V^{daf}, %	$Q^d{}_{s,V}$, MJ/kg
Pine needles of Himalayas in India [31]	45.81	5.38	0.98	46.11	0.01	1.5–1.74	–	16.7–18.5
Pine needles (*Pinus pinaster*) [32]	47.97–48.42	6.84–6.96	0.68–0.75	41.92–42.28	0.83–0.87	1.97	79.4	21.12–22.1
Pine needles (*Pinus sylvestris*) [33]	48.21	6.57	–	43.72	–	1.5	72.38	19.24
Birch leaves (*Betula pendula*) [34]	50.1–51.1	5.8–6.4	1.1–1.3	41.6–42.4	0.09–0.11	5.2–5.8		16.05–19.12
Leaves of various trees [35]	41.1–59.6	5.3–9.7	1.03–3.04	27.8–50.6	0.19–0.77	3.8–16	66.8–89.9	14.4–20.7

Table 2. Typical ultimate composition of birch leaves and pine needles (on the base of data [34,36–39]).

Additive	Elements, mg/kg							
	Na	Mg	K	Ca	Mn	Fe	Zn	P
Pine needles (*Pinus sylvestris*) [36–38]	15,388 38,200 –	859 540–760 890–1540	6101 3840–5060 3430–4550	2995 2780–3540 2590–6160	125.3 920–1240 151–370	0.6 – 40–100	24.5 – 33.6–73.3	1399 1010–1290 1360–3040
Birch leaves (*Betula pendula*) [34,39]	51–75 –	2748–4103 5890	6622–15,232 32,200	7521–24,897 29,000	159–1323 1470	172–243 790	124–185 280	2072–2915 3420

Table 3. Properties of wood components (based on data from [37,40–56]).

Samples	Pine Sawdust			Charcoal		Oak Bark	
	Pinus sylvestris from Siberia, Russia	*Pinus sylvestris* [37,40–44]	*Pinus tabulaeformis* [45,46]	Charcoal from Chernogolovka, Russia	Charcoal from Birch, Norway [47]	*Quercus robur* [48–52]	Bark of Various Oak [53–56]
	Chemical analysis, %						
C	52.32	50.9–52.8	47.21–49.65	83.11	72.7–91.4	46.08–51.2	40.77–48.9
H	6.39	5.95–6.2	6.25–8.09	3.46	1.93–4.35	5.5	5.43–6.11
O	40.70	40.5–42.9	41.58–44.40	12.8	7.09–21.61	46.8	39.3–53.54
N	0.24	0.1–0.5	0.05–0.1	0.6	0.37–1.13	0.2–1.32	0.2–0.56
S	0.02	0.01–0.09	0.04–0.21	0.03	<0.02	0.01–0.33	0.1–0.26
	Proximate analysis						
Moisture content, %	7.0	6.4–7.6	7.3–7.84	0.28		–	6.88–12.9
Volatiles, %	83.4	66.9–70.4	73.52–78.95	22.56	6.6–22.3	–	76.3–81.8
Ash content, %	1.6	0.47–5.5	0.76–1.88	1.49	1.4–5.0	0.3	1.64–3.6
Heat of combustion, MJ/kg	18.6	19.3	19.03–19.73	29.60		18.7	17.8–19.3
	Ultimate analysis, mg/kg						
Na	–	5–25	379	–	13–43	154	0.06–5.47%
K	–	120–780	435	–	1508–2538	1339–3520	4.32–16.7%
Ca	–	487–1000	1236	–	3332–4865	9750–20,100	16.2–37.3%
Mg	–	80–370	153	–	682–1099	230–340	1.77–4.14%
Fe	–	1.9–24	154	–	70–254	28.6–131	0.01–0.63%
Mn	–	25–200	28	–	230–477	280	0.16–3.71%
Cost, $/kg		0.006			0.3		0.006

Table 4. Properties of Coal and Filter Cake.

Coal Rank	Proximate Analysis				Ultimate Analysis				
	W^a, %	A^d, %	V^{daf}, %	$Q^a_{s,V}$, MJ/kg	C^{daf}, %	H^{daf}, %	N^{daf}, %	S_t^d, %	O^{daf}, %
Filter cake of coking coal (C) (dry)	–	26.46	23.08	24.83	87.20	5.09	2.05	1.022	4.46
Coking coal (C)	2.05	14.65	27.03	29.76	79.79	4.47	1.84	0.87	12.70

Table 5. Properties of liquid component.

Sample	Density, kg/m³	A^d, %	T_f, °C	T_{ign}, °C	$Q^a_{s,V}$, MJ/kg
Used turbine oil	868	0.03	175	193	44.99

2.2. Experimental Setups

The main elements of the experimental setup included a rotary muffle furnace and a gas analyzer [57]. A muffle furnace can create an air environment with temperatures 700–1000 °C. This temperature range is especially typical of CWS and CWSP combustion in the industrial conditions. Currently, there are several concepts of CWS and CWSP combustion in boilers of different capacity. These include the conventional fluidized bed combustion, vortex combustion of atomized flow, as well as co-firing with other types of fuels, for instance, with coal. Vortex combustion of coal-water fuel is the most widespread one. Combustion of soaring fine aerosol flow makes it possible to prolong the lifetime of particles in a combustion chamber. It also provides rapid mixing of the fuel and oxidizer, which, in turn, ensures more complete burnout of slurry fuel droplets. When it comes to laboratory research, the concept of soaring particles provides a way to gain a deeper insight into the typical mechanisms and stages of CWSP combustion: inert heating of a droplet, evaporation of moisture from its near-surface layer, evaporation of the liquid fuel component in heated air and thermolysis of the organic matter of coal, oxidation of volatiles and vapors of liquid fuel component in heated air, as well as heating and heterogeneous ignition of coke residue. However, using this technology to estimate the anthropogenic impact and measure the concentration of emissions from the combustion of a slurry fuel droplet is very cost-intensive. Vortex and fluidized bed combustion may have different conditions of fuel ignition and combustion, but the portions of fuel being burned are identical, so specific emissions can be considered comparable. To measure the emissions from fluidized bed combustion, it is enough to burn even a small portion of fuel, unlike with vortex combustion.

For fluidized bed combustion, boiler furnaces must also provide quite a long fuel lifetime in the combustion chamber and maintain the required high temperature throughout the said chamber. A thermally insulated rotary muffle furnace in the experimental setup can provide such conditions and make them near-real. The ceramic tube of the muffle furnace protects it without sharply reducing the temperatures in the near-wall area and in the active combustion zone. This is necessary for the stable CWS and CWSP combustion.

For the experiments, the fuel batch was weighed on an analytical balance with 0.01 g increments. The mass of the batch ranged within 0.5–1.5 g in each experiment. The gaseous products released during combustion of fuel in the muffle furnace were recorded and analyzed by the gas analyzer. Its main properties can be seen in Table 6. When averaging, only those results of the experiments were taken into account that did not differ by more than 2.5%.

Table 6. Gas analyzer sensors.

Process	Measurement Range	Accuracy
O_2	0–25 vol%	±0.2 vol%
CO	0–10,000 ppm	±10 ppm or ±10% of value (0–200 ppm) ±20 ppm or ±5% of value (201–2000 ppm) ±10% of value (2001–10,000 ppm)
NO_x	0–4000 ppm	±5 ppm (0–99 ppm) ±5% of value (100–1999 ppm) ±10% of value (2000–4000 ppm)
SO_2	0–5000 ppm	±10 ppm (0–99 ppm) ±10% of value (beyond this range)
CO_2 (derived from O_2 measurement)	0–CO_2	±0.2 vol%

2.3. Research Procedures

The following main components were used: filter cakes, used turbine oil and wood components (needles, leaves, sawdust, charcoal, and bark). Grinding the solid fuel component and plant additives to dust. Rotor Mill Pulverisette 14 was used for grinding. Then, a sample with an average particle size of 80–100 μm was collected using plansifter RL-1. Filter cakes from coal washing plants contain coal particles with a size of 60–80 μm. Therefore, their grinding is not necessary. Batches of slurry components were prepared using the ViBRA HT 84RCE (increment 10^{-5} g). The mass of the batches was calculated from the mass of the resulting composition and corresponding mass fractions of the components: filter cakes 75–100%, used turbine oil 10%, needles 7–15%, leaves 7–15%, mixture of needles and leaves 15%, bark 10%, sawdust 10% and charcoal 10%. The components were mixed in two stages by a homogenizer MPW-324 with a disperser in a metal container, which took 10 minutes.

The procedure of determining the amount of emissions from the fuel combustion comprised the following stages. The fuel sample was placed into a substrate made of stainless steel mesh which was fixed with fasteners at the end of the modular probe of the gas analyzer. The minirobotic arm moved the fuel sample and the gas analyzer probe to the combustion chamber. One experiment lasted 30–60 s, depending on the temperature in the combustion chamber. The flue gases from the combustion of the fuel moved towards the sensor. The sample went to the measuring sensors of the gas analyzer through a gas sampling hose. The EasyEmisson software (version 2.7, Lenzkirch, Germany) performed a continuous monitoring of flue gases. The values of CO, CO_2, NO_x, SO_x were recorded.

3. Results and Discussion

Figures 1 and 2 present the SO_x, NO_x from the combustion of CWS and CWSP containing forest fuels such as birch leaves, pine needles and their mixtures (with an equal proportion of leaves and needles). Adding forest fuels to CWSP significantly reduces the gaseous emissions of sulfur oxides (Figure 1). The values of SO_x emissions for such slurries (based on filter cake C) range from 9 to 117 ppm versus 62–360 ppm for coal of the same grade, depending on the combustion chamber temperature (Figure 1).

The decrease in the share of emissions (from 33 to 86%) was due to the chemical composition of the components introduced in the slurry. Alkaline and alkaline-earth metals (Ca, Na, K) present in forest fuel (Table 1) can form substances that remain in the coal ash ($2CaO + 2SO_2 + O_2 = 2CaSO_4$), preventing the formation of SO_x. The addition of a 15% forest fuel mixture had the most noticeable impact on sulfur oxide release. The SO_x concentrations in the temperature range under consideration were 17–90 ppm.

According to the data presented in Figure 2, the lowest concentrations of oxides and nitrogen are typical of filter cake C with a 15% addition of forest fuel mixture (88–218 ppm). The latter suggests that a synergistic effect emerges when a mixture of leaves and pine needles is used, especially at high temperatures (900–1000 °C). During forest fuel thermolysis, a part of metals (for instance, iron) remains in solid pyrolysis products. At high temperatures typical of the late pyrolysis, iron reacts with sulfur and nitrogen oxides ($3CO + Fe_2O_3 = 3CO_2 + 2Fe$; $2Fe + 3NO = 3/2N_2 + Fe_2O_3$). The synergism between pine needles and leaves reduces the concentration of sulfur oxide (Figure 1) and nitrogen oxide (Figure 2). In addition, there is no need to sort the incoming forest fuel for slurry preparation.

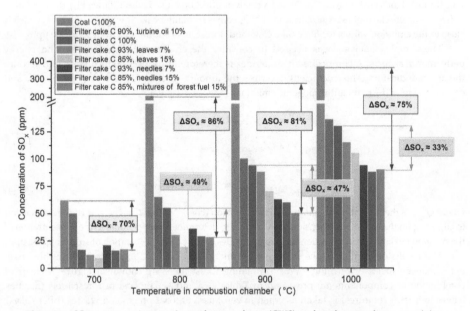

Figure 1. SO_x concentrations at the coal-water slurry (CWS) and coal-water slurry containing petrochemicals (CWSP) (with leaves, needles, or their mixtures) combustion.

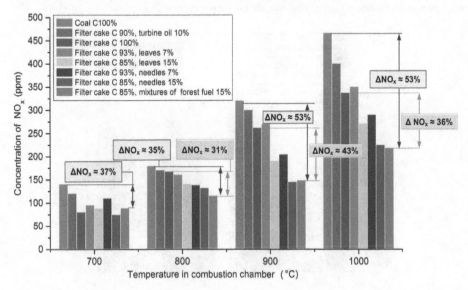

Figure 2. NO_x concentrations at the CWS and CWSP (with leaves, needles, or their mixtures) combustion.

The experimental results (Figure 2) have shown that using forest fuels as additives to coal-water slurries also reduces NO_x emissions by 35–53% and 5–43%, as compared to coal or CWSP based on used turbine oil without any additives, respectively. First, due to a quick release of volatile particles of forest fuels and their subsequent burning, the amount of O_2 in the combustion chamber decreases. Therefore, the reactions involving fuel nitrogen and oxygen produce a lower amount of NO and NO_2. Second, a low nitrogen content in plant additives also contributed to lower amounts of NO_x emissions [58]. Thirdly, metal ions such as Mn, Cu, Fe had a catalytic effect on NO_x oxides followed by the formation of free nitrogen. The highest concentration of NO_x comes from filter cake C with 7% of leaves, which is close to the emission values for filter cake C without forest fuels and ranges from 95 to 350 ppm.

The complex analysis was needed to consider the environmental, economic and energy performance aspects. A composite integral index is therefore introduced, which takes into account the above indicators. This coefficient describes the amount of energy per cost of fuel slurry and concentration of the main anthropogenic emissions [33]:

$$D_{cwsp}{}^{NOx} = Q^a{}_{s,V\_cwsp}/(C_{cwsp} \cdot NO_{x\_cwsp}); \tag{1}$$

$$D_{cwsp}{}^{SOx} = Q^a{}_{s,V\_cwsp}/(C_{cwsp} \cdot SO_{x\_cwsp}); \tag{2}$$

$$D_{cwsp}{}^{NOx\&SOx} = D_{cwsp}{}^{NOx} \cdot D_{cwsp}{}^{SOx}; \tag{3}$$

$$D_{relative} = D_{cwsp}{}^{NOx\&SOx}/D_{coal}{}^{NOx\&SOx}. \tag{4}$$

where $Q^a{}_{s,V}$ is the heat of combustion of the suspension (coal), MJ/kg; C is the cost of the suspension (coal), \$/kg (in the case of a suspension $Q^a{}_{s,V}$ and C are determined proportional to the concentration of the components); NO_x concentration of nitrogen oxides, ppm; SO_x concentration of sulfur oxides, ppm.

The results of calculating the relative performance indicators ($D_{relative}$) considering the main performance aspects of burning CWSP containing forest fuels are shown in Figure 3. Heats of combustion of components are presented in Tables 1, 3 and 4. The cost of forest fuels (needles, leaves and their mixtures) is taken as equal to zero; as with coal processing wastes (filter cakes), transportation expenses making up 0.0058 \$/kg were the only expenses accounted. The costs of slurries were determined in proportion to the components' concentrations assuming zero water cost, since process and waste water may be used for the preparation of CLFs.

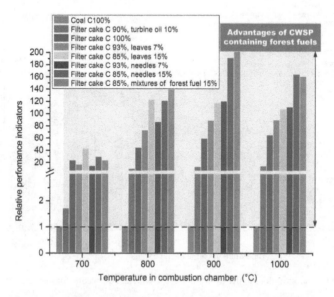

Figure 3. Relative performance indicators ($D_{relative}$) of burning high-potential CWSP fuels containing leaves, needles, or their mixture vs. coal at varying temperatures in the combustion chamber.

The resulting dependences have shown that adding pine needles, leaves and their mixtures appears attractive in terms of environmental friendliness, energy, and cost efficiency. Plant-based slurries, having the same cost as a filter cake of coking coal, are marked by lower concentrations of the main anthropogenic emissions (SO_x, NO_x), and in terms of heat of combustion, are highly competitive with coal-water fuels.

In the preparation of slurry fuels before the experiments with burning their batches, it was established that all the wood additives under study can significantly affect the rheology. Being added to CWSP fuels, these components adsorb some of the fuel moisture, thus preventing its lamination. The maximum allowable relative mass fraction for the additives and dopants under study should equal 15%.

The reduction of sulfur dioxide emission in CWSP fuels with wood components (Figures 4 and 5) can be attributed to a low sulfur fraction in the latter, which has direct impact on the overall sulfur content in the slurry. It was established that a fuel based on filter cakes, used turbine oil and 10% of tree bark or 10% charcoal has the lower environmental indicators for sulfur oxides (10–108 ppm). Charcoal can rapidly adsorb many substances, including sulfur, from a fluid or gaseous medium. These substances, sulfur in particular, are present as oxides. Therefore, it is safe to conclude that charcoal adsorbs sulfur and nitrogen from the pyrolysis of coal or liquid fuel component of CWSP.

Although the combustion heat of tree bark is comparable with that of a filter cake, its presence in the slurry can increase the energy performance and improve the ignition and combustion characteristics due to a high content of highly volatile substances in the bark particles. Ignition delay and combustion times decrease. In terms of NO_x emission (Figure 5), a 10% addition of tree bark did slight compensate for the presence of a liquid fuel component in the slurry, which is largely responsible for the formation of fuel oxides [59].

The 10% of sawdust in the CWSP reduces of NO_x emission in by more than 1.5 times (209–231 ppm vs. 320–466 ppm for coal) at 900–1000 °C. Sawdust intensifies ignition and increases the yield of carbon monoxide ($NO_x + XCO = 1/2N_2 + XCO_2$).

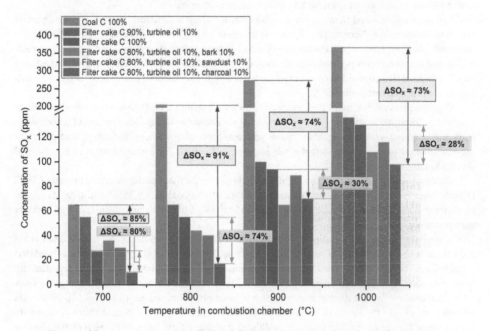

Figure 4. SO_x concentrations at the CWS and CWSP (with bark, sawdust, charcoal) combustion.

9

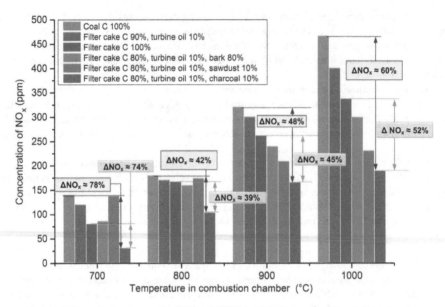

Figure 5. NO$_x$ concentrations at the CWS and CWSP (with bark, sawdust, charcoal) combustion.

Adding even 10% of sawdust to CWSP based on flotation waste lowers the ignition temperature (70–80 °C) and increases the combustion rate. The sawdust enhances the effects of fuel droplet micro-explosions [60]. This shortens the ignition delay time and overall reduces the energy consumption at the firing stage. The optimal concentrations of nitrogen oxides are also reached with a 10% concentration of charcoal and do not exceed 190 ppm.

The pyrolysis of wood biomass leaves solid residue, which is similar to charcoal in its properties. As a confirmation, the thermal decomposition of forest fuel as part of CWSP occurs under oxygen deficiency (oxygen cannot reach the surface of wood particles, since the gaseous combustion products of coal and liquid fuel components oust air from the combustion zone). Therefore, it is safe to assume that the thermal decomposition of wood as part of slurry fuels produces charcoal, which can adsorb sulfur and nitrogen compositions.

Figure 6 presents the main performance aspects of burning CWSP fuels with wood additives. The heat of combustion and the cost of components are shown in Tables 3–5. The cost of wood waste (bark and sawdust) is taken as equal to zero; as with coal processing wastes (filter cakes), transportation expenses making up 0.006 $/kg were the only expenses accounted. The average market cost of charcoal was 0.3 $/kg.

Charcoal, sawdust, and bark can significantly improve the main performance of burning CWSP. Despite high environmental performance indicators of charcoal as a CWSP additive, its global production is not enough. The use of sawdust and bark as additives to composite coal fuel appears very promising.

The results of numerous studies and the industrial experience of co-firing coal fuels with biomass indicate that this technology is associated with several issues [61–64]. Among them, ash-related problems are the key issues. The chemical composition of lignocellulosic fuels differs significantly from traditional coal. Biomass commonly contains large amounts of water-soluble inorganic salts, which can easily volatilize during combustion and become part of the gas phase [61,62]. This leads to high level of activity for alkali materials in the ash and, consequently, high dirtiness propensity during the co-firing process. The quantity volatilized depends on the property of the fuel, the ambient air, and the combustion technology. SiO$_2$ and CaO dominate in the biomass ashes, oxides of Mg, Al,

K, Na and P is much lower in the ash. The ash from hard biomass (wood) includes large amount of oxides with low melting points, primarily K and P. In addition, they keep substantially lower levels of heavy metals, then soft biomass (straw) [61]. At high temperatures, metals and their oxides partially evaporate and amounted an active part of the reactions in the gaseous phase. During the gases motion in the boiler's channel, they precipitate on its elements at low temperatures and form small particles on the surface, for example CaO. Then these particles become part of the gases in so-termed "fly ash" (<1 μm). Because of a reoxidation and coagulation, particles agglomerate, forming ash size more than 10 μm (coarse fly ash). Non-volatile ash compounds melt and coalesce on or in the surface of the particle, contingent on the temperature and chemical compound of the particle and the ambient gases [61,62]. Depending on the particle's density and type, the technology used and the gas speed, these ash fractions could be entrained by the gases, but the majority is deposited. For combustion chamber, heavy ash deposition leads to contamination, corrosion, and defluidization. This can reduce the efficiency of the combustion chamber and damage its equipment, as well as increase maintenance costs [61–64].

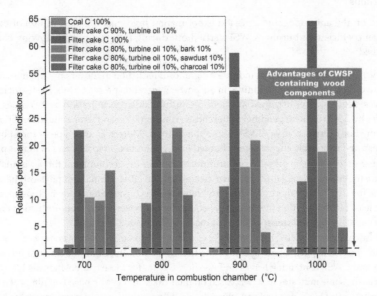

Figure 6. Relative performance indicators ($D_{relative}$) of burning high-potential CWSP fuels containing bark, sawdust, or charcoal vs. coal at varying temperatures in the combustion chamber.

During combustion of slurry fuel with forest biomass, the chemical composition of which contributes to the corrosive effect on boilers, it is important to realize that ash level is one significant factor in the design stage of the equipment. The combustion process of slurry fuels should be adapted to the requirements for fuels, especially when it comes to industrial waste or forest biomass that are chemically different from traditional fuels to achieve sustainable and stable development of such technologies. There are an amount of decisions that can be taken to avert and decrease corrosion, such as controlling the steam temperature in the boiler design to a level at which the corrosion ratio is agreeable, the choice of more noncorrosive alloy of heat exchangers and other boiler elements and use of additive component that modify the combustion gases chemistry and prevent the ash deposition [61,65].

Dmitrienko et al. [57] used two approaches (one considering solely the environmental performance and the other, the combustion heat, fuel cost and anthropogenic emissions) to show that CWSP and CWS fuels are high-potential solutions to many environmental, economic and energy problems of the modern coal-fired power industry by choosing the necessary components and their concentration. The main conclusion of Dmitrienko et al. [57] is that it is worthwhile to involve numerous coal and oil processing wastes (filter cakes, oils, sludges, etc.) in the fuel cycle. The experimental data presented in this research highlight great prospects of solving tasks set on a brand-new environmental level [57]. In particular, the use of forest fuels makes it possible to reduce anthropogenic emissions to such low levels that it makes sense to simplify flue gas purification cycles at power plants. Thus, the economic benefit from CWS and CWSP technology implementation can be even greater than described by Dmitrienko et al. [57]. Moreover, forest biomass reduces not only the anthropogenic gaseous emissions but also ash residue. This is a very important point for coal enterprises. Using CWS and CWSP fuels results in the lower volume of ash formation as compared to that from coal combustion, as well as longer service life of heat and power equipment with high energy performance indicators [66]. This is explained by less ash sticking to heat exchange surfaces, so the heat absorption remains rather high over a long period of time.

4. Conclusions

Based on the anthropogenic emission experimental investigations and results of calculating the integral coefficient of burning CWSP carried out during this paper, the following conclusions were reached:

(i)   The vast majority of the results indicate a significant reduction in the amount of emissions due to the involvement of biomass additives in the power generation process. Thus, the use of forest fuels (leaves, needles, sawdust, and bark) reduces sulfur oxide concentration by 2–5 times, nitrogen oxides by 1.5–2 times (depending on the chosen concentration and temperature conditions of fuel combustion) versus coal or CWSP without additives. Moreover, this type of forest biomass is a cheap and renewable energy source formed in large amounts in forests and in timber processing.

(ii)   The use of wood waste gives an opportunity to recover the accumulated timber industry waste, reduce the environmental load, improve rheological as well as thermal and physical characteristics of the fuel. However, there are other problems worth considering. They include modification or at least reconfiguration of fuel control equipment of boiler units, fuel production and supply systems, slagging, and transportation of the components to the station.

(iii)   The calculated complex indicators $D_{relative}$ takes into account the energy, economic and environmental aspects of using composite fuel liquids. It illustrates the obvious advantages of CWS and CWSP containing leaves and needles. $D_{relative}$ for these compositions is 1.2–10 times as high as the same indicator for CWS based on filter cakes and CWSP based on filter cakes and 10% of turbine oil. $D_{relative}$ for fuel samples with sawdust or bark also demonstrates the benefits of biomass additives and exceeds the values of normal CWSP without dopants by 1.2–2.5 times.

(iv)   The main way to further develop this research is to analyze and specify effective CWS and CWSP fuel compositions from numerous components, additives, and dopants. A compiled database of experimental information with the main energy, economic and environmental performance indicators of burning coal, CWS and CWSP with different dopants (an experimental setup and method from study [67] can be used) will make it possible to develop a predictive model. It can be-based, for example, on statements and numerical solution methods of partial differential equations described in studies [68–70] when studying slurry fuel heating, evaporation, thermal decomposition, combustion processes. This model will enable choosing a relevant component composition to reach high-performance indicators of power equipment.

Author Contributions: P.S. wrote the paper; G.N. performed the experiments.

Funding: This research received no external funding.

Acknowledgments: Research was founded by National Research Tomsk Polytechnic University (project VIU-ISHFVP-184/2018).

Conflicts of Interest: The authors declare no conflict of interest.

References

1.   Liu, F.; Lyu, T.; Pan, L.; Wang, F. Influencing factors of public support for modern coal-fired power plant projects: An empirical study from China. *Energy Policy* **2017**, *105*, 398–406. [CrossRef]
2.   Li, H.; Yang, S.; Zhang, J.; Qian, Y. Coal-based synthetic natural gas (SNG) for municipal heating in China: Analysis of haze pollutants and greenhouse gases (GHGs) emissions. *J. Clean. Prod.* **2016**, *112*, 1350–1359. [CrossRef]
3.   Su, F.; Itakura, K.; Deguchi, G.; Ohga, K. Monitoring of coal fracturing in underground coal gasification by acoustic emission techniques. *Appl. Energy* **2017**, *189*, 142–156. [CrossRef]
4.   Pearse, R. The coal question that emissions trading has not answered. *Energy Policy* **2016**, *99*, 319–328. [CrossRef]
5.   Guttikunda, S.K.; Jawahar, P. Atmospheric emissions and pollution from the coal-fired thermal power plants in India. *Atmos. Environ.* **2014**, *92*, 449–460. [CrossRef]
6.   Zhao, C.; Luo, K. Sulfur, arsenic, fluorine and mercury emissions resulting from coal-washing byproducts: A critical component of China's emission inventory. *Atmos. Environ.* **2017**, *152*, 270–278. [CrossRef]
7.   *World Energy Outlook Special Report 2016: Energy and Air Pollution*; International Energy Agency: Paris, France, 2016.
8.   Glushkov, D.O.; Strizhak, P.A.; Chernetskii, M.Y. Organic coal-water fuel: Problems and advances (Review). *Therm. Eng.* **2016**, *63*, 707–717. [CrossRef]
9.   Khodakov, G.S. Coal-water suspensions in power engineering. *Therm. Eng.* **2007**, *54*, 36–47. [CrossRef]
10. Strizhak, P.A.; Vershinina, K.Y. Maximum combustion temperature for coal-water slurry containing petrochemicals. *Energy* **2017**, *120*, 34–46. [CrossRef]
11. Bhuiyan, A.A.; Blicblau, A.S.; Islam, A.K.M.S.; Naser, J. A review on thermo-chemical characteristics of coal/biomass co-firing in industrial furnace. *J. Energy Inst.* **2018**, *91*, 1–18. [CrossRef]
12. Badour, C.; Gilbert, A.; Xu, C.; Li, H.; Shao, Y.; Tourigny, G.; Preto, F. Combustion and air emissions from co-firing a wood biomass, a canadian peat and a Canadian lignite coal in a bubbling fluidised bed combustor. *Can. J. Chem. Eng.* **2012**, *90*, 1170–1177. [CrossRef]
13. Liu, G.; Liu, Q.; Wang, X.; Meng, F.; Ren, S.; Ji, Z. Combustion Characteristics and Kinetics of Anthracite Blending with Pine Sawdust. *J. Iron Steel Res. Int.* **2015**, *22*, 812–817. [CrossRef]
14. Gil, M.V.; Pevida, C.; Pis, J.J.; Rubiera, F. Thermal behaviour and kinetics of coal/biomass blends during co-combustion. *Bioresour. Technol.* **2010**, *101*, 5601–5608. [CrossRef] [PubMed]
15. Maj, G. Emission Factors and Energy Properties of Agro and Forest Biomass in Aspect of Sustainability of Energy Sector. *Energies* **2018**, *11*, 1516. [CrossRef]
16. Fitzpatrick, E.M.; Bartle, K.D.; Kubacki, M.L.; Jones, J.M.; Pourkashanian, M.; Ross, A.B.; Williams, A.; Kubica, K. The mechanism of the formation of soot and other pollutants during the co-firing of coal and pine wood in a fixed bed combustor. *Fuel* **2009**, *88*, 2409–2417. [CrossRef]
17. Lei, K.; Ye, B.; Cao, J.; Zhang, R.; Liu, D. Combustion characteristics of single particles from bituminous coal and pine sawdust in O_2/N_2, O_2/CO_2, and O_2/H_2O atmospheres. *Energies* **2017**, *10*, 1695. [CrossRef]
18. Hu, W.; Liang, F.; Xiang, H.; Zhang, J.; Yang, X.; Zhang, T.; Mi, B.; Lui, Z. Investigating co-firing characteristics of coal and masson pine. *Renew. Energy* **2018**, *126*, 563–572. [CrossRef]
19. Zhu, M.; Zhang, Z.; Zhang, Y.; Liu, P.; Zhang, D. An experimental investigation into the ignition and combustion characteristics of single droplets of biochar water slurry fuels in air. *Appl. Energy* **2017**, *185*, 2160–2167. [CrossRef]
20. Li, W.; Li, W.; Liu, H. The resource utilization of algae—Preparing coal slurry with algae. *Fuel* **2010**, *89*, 965–970. [CrossRef]
21. *Global Forest Products Facts and Figures*; Food and Argiculture Organization of the United Nation: Rome, Italy, 2016; ISBN I666EN/1/12.16.
22. Sathre, R.; Gustavsson, L.; Truong, N.L. Climate effects of electricity production fuelled by coal, forest slash and municipal solid waste with and without carbon capture. *Energy* **2017**, *122*, 711–723. [CrossRef]

23. Lehtonen, A.; Heikkinen, J.; Makipa, R.; Sievanen, R.; Lisk, J. Biomass expansion factors (BEFs) for Scots pine, Norway spruce and birch according to stand age for boreal forests. *For. Ecol. Manag.* **2004**, *188*, 211–224. [CrossRef]

24. Ratajczak, E.; Bidzińska, G.; Szostak, A.; Herbeć, M. Resources of post-consumer wood waste originating from the construction sector in Poland. *Resour. Conserv. Recycl.* **2015**, *97*, 93–99. [CrossRef]

25. McKendry, P. Energy production from biomass (part 1): Overview of biomass. *Bioresour. Technol.* **2002**, *89*, 37–46. [CrossRef]

26. Vicente, E.D.; Alves, C.A. An overview of particulate emissions from residential biomass combustion. *Atmos. Res.* **2018**, *199*, 159–185. [CrossRef]

27. Parikka, M. Global biomass fuel resources. *Biomass Bioenergy* **2004**, *27*, 613–620. [CrossRef]

28. Huron, M.; Oukala, S.; Lardière, J.; Giraud, N.; Dupont, C. An extensive characterization of various treated waste wood for assessment of suitability with combustion process. *Fuel* **2017**, *202*, 118–128. [CrossRef]

29. Ramage, M.H.; Burridge, H.; Busse-Wicher, M.; Fereday, G.; Reynolds, T.; Shah, D.U.; Wu, G.; Yu, L.; Fleming, P.; Densley-Tingleye, D.; et al. The wood from the trees: The use of timber in construction. *Renew. Sustain. Energy Rev.* **2017**, *68*, 333–359. [CrossRef]

30. Junginger, M.; Goh, C.S.; Faaij, A. *International Bioenergy Trade: History, Status & Outlook on Securing Sustainable Bioenergy Supply, Demand and Markets*; Springer Science Business Media Dordrecht: Dordrecht, The Netherlands, 2014; ISBN 978-94-007-6982-3.

31. Safi, M.J.; Mishra, I.M.; Prasad, B. Global degradation kinetics of pine needles in air. *Thermochim. Acta* **2004**, *412*, 155–162. [CrossRef]

32. Viana, H.F.S.; Rodrigues, A.M.; Godina, R.; Matias, J.C.O.; Nunes, L.J.R. Evaluation of the Physical, Chemical and Thermal Properties of Portuguese Maritime Pine Biomass. *Sustainability* **2018**, *10*, 2877. [CrossRef]

33. Nizamuddin, S.; Baloch, H.A.; Griffin, G.J.; Mubarak, N.M.; Bhutto, A.W.; Abro, R.; Mazari, S.A.; Ali, B.S. An overview of effect of process parameters on hydrothermal carbonization of biomass. *Renew. Sustain. Energy Rev.* **2017**, *73*, 1289–1299. [CrossRef]

34. Pňakovič, Ľ.; Dzurenda, L. Combustion characteristics of fallen fall leaves from ornamental trees in city and forest parks. *BioResources* **2015**, *10*, 5563–5572. [CrossRef]

35. Fernandes, E.R.K.; Marangoni, C.; Souza, O.; Sellin, N. Thermochemical characterization of banana leaves as a potential energy. *Energy Convers. Manag.* **2013**, *75*, 603–608. [CrossRef]

36. Giertych, M.J.; de Temmerman, L.O.; Rachwal, L. Distribution of elements along the length of Scots pine needles in a heavily polluted and a control environment. *Tree Physiol.* **1997**, *17*, 697–703. [CrossRef] [PubMed]

37. Skonieczna, J.; Małek, S.; Polowy, K.; Węgiel, A. Element content of scots pine (*Pinus sylvestris* L.) stands of different densities. *Drewno* **2014**, *57*, 77–87. [CrossRef]

38. Mikhailova, T.A.; Kalugina, O.V.; Afanaseva, L.V.; Nesterenko, O.I. Trends of chemical element content in needles of scots pine (*Pinus sylvestris* L.) under various natural conditions and emission load. *Contemp. Probl. Ecol.* **2010**, *3*, 173–179. [CrossRef]

39. Khanina, M.A.; Guselnikova, E.N.; Rodin, A.I.; Ivanova, V.V. The influence of ecological factors on element structure of leaves of the birch. *Meditsina i Obrazovaniye v Sibiri* **2015**, *6*, 1–11.

40. Saarela, K.E.; Harju, L.; Rajander, J.; Lill, J.O.; Heselius, S.J.; Lindroos, A.; Mattsson, K. Elemental analyses of pine bark and wood in an environmental study. *Sci. Total Environ.* **2005**, *343*, 231–241. [CrossRef] [PubMed]

41. Vassilev, S.V.; Baxter, D.; Andersen, L.K.; Vassileva, C.G. An overview of the chemical composition of biomass. *Fuel* **2010**, *89*, 913–933. [CrossRef]

42. Filbakk, T.; Jirjis, R.; Nurmi, J.; Høibø, O. The effect of bark content on quality parameters of Scots pine (*Pinus sylvestris* L.) pellets. *Biomass Bioenergy* **2011**, *35*, 3342–3349. [CrossRef]

43. Dong, Y.; Haverinen, J.; Tuuttila, T.; Jaakkola, M.; Holm, J.; Levequee, J.M.; Lassi, U. Rapid one-step solvent-free acid-catalyzed mechanical depolymerization of pine sawdust to high-yield water-soluble sugars. *Biomass Bioenergy* **2017**, *102*, 23–30. [CrossRef]

44. Krutul, D.; Zielenkiewicz, T.; Radomski, A.; Zawadzki, J.; Antczak, A.; Drożdżek, M.; Makowski, T. Metals accumulation in scots pine (*Pinus sylvestris* L.) wood and bark affected with environmental pollution. *Wood Res.* **2017**, *62*, 353–364.

45. Liao, C.; Wu, C.; Yan, Y.; Huang, H. Chemical elemental characteristics of biomass fuels in China. *Biomass Bioenergy* **2004**, *27*, 119–130. [CrossRef]

46. Mei, Y.; Liu, R.; Zhang, L. Influence of industrial alcohol and additive combination on the physicochemical characteristics of bio-oil from fast pyrolysis of pine sawdust in a fluidized bed reactor with hot vapor filter. *J. Energy Inst.* **2016**, *90*, 923–932. [CrossRef]

47. Bui, H.; Wang, L.; Tran, K.; Skreiberg, O. CO_2 gasification of charcoals produced at various pressures. *Fuel Process. Technol.* **2016**, *152*, 207–214. [CrossRef]

48. Balboa-Murias, M.A.; Rojo, A.; Álvarez, J.G.; Merino, A. Carbon and nutrient stocks in mature *Quercus robur* L. stands in NW Spain. *Ann. For. Sci.* **2006**, *63*, 557–565. [CrossRef]

49. Gómez-García, E.; Diéguez-Aranda, U.; Cunha, M.; Rodríguez-Soalleiro, R. Comparison of harvest-related removal of aboveground biomass, carbon and nutrients in pedunculate oak stands and in fast-growing tree stands in NW Spain. *For. Ecol. Manag.* **2016**, *365*, 119–127. [CrossRef]

50. Krutul, D.; Zielenkiewicz, T.; Zawadzki, J.; Radomski, A.; Antczak, A.; Drozdzek, M. Influence of urban environment originated heavy metal pollution on the extractives and mineral substances content in bark and wood of oak (*Quercus robur* L.). *Wood Res.* **2014**, *59*, 177–190.

51. De Visser, P.H.B. The relations between chemical composition of oak tree rings, leaf, bark, and soil solution in a partly mixed stand. *Can. J. For. Res.* **1992**, *22*, 1824–1831. [CrossRef]

52. Telmo, C.; Lousada, J.; Moreira, N. Proximate analysis, backwards stepwise regression between gross calorific value, ultimate and chemical analysis of wood. *Bioresour. Technol.* **2010**, *101*, 3808–3815. [CrossRef] [PubMed]

53. Channiwala, S.A.; Parikh, P.P. A unified correlation for estimating HHV of solid, liquid and gaseous fuels. *Fuel* **2002**, *81*, 1051–1063. [CrossRef]

54. Jin, W.; Singh, K.; Zondlo, J. Pyrolysis Kinetics of Physical Components of Wood and Wood-Polymers Using Isoconversion Method. *Agricultureic* **2013**, *3*, 12–32. [CrossRef]

55. Ozbay, N.; Yargic, A.S. Liquefaction of Oak Tree Bark with Different Biomass/Phenol Mass Ratios and Utilizing Bio-based Polyols for Carbon Foam Production. *AIP Conf. Proc.* **2017**, *1809*, 020039. [CrossRef]

56. Ruiz-Aquino, F.; González-Pena, M.M.; Valdez-Hernández, J.I.; Revilla, U.S.; Romero-Manzanares, A. Chemical characterization and fuel properties of wood and bark of two oaks from Oaxaca, Mexico. *Ind. Crops Prod.* **2015**, *65*, 90–95. [CrossRef]

57. Dmitrienko, M.A.; Nyashina, G.S.; Strizhak, P.A. Environmental indicators of the combustion of prospective coal water slurry containing petrochemicals. *J. Hazard. Mater.* **2017**, *338*, 148–159. [CrossRef] [PubMed]

58. Xie, J.-J.; Yang, X.-M.; Zhang, L.; Ding, T.-L.; Song, W.-L.; Lin, W.-G. Emissions of SO_2, NO and N_2O in a circulating fluidized bed combustor during co-firing coal and biomass. *J. Environ. Sci.* **2007**, *19*, 109–116. [CrossRef]

59. Rokni, E.; Panahi, A.; Ren, X.; Levendis, Y.A. Curtailing the generation of sulfur dioxide and nitrogen oxide emissions by blending and oxy-combustion of coals. *Fuel* **2016**, *181*, 772–784. [CrossRef]

60. Liu, J.; Wang, R.; Xi, J.; Zhou, J.; Cen, K. Pilot-scale investigation on slurrying, combustion, and slagging characteristics of coal slurry fuel prepared using industrial waste liquid. *Appl. Energy* **2014**, *115*, 309–319. [CrossRef]

61. Nunes, L.J.R.; Matias, J.C.O.; Catalão, J.P.S. Biomass combustion systems: A review on the physical and chemical properties of the ashes. *Renew. Sustain. Energy Rev.* **2016**, *53*, 235–242. [CrossRef]

62. Shao, Y.; Xu, C.; Zhu, J.; Preto, F.; Wang, J.; Li, H.; Badour, C. Ash Deposition in Co-firing Three-Fuel Blends Consisting of Woody Biomass, Peat, and Lignite in a Pilot-Scale Fluidized-Bed Reactor. *Energy Fuels* **2011**, *25*, 2841–2849. [CrossRef]

63. Vassilev, S.V.; Vassileva, C.G.; Baxter, D. Trace element concentrations and associations in some biomass ashes. *Fuel* **2014**, *129*, 292–313. [CrossRef]

64. Williams, A.; Jones, J.M.; Ma, L.; Pourkashanian, M. Pollutants from the combustion of solid biomass fuels. *Prog. Energy Combust.* **2012**, *38*, 113–137. [CrossRef]

65. Amir, M.K.; Samad, M.; Behzad, S.M.; Mehdi, S.M. Corrosion Prevention in Boilers by Using Energy Audit Consideration. *Appl. Mech. Mater.* **2014**, *532*, 307–310. [CrossRef]

66. Nyashina, G.; Legros, J.C.; Strizhak, P. Environmental potential of using coal-processing waste as the primary and secondary fuel for energy providers. *Energies* **2017**, *10*, 405. [CrossRef]

67. Glushkov, D.O.; Strizhak, P.A. Ignition of composite liquid fuel droplets based on coal and oil processing waste by heated air flow. *J. Clean. Prod.* **2017**, *165*, 1445–1461. [CrossRef]

68. Syrodoy, S.V.; Kuznetsov, G.V.; Salomatov, V.V. The influence of heat transfer conditions on the parameters characterizing the ignition of coal-water fuel particles. *Therm. Eng.* **2015**, *62*, 703–707. [CrossRef]
69. Salomatov, V.V.; Kuznetsov, G.V.; Syrodoy, S.V.; Gutareva, N.Y. Ignition of coal-water fuel particles under the conditions of intense heat. *Appl. Therm. Eng.* **2016**, *106*, 561–569. [CrossRef]
70. Syrodoy, S.V.; Kuznetsov, G.V.; Zhakharevich, A.V.; Gutareva, N.Y.; Salomatov, V.V. The influence of the structure heterogeneity on the characteristics and conditions of the coal–water fuel particles ignition in high temperature environment. *Combust. Flame* **2017**, *180*, 196–206. [CrossRef]

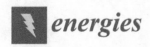

Article

The Main Elements of a Strategy for Combined Utilization of Industrial and Municipal Waste from Neighboring Regions by Burning it as Part of Composite Fuels

Dmitrii Glushkov, Geniy Kuznetsov *, Kristina Paushkina and Dmitrii Shabardin

National Research Tomsk Polytechnic University, Tomsk 634050, Russia; dmitriyog@tpu.ru (D.G.);
kkp1@tpu.ru (K.P.); dpshabardin@mail.ru (D.S.)
* Correspondence: kuznetsovgv@tpu.ru; Tel.: +7-(3822)-701-777 (ext. 1615)

Received: 26 August 2018; Accepted: 21 September 2018; Published: 22 September 2018

Abstract: An experimental study has been conducted into the ignition and combustion processes of composite fuel droplets fed into a heated muffle furnace on a holder. Consistent patterns and characteristics of physical and chemical processes have been established for a group of fuel compositions: wet coal processing waste (a mixture of fine coals and water) 85% + municipal solid waste (wood, or plastic, or rubber) 10% + used oil 5%. Burning a coal-water slurry instead of dry coal dust is characterized by a positive environmental effect. Adding used oil to a coal-water slurry results in better energy performance characteristics of the composite fuel during combustion. Adding fine municipal solid waste (MSW) to the fuel composition makes it possible to effectively recover it by burning in boiler furnaces with energy performance characteristics of combustion and environmental characteristics of flue gases that are as good as those of composite fuel compositions without MSW. Sustainability of the composite fuel ignition process and complete burnout of liquid and solid combustible components have been determined. The values of the guaranteed ignition delay times for droplets with a size (diameter) of about 2 mm have been established for the composite fuel compositions under study in the ambient temperature range 600–1000 °C. The minimum values of ignition delay times are about 3 s, the maximum values are about 15 s under the near-threshold ignition conditions. The obtained findings enabled to elaborate the main elements of the strategy for combined recovery of industrial and municipal waste by burning it as part of composite fuels.

Keywords: municipal solid waste; coal processing waste; oil refining waste; waste management; composite fuel; energy production

---

1. Introduction

Nowadays the problem of municipal solid waste (MSW) and industrial waste utilization [1] is very acute in the modern world and the Russian Federation is not an exception. One area of great interest is the design of management systems for waste handling by processing it into energy and valuable products [2]. According to the Russian Federal State Statistics Service [3], about 5441 Mt of industrial waste were produced in country in 2016, which was 7.5% more than in 2015. More than 40,000 Mt of waste have been accumulated in the Russian Federation so far, of which 86% comes from solid and liquid fossil fuel extraction. In terms of the geographic distribution, the Siberian Federal District accounts for the majority of the waste. It makes up about 60% of the total amount of industrial waste in the country.

In 2016, the level of industrial waste recovery was 2685 Mt, or 49.3% of the total volume of its production (Figure 1). The rest of the waste was buried (504 Mt, or 9.3%) or stockpiled at open-air disposal sites for temporary storage (2253 Mt, or 41.4%). Industrial waste disposal sites occupy large amounts of agricultural land, damaging the land cover, soil, and landscape. Moreover, coal industry waste is not only a fire hazard but also contains acid-forming substances, heavy metals, and other elements dangerous for the environment. When exposed to the intensive physical and chemical effects of natural factors (air, water, and solar energy), it becomes a source of integrated pollution of the environment. Coal dust from coal mining and processing pollutes the air and water. Waste containing used oils and petroleum products is toxic. Storing oil waste has the following detrimental effects on the environment: it enhances the greenhouse effect, causes acid rain, decreases the quality of water, and contaminates the groundwater. One liter of used oil can contaminate about 7 Ml of ground waters. The pollution of water with petroleum products reduces the amount of dissolved oxygen and causes many marine species to die [4]. The contamination of soil with hydrocarbons makes its further use for agricultural purposes impossible.

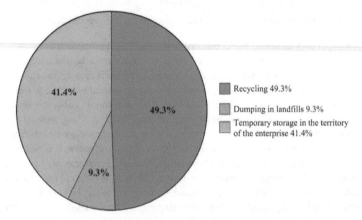

Figure 1. Industrial waste recovery structure in the Russian Federation [3].

The volume of MSW (paper, cardboard, plastic, wood, and textiles) production in the Russian Federation in 2016 made up 52.4 Mt. About 3.9 Mt of MSW, or 7.4%, were destined for re-use (Figure 2). About 1.0 Mt of MSW, or 1.9%, were transferred for decontamination and destruction, including by burning it at trash incineration plants. The overwhelming majority of waste (47.5 Mt, or 90.7%) went to waste disposal and landfill sites for burial and temporary storage [3].

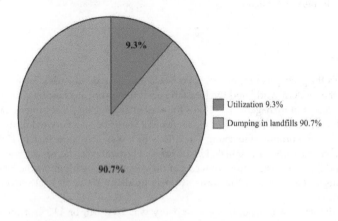

Figure 2. Municipal solid waste (MSW) recovery structure in Russian Federation [3].

The extremely low MSW recovery indicators are explained by the low level of waste management development. There are only 243 MSW recovery complexes, 53 sorting complexes, and 10 trash incineration plants in the whole Russian Federation. The number of specially equipped facilities for waste disposal (MSW disposal sites) in the country is about 1399, which is several times lower than the number of authorized landfill sites (there are 7153 of them). Unauthorized landfill sites number more than 17,500. They pose the biggest environmental threat. In the waste management structure, they are regarded as environmental damage that has already been built up over the last years. Dedicated waste disposal sites, authorized and unauthorized MSW landfill sites take up huge amounts of land with a total area of more than 50,000 ha. Landfill storage of non-recovered waste entails the following negative factors: a wide spread of substances and bacterial flora that are dangerous for people's health, including when they find their way into the air and ground waters; formation of dioxides if ignition occurs; low economic performance indicators taken the environmental risks, the costs of land and waste disposal site maintenance.

Low level of waste management development in the Russian Federation does not conform to the "Basic Principles of National Policy for Environmental Development of the Russian Federation for the period until 2030" [5]. It is a current objective to develop routines that will effect the transition from waste stockpiling towards recovery and re-use. Additionally, minimizing the adverse effects on the environment of the waste already accumulated is a priority.

According to the global experience, discontinuing waste stockpiling and burial in favor of re-use requires the implementation of an intermediate stage in the medium term, namely waste disposal by burning to produce heat and electricity [6]. Such routines will reduce the annual growth rates of industrial and municipal waste, and, in some cases, will allow for partial or complete recovery of the accumulated waste that is unsuitable for re-use. It is an important task to develop routines for using industrial waste and MSW to reduce the load of landfill sites and improve the environmental situation around them. Normally, such problems are solved by simply burning waste [7] or by burning waste with energy generation [8]. However, the low calorific value of MSW (about 10 MJ/kg [9]) versus that of conventional coal fuels (20–30 MJ/kg [10]), as well as a relatively high concentration of harmful gases make complete replacement of coal with combustible waste not economically, environmentally or technically viable.

An alternative approach to solving this problem is to use fine MSW as a composite liquid fuel component consisting of coal processing waste (or a mixture of low-quality brown or bituminous coals with water) and a used combustible liquid (transformer, turbine, engine oils, etc.). Based on the assessments in [11], it is possible to predict that the introduction of 10–20% of typical municipal solid waste into composite fuels will decrease the territories of new landfill sites built for MSW burial by 25%. Besides it will provide for more economical use of non-renewable hydrocarbon fuels that nowadays are directly burned to produce heat and electricity.

Based on the above, the purpose of this research is to experimentally study the patterns and characteristics of composite fuel ignition and combustion, as well as to analyze the prospects of joint implementation of industrial and municipal waste recovery strategy with power generation at local thermal power plants by several neighboring regions of the Russian Federation.

The relevance of the present study is explained by the following: expanding the scope of raw materials in thermal power engineering by using relatively cheap composite fuels from waste rather than high-grade coals saves solid fossil fuels and financial means for their acquisition. As a rule, a limited amount of accumulated or annually produced combustible waste in a separate region does not allow to implement a strategy of efficient waste recovery with energy production on an industrial scale. The authors of the manuscript provide an example of several neighboring regions of the Russian Federation with different levels of social development and different industrial structures to validate the prospects of combined implementation of the strategy for efficient recovery of industrial and municipal waste with energy production at local thermal power plants.

The novelty of the experimental investigation and theoretical analysis consists in the prospect of practical application of the proposed waste management strategy. The latter would be a temporary strategy while transition is being made from waste burial to re-use or recycling. During the 15–25 years of transition, the waste management strategy will reduce the amount of waste to be buried, on the one hand, and eliminate interim steps involving the construction of costly incineration plants which will be in demand for a relatively short period of time, on the other hand.

2. Experimental Investigation

2.1. Materials Preparation

The research was performed on five fuel compositions based on filter cakes of coking coal (filter cake). Filter cake was produced in the Severnaya coal washing plant that is situated in Kemerovo region of the Russian Federation. Combustible waste is a by-product of coal processing—preparation for long hauls to the consumer. The initial coal is washed with water to remove fine fractions (5–15% of the initial amount of coal). This reduces the level of environmental contamination with coal dust during transportation by trains and sea ship; it also improves fire safety of the coal dust. After the coal has been washed, the liquid containing fine particles is left in tanks. These particles (up to 80 μm in size) settle on the bottom. The upper layer of water is pumped for re-use. The liquid residue is passed through press filters. The moist residue is the filter cake with water content about 40%. At the locations of the coal washing plants, filter cakes are stockpiled on open sites. Large areas are contaminated not only by stockpiled filter cake, but also by a fine coal dust that pollutes the surrounding lands due to the wind.

Previous research findings of composite fuel ignition and combustion [12] processes (filter cakes + used oil) testify to the fact that filter cakes can be used as the main combustible component when preparing composite fuels based on industrial waste and MSW. Taking into account the high level of coal mining (about 7.5 Mt) and consumption in the thermal power engineering (about 40% of the total power generation), and consequently, rather large production volumes of filter cakes from coal processing, it can be concluded that filter cakes are a promising component for composite fuels [13].

Wet filter cakes (with a coal mass fraction of about 60%) were used in this study as the main component of our composite fuels. Five different compositions are used in experiments: No. 1—filter cake 100%; No. 2—filter cake 95% + used engine oil 5%; No. 3—filter cake 85% + wood 10% + used engine oil 5%; No. 4—filter cake 85% + rubber 10% + used engine oil 5%; No. 5—filter cake 85% + plastic 10% + used engine oil 5%. The mass fractions are listed here. The particle size of typical municipal solid waste is comparable to that of coal particles (about 100 μm). The main characteristics of the fuel components are listed in Tables 1–3. The characteristics of the filter cake are presented for dry samples. The filter cake has been dried at about 105 °C until the moisture has fully evaporated.

Fuel droplets were generated by an electronic dispenser (limit dosage volumes were 1 μL and 10 μL, pitch variation was 0.1 μL). The fuel droplet size (diameter) was about 2 mm (the volume was about 4.2 μL, the mass about 3.9 mg).

Table 1. Properties of fuel components. Proximate analysis.

Component	W^a, %	A^d, %	V^{daf}, %	$Q^a_{s,v}$, MJ/kg	References
Filter cake	-	26.5	23.1	24.83	[14]
Wood	20.0	2.0	83.1	16.45	[15]
Rubber	2.0	1.8	67.4	33.50	[9]
Plastic	2.0	0.2	99.5	22.00	[9]
Motor oil	0.3	0.8	100.0	44.02	[15]

Table 2. Properties of fuel components [9,14,15]. Ultimate analysis.

Component	C^{daf}, %	H^{daf}, %	N^{daf}, %	S^{daf}, %	O^{daf}, %	References
Filter cake	87.2	5.1	2.1	1.1	4.5	[14]
Wood	50.3	6.0	0.2	0.1	43.4	[15]
Rubber	97.9	1.2	0.3	0.6	-	[9]
Plastic	66.7	7.9	-	-	25.4	[9]

Table 3. Flash temperature and ignition temperature of fuel components.

Component	Flash Temperature, °C	Ignition Temperature, °C
Filter cake	-	450
Wood	230	340
Rubber	-	350
Plastic	306	415
Motor oil	132	218

2.2. Experimental Setup and Procedure

Ignition and combustion processes of composite liquid fuel droplets of different compositions have been researched using the experimental setup shown in Figure 3. The setup enables one to simulate the conditions of fuel droplet heating that are identical to heating conditions in a boiler furnace. The setup is based on a rotary muffle furnace R 50/250/13 (Nabertherm GmbH, Lilienthal, Germany). The inner diameter of the ceramic tube is 0.04 m and the tube length is 0.45 m; the temperature variation range is 20–1200 °C; the temperature is controlled by the signal of an integrated type S thermocouple. The experiments were conducted after the furnace was heated up to the given temperature. Fuel droplets generated by the dispenser were placed on the holder and introduced to the furnace by the robotic arm through one of the side apertures of the ceramic tube along the tube symmetry axis. The linear velocity of the robotic arm movements did not exceed 0.5 m/s to prevent the droplet deformation and sliding off the holder. A high-speed video camera was fixed on the robotic arm and moved with the holder with a fuel droplet on it. Such scheme made it possible to record all the physical and chemical processes occurring from the moment a fuel droplet was introduced into the furnace up to its burnout. The droplet remained in the video camera focus throughout the whole period of process recording.

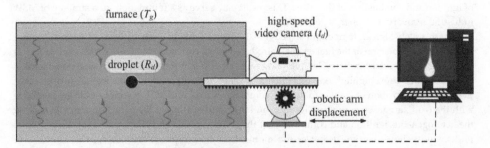

Figure 3. Scheme of experimental setup.

The following parameters were controlled when a series of five experiments under identical initial conditions were conducted: the temperature (T_g) of the heated air in the furnace, the initial diameter (D_d) of the droplet, and ignition delay times (t_d) of the fuel. The air temperature in the furnace was controlled by the readings of a built-in thermocouple. The processes taking place during the fuel droplet heating were recorded by a Phantom v411 high-speed color camera (Vision Research Inc, Wayne, NJ, USA). Its main specifications are as follows: 12 Gb memory; filming rate 4200 frames/s at maximum resolution 1280 × 800 pixels; pixel size 20 μm; 12 bit depth; minimal exposure 1 μs. The Tema Automotive Video software (Image Systems AB, Linköping, Sweden) was used to analyze the video recordings of the experiments. The software and high-speed video recording hardware system made it possible to record fuel droplet characteristics and conduct an analysis of consistent patterns in their combustion. The diameter of the droplet was recorded before it was placed into the muffle furnace. Using the Tema Automotive algorithms, four droplet diameters were automatically measured in four sections. The obtained values were arithmetically averaged and the droplet diameter D_d was then calculated. The systematic error of D_d determination did not exceed 4%. The ignition delay time (t_d) was also calculated automatically as a time interval between two events [16]: the introduction of the fuel droplet into the furnace (heating initiation) and the emergence of luminance around the droplet that corresponds to the moment of gas-phase ignition [17]. The values of t_d were determined by the Threshold algorithm of the Tema Automotive software. The systematic error of t_d did not exceed 3%. Random errors did not exceed 10% for sets of five experiments under identical conditions.

2.3. Experimental Results and Discussion

The results of the experimental research (Figure 4) allow us to conclude that the fuel compositions under study, based on coal processing waste and typical MSW, are ignited consistently when heated in an oxidizing environment (air). The fuel droplet combustion processes continues until the liquid and solid combustible components burn out. It can be concluded that composite fuels based on coal processing waste and typical MSW can be used in thermal power engineering instead of conventional coals.

It has been established that the ignition and combustion patterns of all the composite fuel compositions (Figure 4) with 5–15% of a combustible liquid and MSW are similar to those of the initial filter cakes (composition No. 1). The experimental result can be explained by the dominant influence of this component on the physical and chemical processes during the fuel droplet heating. A possibility of varying the component composition and concentration of separate components of composite fuels in wide ranges creates favorable conditions of using such fuels with predictable characteristics in real life. Wet filter cakes in their initial state and used oils (transformer, turbine, engine, etc.), or filter cakes with a relatively low moisture content (moisture evaporates in the process of storing them at disposal sites) with industrial or municipal sewage water containing combustible liquids can be used to prepare composite fuels. According to the results (Figure 4), adding about 10% of typical MSW to composite fuels does not change the consistent patterns and characteristics of ignition and combustion of the latter. This result may serve as a foundation for a strategy of MSW industrial recovery with energy generation.

At relatively high air temperatures (above 700 °C) in the muffle furnace, the conditions of heating, ignition and combustion of the fuel droplet introduced into it correspond to those of the same processes occurring in the boiler furnace. The following main stages of interdependent physical and chemical processes have been highlighted: inert heating; moisture evaporation from the subsurface layer; thermal decomposition of solid combustible components (coal and MSW); combustible gases mixing with the oxidizer; gaseous mixture ignition and burnout; heating of the solid combustible residue; the heterogeneous ignition and combustion of the solid residue. The video frames of the processes under study show (Figure 4) that at the moment of the gas-phase ignition t_d, the combustible gas mixture forming around the droplet is spherical. The size of this mixture equals 2–3 sizes of the fuel droplet. The more components with a high content of volatiles in the fuel composition and the lower the ambient temperature, the larger the size of the forming gas zone (up until the ignition) around the fuel droplet. This result is explained by the following. At the ignition moment (less than 5 s after heating), the fuel droplet is heated unevenly in conditions of relatively high ambient temperatures (800 °C and above). The evaporation and thermal decomposition processes are not complete. They only take place in the near-surface layer of the fuel droplet. The gas-phase ignition of vapors occurs in the immediate vicinity of the droplet. At relatively low ambient temperatures

(600 °C), the ignition delay time t_d is substantially higher (more than 12 s) than it is at 800 °C and above. During such a long time period, the droplet is warmed up evenly. In such conditions, the intensity of fuel evaporation and thermal decomposition is higher than it is at the temperatures above 800 °C. As a result, the concentration of vapors and the size of the gas zone in the vicinity of the droplet are bigger at the moment of ignition.

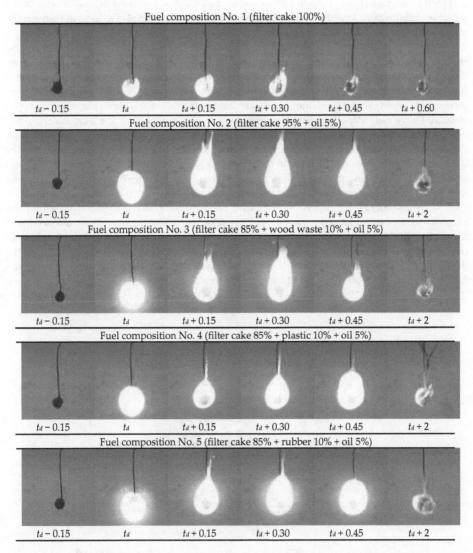

Figure 4. Ignition and combustion of composite liquid fuel droplet at T_g = 800 °C (t_d, s).

After the ignition, the gas-phase combustion process of the volatiles takes place predominantly above the droplet (Figure 4) where a stoichiometric ratio occurs between the components (oxidizer and fuel) of the gaseous mixtures. The intensity of the flame in the frame with $t = (t_d + 0.30)$ s (Figure 4) is lower in the lower regions of the droplet. The intensity increases when going vertically (upwards from the droplet bottom). It means that a rich premixed flame zone is present in this region, where the oxidizer concentrations are relatively higher than the fuel concentrations [18]. The intensity of evaporation and thermal decomposition of the composite fuel components is quite low. This causes the flame to appear nearer to the droplet surface. As a result, the solid combustible residue is heated by the energy released from the process of the gaseous mixture burnout, and then it ignites. The combustion front is spread inside the droplet from its surface to the center. For the municipal waste investigated in this research, the share of a non-combustible solid residue is much smaller than that for the filter cake (Table 1). Adding MSW into the fuel composition causes not only a pronounced gas-phase combustion of volatiles around the droplet to take place, but also reduces the ash residue after the burnout of combustible components.

Figure 5 contains a crosshatched region that illustrates the range of changing the main process characteristics (t_d) under guaranteed ignition conditions of different composite fuel compositions when the air temperature T_g was varied from 600 to 1000 °C. The upper limit of the region (Figure 5) is a curve characterizing ignition delay times of the initial filter cake (without adding MSW and combustible liquid) vs. temperature. The lower limit of the region (Figure 5) is a curve $t_d = f(T_g)$ obtained for the fuel composition based on filter cakes, plastic, and used oil. The ignition delay times of other compositions lie within the boundaries of the crosshatched region (Figure 5). The results from Figure 5 make it possible to conclude that the air temperature 600 °C is the minimum necessary for the ignition process of composite fuels based on coal processing waste, MSW, and a combustible liquid to initiate. The maximum difference in t_d for compositions with different components makes up less than 20% at the air temperatures 600–1000 °C. It was also established that at the air temperature above 1000 °C, the intensity of processes occurring during the induction period is so high that the heat and mass transfer in the droplet and around it has a less significant impact on the ignition characteristics than at T_g = 600–1000 °C. When T_g > 1000 °C, the ignition delay times for the identical fuel droplet with the same composition are only marginally different. For example, at T_g = 1100 °C and T_g = 1200 °C the difference in t_d is less than 5%. For a rapid evaluation of mean ignition delay time values of composite fuels (filter cakes 85% + MSW 10% + oil 5%), an approximate equation can be used:

$$t_d \approx 203.1 \exp\left(-0.0044T_g\right) \text{ at } 600 \leq T_g \leq 1000 \text{ °C.}$$

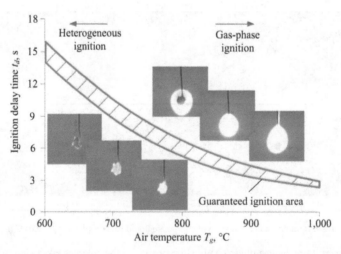

Figure 5. Crosshatched area shows ignition delay times of droplets sized 2 mm of different compositions of composite fuels based on industrial and municipal waste at various air temperatures.

The analysis of the high-speed video recordings of composite fuel ignition and combustion processes under the conditions of low ($T_g \approx$ 600–700 °C), moderate ($T_g \approx$ 700–800 °C) and high (T_g > 800 °C) intensity of the droplet heating made to possible to determine consistent patterns of the occurring physical and chemical processes, which can be used in the development of a relevant mathematical model. At relatively low temperatures, the duration of the induction period is rather long (14–16 s) (Figure 5). Under such conditions, the intensity of the occurring thermal decomposition of combustible components is not high. Due to the heat and mass transfer processes in the vicinity of the fuel droplet, the concentration of combustible gases is not high enough for the combustion process to initiate. The heterogeneous ignition and subsequent combustion with a relatively low intensity is typical of the temperatures $T_g \approx 600$–$700\ °C$. In the range of high temperatures (above $700\ °C$), the intensities of heating the subsurface droplet layer, thermal decomposition of combustible components and formation of the gas mixture are high (Figure 5). Figure 4 illustrates the above regularity of physical and chemical processes under such conditions.

The research findings (Figures 4 and 5) lead to an important practical conclusion that it is possible to efficiently recover typical MSW at thermal power engineering facilities. Besides, adding MSW into the main fuel will allow to reduce fuel consumption without worsening the energy performance characteristics of the process (see combustion heat values in Table 1).

3. Prospects for Utilization of Industrial and Municipal Waste from Several Regions by Burning with Energy Generation

In most countries with a developed primary economic sector (primarily due to fossil fuel mining), regions where fossil fuels are extracted tend to be surrounded by other regions with a high level of industrial and social development (Figure 6). This neighborhood creates favorable conditions for increasing the volumes of fossil fuel extraction, on the one hand, and for the development of industrial enterprises and population growth, on the other hand. In such conditions, regions with a developed primary sector of the economy, e.g., due to coal mining and exporting, have to deal with a major problem of reducing the negative environmental impact from coal washing plants storing filter cakes at open-air disposal sites. The main problem of regions with a high level of industrial and social development is recycling and recovery of MSW whose annual production volume is comparable to that of industrial waste of large coal mining and coal processing enterprises, which reaches millions of tons a year.

Figure 6. Location of industrial waste and MSW sources [5].

It appears promising for the neighboring regions to solve these problems in a comprehensive way by implementing a strategy of combined recovery of industrial and municipal waste by burning it as part of composite fuels at local thermal power plants. It will reduce the negative impact of waste on the environment, on the one hand, and diminish the consumption volume of high-grade coals for heat and electricity generation, on the other hand. In this research, the main elements of the proposed strategy have been elaborated using three neighboring regions of the Russian Federation located in Western Siberia as an example: Kemerovo region (No. 1), Novosibirsk region (No. 2), and Tomsk region (No. 3). The obtained results will serve as a foundation for developing similar waste management strategies in other regions of the world, considering their peculiar features.

It is believed that in practice, the new waste management strategy will be implemented in a rather short period of time (15–25 years). This strategy is likely to be in great demand among countries with no MSW recycling but burial at disposal sites, so the proposed waste management strategy will be a temporary solution in the transition from waste burial to re-use or recycling. During this period, it is essential to minimize the volumes of non-recycled waste disposal. Additionally, organizing complicated technologies of MSW re-use or recycling requires time to streamline the processes at each stage—from separate collection of waste by citizens to re-use as raw materials by industrial enterprises. Therefore, within this time, the proposed waste management strategy will reduce the amount of waste to be buried, on the one hand, and eliminate interim steps involving the construction of costly incineration plants which will be in demand for a relatively short period of time, on the other hand.

3.1. General Information

In the Kemerovo region (No. 1), 2801 Mt of waste are annually produced, MSW making up 0.9 Mt of it. Overall in the region, there are 173,159 MSW sources, 139 fossil fuel extraction sources, 538 manufacturing waste sources, 1517 sources of production and non-production waste (materials and goods that lost their consumer properties), 497 sources of waste related to power, gas, and vapor supply, 294 sources of water supply and sewage water disposal waste, 176 sources of construction and repair waste, 72 sources of agricultural, forestry, fish-farming, and fishing waste, as well as 1073 sources of other waste. The MSW breakdown is presented in Figure 7.

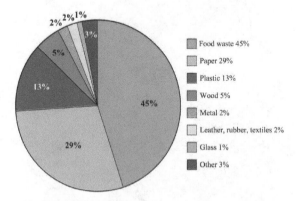

Figure 7. MSW structure of the Kemerovo region (No. 1).

About 98% of all the MSW are stockpiled and buried at disposal sites, and just about 2% are recovered. The Kemerovo region is characterized by an uneven distribution of its MSW production areas: the population density is about 28 people per 1 km^2, of which 85% are concentrated in urban areas. The total population of the region is about 2.7 million people. In each of the five large cities (Kemerovo, Novokuznetsk, Prokop'yevsk, Belovo, and Mezhdurechensk), there are more than 100,000 citizens. Over 80% of the MSW are produced in large cities.

In Kemerovo region, there are deposits of coal, iron ore, gold, silver, manganese, zinc, lead, copper, etc. On an industrial scale, primarily coal (brown and bituminous) is mined. This coal is not only used for thermal power engineering purposes, but is also exported abroad. Coal mining and coal processing enterprises are the main sources of combustible industrial waste. The greatest amount of waste (filter cakes) is produced where major surface mines are located around the city of Kemerovo (39.0%), as well as in "Mezhdurechenskiy" (12.4%), "Kiselyovskiy" (8.9%), "Berezovskiy" (8.6%), and "Prokopievskiy" (7.5%) surface mines.

In the Novosibirsk region (No. 2), 3.9 Mt of waste, of which 1.3 Mt are MSW, are annually produced. The bulk of MSW (1.2 Mt—92.3%) is produced in the regional center (the city of Novosibirsk) and around it (the city of Berdsk). This is conditioned by the fact that 2.2 million people out of the total 2.8 million population live there. About 78% of the MSW is destined for burial. Only a third of MSW disposal sites comply with the current safety requirement regulations. The volumes of MSW stockpiled at disposal sites is growing every year. Near the regional center (the city of Novosibirsk), the volume of waste stockpiled at disposal sites has grown 2.5 times over the last 15 years, and makes up about 70 Mt. The breakdown of the MSW is presented in Figure 8.

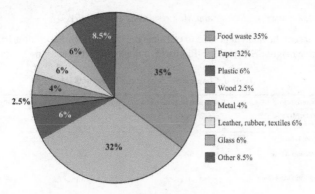

Figure 8. MSW structure of Novosibirsk region (No. 2).

The Novosibirsk region is one of the largest industrial districts of Siberia. The industrial complex is composed of large and medium-sized enterprises producing more than 80% of all the output: agricultural products, foodstuffs, electrical appliances, electronic and optical equipment, metallurgical equipment, metalware and non-metallic products, and construction materials (reinforced concrete). Novosibirsk region annually produces about 2.6 Mt of industrial waste, 43.8% of which is fossil fuel extraction waste, in particular, coal mining waste.

In the Tomsk region (No. 3), 1.3 Mt of waste are annually produced, of which 0.4 Mt are MSW, and 0.9 Mt are industrial waste. The bulk of MSW comes from major cities: Tomsk (0.3 Mt—74.7%) and Seversk (0.03 Mt—7.5%) This is explained by the fact that 0.8 million people out of the total 1.1 million live there. About 98% of MSW are destined to be buried at dedicated waste disposal sites. The MSW breakdown is presented in Figure 9.

The bulk of industrial waste is produced where large industrial facilities are located: in Tomsk (0.3 Mt—26.8%), Seversk (0.3 Mt—24.3%), and near them (0.1 Mt—12.3%). There are about 3500 industrial enterprises in Tomsk region. The structure of industrial production is multisectoral. The main industry branches are: fossil fuel extraction, electric power, non-ferrous mining, chemical, petrochemical, wood, wood working, and food industries, as well as mechanical engineering and metal working. The oil and petrochemical industries, mechanical engineering and metal working take the dominant position (over 60% of the total industrial output). About 0.2 out of 0.9 Mt of the total waste volumes annually produced are recovered at enterprises, 0.2 Mt are delivered to third parties as recyclable materials, and 0.5 Mt are stockpiled at temporary storage sites in the facilities of enterprises or buried at dedicated disposal sites.

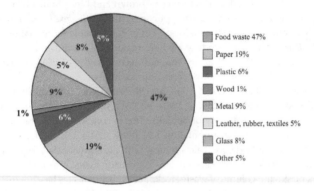

Figure 9. MSW structure of Tomsk region (No. 3).

3.2. Energy Potential of Industrial Waste and Municipal Solid Waste (MSW)

Tables 4–6 present the main data used to calculate the energy potential of industrial and municipal solid waste in three neighboring regions (Figure 6) of the Russian Federation.

Table 4. Characteristics of waste by regions [5].

Characteristics	Kemerovo Region (No. 1)	Novosibirsk Region (No. 2)	Tomsk Region (No. 3)
Total volume of waste, Mt/y	2801	3.9	1.3
Industrial waste, Mt/y	2800	2.6	0.9
Structure of industrial waste:			
• fossil fuel extraction waste	99.5%	43.8%	17.2%
• manufacturing industry waste	0.2%	8.1%	32.8%
• waste related to generation and distribution of power, gas, and water	0.1%	24.5%	7.2%
• construction and repair waste	<0.01%	0.1%	1.0%
• agricultural, forestry, fish-farming, and fishing waste	0.04%	16.0%	32.9%
• other waste	0.15%	7.5%	8.9%
Re-use/recovery, Mt/y	1876.0 (67.0%)	0.9 (34.1%)	0.3 (28.0%)
Burial/stockpiling at disposal sites, Mt/y	924.0 (33.0%)	1.7 (65.9%)	0.6 (72.0%)
Municipal solid waste (MSW), Mt/y	0.9	1.3	0.4
The structure of MSW:			
• food waste	45%	35%	47%
• paper	29%	32%	19%
• plastic	13%	6%	6%
• metal	2%	4%	9%
• leather, rubber, textiles	2%	6%	5%
• glass	1%	6%	8%
• wood	5%	2.5%	1%
• other	3%	8.5%	5%
Re-use/recovery, Mt/y	0.02 (1.9%)	0.3 (22%)	0.008 (2.0%)
Burial/stockpiling at disposal sites, Mt/y	0.88 (98.1%)	1.0 (78%)	0.392 (98.0%)

Table 5. Volume of sewage water produced in oil processing (per 1 t of oil) [3].

Type of Plant	Volume of Sewage Water Produced, m^3/t of Oil	
	of the First System	of the Second System
Fuel refinery plant	0.23–0.25	0.10–0.20
Lube and fuel refinery plant	0.40–1.50	0.10–0.25
Petrochemical plant	2.00–3.00	1.20–2.00

Table 6. Heat of combustion of typical MSW [9].

Type of Waste	Heat of Combustion, MJ/kg
Food waste	5–10
Paper	15–20
Plastic	20–25
Leather, rubber, textiles	20–35
Wood	15–20
Other	<10

The source data to calculate the energy potential of MSW is as follows:

MSW that has been already accumulated and not recovered as of 2017 (stockpiled at disposal sites) (G^0_{MSW}):

- in Kemerovo region—54.0 Mt;
- in Novosibirsk region—70.0 Mt;
- in Tomsk region—21.0 Mt.

The annual production volume of MSW (G_{MSW}):

- in Kemerovo region—0.9 Mt;
- in Novosibirsk region—1.3 Mt;
- in Tomsk region—0.4 Mt.

The share of non-recovered combustible MSW stored at disposal sites (k^{MSW} coefficient) for every region:

- in Kemerovo region—0.5;
- in Novosibirsk region—0.7;
- in Tomsk region—0.6.

The averaged heat of combustion (q^{MSW}) for typical MSW composition is (Table 6):

- in Kemerovo region—14.1 MJ/kg;
- in Novosibirsk region—12.0 MJ/kg;
- in Tomsk region—10.0 MJ/kg.

The energy potential from burning the MSW accumulated by 2017 in every region:

$$Q^0_{MSW} = G^0_{MSW} \times q^{MSW} \qquad (1)$$

- in Kemerovo region $Q^0_{MSW} = 54.0 \times 10^6 \times 10^3 \times 14.1 \times 10^6 = 761$ PJ;
- in Novosibirsk region $Q^0_{MSW} = 70.0 \times 10^6 \times 10^3 \times 12.0 \times 10^6 = 830$ PJ;
- in Tomsk region $Q^0_{MSW} = 21.0 \times 10^6 \times 10^3 \times 10.0 \times 10^6 = 209$ PJ.

The energy potential from burning MSW accumulated within a year:

$$Q^{MSW} = k^{MSW} \times G^{MSW} \times q^{MSW} \tag{2}$$

- in Kemerovo region $Q^{MSW} = 0.5 \times 0.9 \times 10^6 \times 10^3 \times 14.1 \times 10^6 = 13$ PJ;
- in Novosibirsk region $Q^{MSW} = 0.7 \times 1.3 \times 10^6 \times 10^3 \times 12.0 \times 10^6 = 15$ PJ;
- in Tomsk region $Q^{MSW} = 0.6 \times 0.4 \times 10^6 \times 10^3 \times 10.0 \times 10^6 = 2$ PJ.

3.3. Energy Potential of Coal Processing and Oil Refining

The source data to calculate the energy potential of filter cakes and low rank coals is as follows:
Filter cakes ($G^0{}_{fc}$) that have been accumulated and not recovered (are stored at disposal sites) by 2017 in every region:

- in Kemerovo region—368.6 Mt;
- in Novosibirsk region—5.8 Mt;
- in Tomsk region, coal is not extracted or processed.

The annual production volume of filter cakes (G_{fc}) makes up 10–15% of the volume of processed coal:

- in Kemerovo region—8.1 Mt;
- in Novosibirsk region—0.4 Mt;
- in Tomsk region, coal is not extracted or processed.

The heat of combustion (q_{fc}) of typical filter cakes (Table 1) is about 24.8 MJ/kg.
The energy potential from burning the filter cakes accumulated by 2017:

$$Q^0{}_{fc} = G^0{}_{fc} \times q_{fc} \tag{3}$$

- in Kemerovo region $Q^0{}_{fc} = 368.6 \times 10^6 \times 10^3 \times 24.8 \times 10^6 \times 0.6 = 5485$ PJ;
- in Novosibirsk region $Q^0{}_{fc} = 5.8 \times 10^6 \times 10^3 \times 24.8 \times 10^6 \times 0.6 = 86$ PJ;
- in Tomsk region $Q^0{}_{fc} = 0$ J.

The energy potential from burning filter cakes accumulated within a year:

$$Q_{fc} = G_{fc} \times q_{fc} \tag{4}$$

- in Kemerovo region $Q_{fc} = 8.1 \times 10^6 \times 10^3 \times 24.8 \times 10^6 \times 0.6 = 201$ PJ;
- in Novosibirsk region $Q_{fc} = 0.4 \times 10^6 \times 10^3 \times 24.8 \times 10^6 \times 0.6 = 10$ PJ;
- in Tomsk region $Q_{fc} = 0$ J.

The source data to calculate the energy potential of liquid combustible waste (used oils). Used oils, combustible oil production and refining waste ($G^0{}_{oil}$) that have been accumulated and not recovered (are stored at disposal sites) by 2017 in every region:

- in Kemerovo region, there is no liquid combustible waste;
- in Novosibirsk region—97.3 kt;
- in Tomsk region—2219 kt.

The annual production volume of used oils, combustible oil production and refining waste (G_{oil}):

- in Kemerovo region—0.5 kt;
- in Novosibirsk region—0.7 kt;

- in Tomsk region—11.5 kt.

About 40% of used oils are not recovered (l_{oil} coefficient). They can be used to produce composite fuels. The heat of combustion (q_{oil}) of typical combustible liquids is 44.0 MJ/kg. The energy potential from burning the used oils accumulated by 2017:

$$Q^0_{oil} = G^0_{oil} \times q_{oil} \qquad (5)$$

- in Kemerovo region $Q^0_{oil} = 0$ J;
- in Novosibirsk region $Q^0_{oil} = 97.3 \times 10^3 \times 10^3 \times 44.0 \times 10^6 = 4$ PJ;
- in Tomsk region $Q^0_{oil} = 2219.4 \times 10^3 \times 10^3 \times 44.0 \times 10^6 = 98$ PJ.

The energy potential from burning used oils accumulated within a year:

$$Q_{oil} = l_{oil} \times G_{oil} \times q_{oil} \qquad (6)$$

- in Kemerovo region $Q_{oil} = 0.4 \times 0.5 \times 10^3 \times 10^3 \times 44.0 \times 10^6 = 0.009$ PJ;
- in Novosibirsk region $Q_{oil} = 0.4 \times 0.7 \times 10^3 \times 10^3 \times 44.0 \times 10^6 = 0.012$ PJ;
- in Tomsk region $Q_{oil} = 0.4 \times 11.5 \times 10^3 \times 10^3 \times 44.0 \times 10^6 = 2$ PJ.

The results of the conducted study are presented in Table 7. It can be concluded that industrial waste and MSW already accumulated and annually produced have a rather high level of energy potential. In the medium term, such waste can be used to satisfy the needs for energy resources when local coal-fired thermal power plants generate energy.

Table 7. Energy potential of each region due to accumulated and annually produced waste.

Energy Resources	Accumulated in Total		Annual Growth	
	G^0, Mt	Q^0, PJ	G, Mt	Q, PJ
Kemerovo region (No. 1)				
Filter cakes	368.8	5485	8.1	121
MSW	54.0	761	0.9	13
Petroleum products	-	-	0.0005	0.009
Novosibirsk region (No. 2)				
Filter cakes	5.8	86	0.4	6
MSW	70.0	830	1.3	15
Petroleum products	0.0973	4	0.0007	0.012
Tomsk region (No. 3)				
Filter cakes	-	-	-	-
MSW	21	209	0.4	2
Petroleum products	2.2	98	0.0115	2
Overall data for three regions				
Filter cakes	374.6	5571	8.5	127
MSW	145.0	1800	2.6	30
Petroleum products	2.3	102	0.0127	2

3.4. Need for Energy Resources of Coal-Fired Thermal Power Engineering

The regions under study (Figure 6) receive their auxiliary heat and electricity supply primarily from local thermal power plants. In the Kemerovo and Novosibirsk regions, bituminous coals extracted at local deposits are the main fuel, whereas natural gas is a backup fuel. In the Tomsk region, the natural gas extracted locally is the main energy resource. Despite this, all the boilers were initially designed and constructed to burn coals. Their modernization and conversion to natural gas was performed after the commercial extraction of the latter was started. The main specifications of thermal power plants are presented in Table 8.

Combined thermal and electric power generation in Kemerovo region is realized at three plants. Smaller boiler plants (about 100 of them), whose total installed heat power is 904 Gcal/h, provide heat mainly to housing and utilities sector, as well as smaller industrial enterprises. There are four main sources of heat and electricity supply in Novosibirsk region. The total installed capacity of boiler plants (over 200 of them) is 3733 Gcal/h. About 93% of this power is generated by boiler plants using natural gas, the rest is generated by coal-fired boiler plants and those using fuel oil. The main sources of heat and electricity supply in Tomsk are three thermal power plants. The total installed capacity of boiler plants (52 of them) is 762 Gcal/h. About 81% of this power is generated by boiler plants using natural gas, the rest is generated by coal-fired boiler plants and those using fuel oil. The main characteristics of coals used as fuel at thermal power plants are presented in Table 9. These are primarily bituminous coals of Kuznetsk and Kansk-Achinsk Basins.

The simultaneous analysis of data from Tables 8 and 9 makes it possible to calculate the amount of coal used by thermal power plants and boiler plants of the three regions to produce heat and electricity. The annual consumption of high-grade coal is about 10,216 kt (Kemerovo region—2664 kt, Novosibirsk region—7149 kt, and Tomsk region—403 kt). The combustion of this amount of coal releases about 217 PJ of heat. This energy is converted into electricity and heat with due consideration of the 70% efficiency of thermal power plants. In the conversion process, the low efficiency of thermal power plants causes irrecoverable losses of a rather large amount of energy released during high-grade coal combustion. A high-grade fuel is used inefficiently. Replacing coal with a composite fuel from coal processing waste (or from low-quality coal), MSW, used oils (or combustible oil production and oil refining waste) will reduce the consumption of high-grade non-renewable fossil fuels.

Table 8. Specifications of thermal power plants [5].

Thermal Power Plant (TPP)	Installed Capacity	Installed Heat Power	Electric Power Generation	Thermal Power Generation	Specific Consumption of Fuel Equivalent per Electric Power Supply	Specific Consumption of Fuel Equivalent per Thermal Power Supply	Fuel Balance			Produced Thermal Power from Burning Coals in Boiler Furnace
							Coal	Gas	Fuel Oil	
	MW	Gcal/h	GW·h	Tcal	g of Fuel Equivalent/kW·h	kg of Fuel Equivalent/Gcal	%	%	%	PJ
Kemerovo region (No. 1)										
TPP 1	485	1540	2053	2524	362	152	76.8	22.8	<0.1	25
TPP 2	80	749	175	725	383	168	91.3	8.7	0	5
TPP 3	565	1449	1946	2911	370	160	99.4	0.5	<0.1	35
Novosibirsk region (No. 2)										
TPP 1	345	920	1122	1899	328.5	146.8	91.1	8.8	0.1	17
TPP 2	511.5	1115	2246	2390	293.5	143.6	99.7	0	0.3	29
TPP 3	384	1120	1363	2251	314.8	143.8	80.9	19.1	0	18
TPP 4	1200	2730	7065	4652	292.9	138	99.5	0.1	0.4	79
Tomsk region (No. 3)										
TPP 1	14.7	795	3.3	651	326	138	0	99.9	<0.1	0
TPP 2	140	780	746	1694	268.5	131	0	99.9	<0.1	0
TPP 3	331	815	1104	2154	249	159.6	47.6	52.4	0	9
Overall data for three regions										
-	4056	12,013	17,823	21,851	-	-	-	-	-	217

Table 9. Characteristics of coals used for power generation [5].

Coal-Fired TPP	Coal	$Q^a_{s,V}$, MJ/kg	W^a, %	A^d, %	V^{daf}, %
Kemerovo region (No. 1)					
TPP 1	Low-caking coal	21.8	13.7	17.2	27
TPP 2	Long-flame coal	21.8	14.7	15.4	42.3
TPP 3	Flame coal	21.4	14.1	13.1	41.2
Boiler plants	Flame coal	21.4	14.1	13.1	41.2
Novosibirsk region (No. 2)					
TPP 1	Low-caking and nonbaking coals	24.7	8.3	16.9	23.9
TPP 2	Low-caking and nonbaking coals	22.1	17.5	15.5	19.1
	Brown coal	14.8	35.7	7.6	46.4
TPP 3	Low-caking and nonbaking coals	25	9	15.7	19
TPP 4	Gas coal and flame coal	21.6	14.7	13.1	41.4
Boiler plants	Low-caking and nonbaking coals	24.9	8.3	16.9	23.9
	Brown coal	14.8	35.7	7.6	46.4
	Low-caking and nonbaking coals	21.8	14.7	13.1	41.4
Tomsk region (No. 3)					
TPP 3	Flame coal	21.4	14.1	13.1	41.2
Boiler plants	Flame coal	21.2	13.9	14.5	44.7
	Long-flame coal	21.8	12	15.9	48

3.5. Strategy of Combined Recovery of Industrial Waste and MSW with Power Generation

Tables 7 and 8 present data on the energy potential of industrial waste and MSW, as well as on the amount of thermal energy produced from burning coal in boiler furnaces of thermal power plants. The comparison of respective characteristics enables to draw a conclusion about the prospects of using composite fuels from coal processing waste, low rank coals, MSW, used oils, combustible oil production and oil refining waste, whose annual production and stocks will, in the medium term (20–30 years), satisfy the needs of coal-fired thermal power engineering of the three regions (Figure 6) for energy resources for 100%.

The strategy of combined recovery of industrial and municipal waste by burning it as part of composite fuels implies the following. All the energy (about 217 PJ) (Table 8) generated by coal-fired thermal power plants will be produced by burning a composite fuel. A typical composition of such fuel is 85% of filter cakes (or a mixture of a low rank coal with water) + 10% of MSW + 5% of used oil. The replacement of coal by an amount of a composite fuel with the equivalent energy output will require in the first year (Table 10): 11.13 Mt of filter cakes; 1.31 Mt of MSW; and 0.65 Mt of used oil. The global forecast for the growth of energy consumption presumes that the amount of energy produced from fuel combustion and necessary for heat and electricity generation will rise by 1% every year (Table 10). The consumption of components used to prepare composite fuels will rise accordingly. According to data in Table 10, a combustible liquid is, in the medium term, a limiting component for the preparation of such composite fuel. After the first four years of implementing the proposed energy program, all the accumulated used oil or liquid combustible waste of oil production and oil refining would be completely recovered, whereas its annual production would not cover the necessary requirements for fuel preparation. Starting with the fifth year, composite fuels should include 85% of filter cakes and 15% of MSW. The annual need for components of such fuel will be at least 11.5 Mt of filter cakes and at least 2.0 Mt of MSW. Table 10 presents by year the composition of the fuel and consumption of each of its components in the implementation of the proposed strategy of waste recovery in the conditions of annual energy consumption growth by 1% vs. the previous year.

Table 10. Component consumption for composite fuel preparation.

Year	Energy from Fuel Combustion, PJ	Coal Consumption, Mt	Composite Fuel Consumption (by Component)			
			Filter Cakes, Mt	MSW, Mt	Oil, Mt	Total, Mt
1	217.0	10.22	11.13	1.31	0.65	13.10
2	219.2	10.32	11.24	1.32	0.66	13.23
3	221.4	10.42	11.36	1.34	0.67	13.36
4	223.6	10.53	11.47	1.35	0.67	13.49
5	225.8	10.63	11.58	2.04	-	13.63
6	228.1	10.74	11.70	2.06	-	13.77
7	230.3	10.84	11.81	2.08	-	13.90
8	232.7	10.96	11.94	2.11	-	14.04
9	235.0	11.06	12.05	2.13	-	14.18
10	237.3	11.17	12.17	2.15	-	14.32
11	239.7	11.29	12.30	2.17	-	14.47
12	242.1	11.40	12.42	2.19	-	14.61
13	244.5	11.51	12.54	2.21	-	14.76
14	247.0	11.63	12.67	2.24	-	14.91
15	249.4	11.74	12.79	2.26	-	15.05
16	251.9	11.86	12.92	2.28	-	15.20
17	254.4	11.98	13.05	2.30	-	15.35
18	257.0	12.10	13.18	2.33	-	15.51
19	259.6	12.22	13.32	2.35	-	15.67
20	262.2	12.34	13.45	2.37	-	15.82
21	264.9	12.47	13.59	2.40	-	15.99
22	267.4	12.59	13.72	2.42	-	16.14
23	270.1	12.72	13.86	2.45	-	16.30
24	272.8	12.84	13.99	2.47	-	16.46
25	275.5	12.97	14.13	2.49	-	16.63
Total:	6129	288.55	314.39	52.82	2.66	369.87

The 25-year implementation of the energy program will ensure the recovery (Table 10) of 314.39 Mt of filter cakes, 52.82 Mt of MSW, and 2.66 Mt of used oil. According to data in Table 7, the proposed measures will completely resolve the issue of recovering the used oils, liquid combustible oil production and oil refining waste, accumulated by 2017, and annually produced coal processing waste, as well as reduce the amount of filter cakes accumulated by 2017 by 84%. Moreover, adding MSW to composite fuels will eliminate the problem of its disposal until the transition has been effected to a new system of waste management with a high share of MSW recovery and re-use, and reduce the volumes of MSW that have been accumulated by 2017 at landfill sites by 36%.

The proposed strategy of combined recovery of industrial and municipal waste by burning it as part of composite fuels has several main positive effects:

(1) Saving on high-grade solid fossil fuels (more than 280 Mt over 25 years) due to reducing their consumption by thermal power engineering through the replacement by composite fuels in the amount equivalent by energy performance indicators (about 315 Mt).
(2) Reducing environmental pollution due to solid waste disposal as part of an environmentally friendly electricity and heat production technology.
(3) Reducing the intensity of landfill site area growth due to scheduled disposal of municipal solid waste.
(4) Efficient investment of financial means saved by reducing energy resource acquisition costs into the development of cutting-edge technologies in commercial thermal power engineering and modernization of thermal power plants.

Experimental research findings [19] enable one to conclude that the main environmental characteristics (CO, CO_2, NO_x, SO_x, and micron-sized ash fraction) of flue gases when burning composite fuels prepared from low rank coals and coal processing waste, water and a combustible liquid are not inferior to those of flue gases from coal dust combustion using the conventional technology of coal-fired thermal power plants. A possibility to vary raw materials for composite fuels in a wide range enables to develop fuel compositions with predictable energy, economy and environmental performance characteristics.

4. Conclusions

(1)   The theoretical analysis showed that implementing the strategy of combined recovery of industrial and municipal waste by burning it as part of composite fuels at local thermal power plants is a promising approach for neighboring regions to deal with the waste management issue. One of the regions is characterized by a high level of solid and liquid fossil fuel extraction. In the neighboring regions, fossil fuel mining is underdeveloped, whereas the level of social advancement and industrial output is high. The energy potential of industrial waste and MSW determines the prospects of its recovery by means of burning it as part of composite fuels. Burning technologies of composite fuels is characterized by positive economic and environmental effects. That is why consistent patterns and necessary conditions have been experimentally discovered for the ignition of typical composite fuel consists of filter cakes with 10% of typical MSW (rubber, or wood, or plastic) and 5% of used motor oil under heating conditions similar to those of fuel combustion in boiler furnaces.

(2)   The fuel compositions under study (85% of filter cakes + 10% of MSW + 5% of used oil) were used as examples to experimentally validate sustainable ignition and combustion of composite fuel droplets up to their complete burnout in the conditions typical of conventional boiler furnaces. A software system for high-speed video recording outlined the main interdependent stages of process: heating; moisture evaporation; thermal decomposition of solid combustible components (MSW and coal); formation of combustible mixture; gas-phase ignition and burnout; heating of the solid combustible residue; heterogeneous ignition and combustion of the latter.

(3)   The values of the guaranteed ignition delay times for droplets with a size (diameter) of about 2 mm have been established for the composite fuel compositions under study in a wide range of the ambient temperature variation 600–1000 °C. The minimum values of ignition delay times are about 3 s, the maximum values are about 15 s. The maximum difference in ignition delay times of fuel compositions with different components is less than 20% at the ambient temperature 600–1000 °C. An approximation equation $t_d = f(T_g)$ has been derived for a rapid evaluation of mean ignition delay time values of composite fuel droplets (filter cakes 85% + MSW 10% + oil 5%).

The obtained results serve as a foundation for the development of joint routines by neighboring regions (those with a high level of solid and liquid fossil fuel extraction, and those with a high level of social advancement) to implement high-potential technologies of burning composite fuels prepared from industrial and municipal waste at thermal power plant. One of the benefits of such technology is a possibility to optimize the modes of the main process equipment operation by varying the compositions of composite fuels and the concentration of combustible components. The new technology is the basis for countries not recycling MSW other than by burial at disposal sites to implement a waste management strategy. During the 15–25 years of transition from waste burial to re-use and recycling, it is essential to minimize the volumes of non-recycled waste disposal. Within this time, the proposed waste management strategy will, on the one hand, reduce the amount of waste to be buried and decrease the consumption of high-grade coal for heat and electricity generation, and on the other hand, will eliminate interim steps involving the construction of costly incineration plants which will be in demand for a relatively short period of time. The analysis of three regions of the Russian Federation, taken as an example, established that complete replacement of coal by a composite fuel with an equivalent energy output will save 10 Mt of solid fossil fuels a year in the course of 25 years (during the period of transition to a new system of waste management with a high share of recovery and re-use of MSW). During the same time interval, about 315 Mt of filter cakes, 53 Mt of MSW, and about 2.6 Mt of used oil will have been disposed of. The proposed measures will completely resolve the issue of

Energies **2018**, *11*, 2534

recovering the used oils, liquid combustible oil production and oil refining waste, accumulated by 2017, and annually produced coal processing waste, as well as reduce the amount of filter cakes accumulated by 2017 by 84%. Moreover, adding MSW to the fuel composition will eliminate the problem of its disposal until the transition has been effected to a new system of waste management with a high share of MSW recovery and re-use, and reduce the volumes of MSW that have been accumulated by 2017 at landfill sites by 36%.

Author Contributions: D.G. conceived and designed the experiments and wrote the paper; G.K. analyzed the data and wrote the paper; K.P. performed the calculations; D.S. performed the experiments.

Funding: This research was funded by Russian Foundation for Basic Research and the government of the Tomsk region of the Russian Federation, grant number [18-43-700001].

Conflicts of Interest: The authors declare no conflict of interest.

References

1. Lishtvan, I.I.; Dudarchik, V.M.; Kraiko, V.M.; Anufrieva, E.V.; Smolyachkova, E.A. Utilization of polymer wastes by joint pyrolysis with peat to produce high-calorific gas. *Solid Fuel Chem.* **2017**, *51*, 273–277. [CrossRef]
2. Rizwan, M.; Saif, Y.; Almansoori, A.; Elkamel, A. Optimal processing route for the utilization and conversion of municipal solid waste into energy and valuable products. *J. Clean. Prod.* **2018**, *174*, 857–867. [CrossRef]
3. Ministry of Natural Resources and Environment of the Russian Federation. Available online: www.mnr.gov. ru/docs/o_sostoyanii_i_ob_okhrane_okruzhayushchey_sredy_rossiyskoy_federatsii/gosudarstvennyy_ doklad_o_sostoyanii_i_ob_okhrane_okruzhayushchey_sredy_rossiyskoy_federatsii_v_2016_/ (accessed on 13 February 2018). (In Russian)
4. Anifowose, B.A.; Odubela, M.T. Oil facility operations: A multivariate analyses of water pollution parameters. *J. Clean. Prod.* **2018**, *187*, 180–189. [CrossRef]
5. The Federal Service for Supervision in the Sphere of Nature Management. Available online: rpn.gov.ru/ sites/all/files/users/rpnglavred/filebrowser/docs/doklad_po_tbo.pdf (accessed on 26 February 2018). (In Russian)
6. Touš, M.; Pavlas, M.; Putna, O.; Stehlík, P.; Crha, L. Combined heat and power production planning in a waste-to-energy plant on a short-term basis. *Energy* **2015**, *90*, 137–147. [CrossRef]
7. Sun, R.; Ismail, T.M.; Ren, X.; Abd El-Salam, M. Numerical simulation of gas concentration and dioxin formation for MSW combustion in a fixed bed. *J. Environ. Manag.* **2015**, *157*, 111–117. [CrossRef] [PubMed]
8. Song, J.; Song, D.; Zhang, X.; Sun, Y. Risk identification for PPP waste-to-energy incineration projects in China. *Energy Policy* **2013**, *61*, 953–962. [CrossRef]
9. Faitli, J.; Magyar, T.; Erdélyi, A.; Murányi, A. Characterization of thermal properties of municipal solid waste landfills. *Waste Manag.* **2015**, *36*, 213–221. [CrossRef] [PubMed]
10. Tabakaev, R.; Shanenkov, I.; Kazakov, A.; Zavorin, A. Thermal processing of biomass into high-calorific solid composite fuel. *J. Anal. Appl. Pyrolysis* **2017**, *124*, 94–102. [CrossRef]
11. Dmitrienko, M.A.; Strizhak, P.A.; Tsygankova, Y.S. Technoeconomic analysis of prospects of use of organic coal-water fuels of various component compositions. *Chem. Pet. Eng.* **2017**, *53*, 195–202. [CrossRef]
12. Nyashina, G.S.; Legros, J.C.; Strizhak, P.A. Environmental potential of using coal-processing waste as the primary and secondary fuel for energy providers. *Energies* **2017**, *10*, 405. [CrossRef]
13. Valiullin, T.R.; Vershinina, K.Y.; Glushkov, D.O.; Shevyrev, S.A. Droplet ignition of coal–water slurries prepared from typical coal- and oil-processing wastes. *Coke Chem.* **2017**, *60*, 211–218. [CrossRef]
14. Glushkov, D.O.; Strizhak, P.A.; Chernetskii, M.Y. Organic coal-water fuel: Problems and advances (Review). *Therm. Eng.* **2016**, *63*, 707–717. [CrossRef]
15. Ilinykh, G.V. Evaluation of thermotechnical properties of solid waste from their morphological composition. *Vestnik Permskogo natsionalnogo issledovatelskogo politekhnicheskogo universiteta. Urbanistika* **2013**, *3*, 125–137. (In Russian)
16. Bazyn, T.; Krier, H.; Glumac, N. Combustion of nanoaluminum at elevated pressure and temperature behind reflected shock waves. *Combust. Flame* **2006**, *145*, 703–713. [CrossRef]

17. Glushkov, D.O.; Shabardin, D.P.; Strizhak, P.A.; Vershinina, K.Yu. Influence of organic coal-water fuel composition on the characteristics of sustainable droplet ignition. *Fuel Process. Technol.* **2016**, *143*, 60–68. [CrossRef]

18. Mishra, D.P.; Patyal, A.; Padhwal, M. Effects of gellant concentration on the burning and flame structure of organic gel propellant droplets. *Fuel* **2011**, *90*, 1805–1810. [CrossRef]

19. Glushkov, D.O.; Paushkina, K.K.; Shabardin, D.P.; Strizhak, P.A. Environmental aspects of converting municipal solid waste into energy as part of composite fuels. *J. Clean. Prod.* **2018**, *201*, 1029–1042. [CrossRef]

Article

Activation of the Fuels with Low Reactivity Using the High-Power Laser Pulses

Roman I. Egorov [1,†], Alexandr S. Zaitsev [1,†] and Eugene A. Salgansky [2,*,†]

1   National Research Tomsk Polytechnic University, 30 Lenin Avenue, 634050 Tomsk, Russia;
     rommel@tpu.ru (R.I.E.); alexzaitsev@tpu.ru (A.S.Z.)
2   Institute of Problems of Chemical Physics, Russian Academy of Sciences, 1 Acad. Semenov Avenue,
     142432 Chernogolovka, Russia
*   Correspondence: sea@icp.ac.ru; Tel.: +7-496-522-1368
†   These authors contributed equally to this work.

Received: 8 October 2018; Accepted: 13 November 2018; Published: 15 November 2018

Abstract: In this paper we have proposed the simple and effective approach to activation of the low reactivity industrial fuel which can be used immediately inside the furnace. The high-power laser pulses initiates partial gasification of the fuel together with its ultra-fine atomization. The gas-aerosol cloud surrounding the initial coal-water slurry droplet can consist of approximately 10% (after absorption of hundred pulses) of the initial droplet weight. The ratio of the syngas and aerosol weights is like 1:2 when pulse intensity is higher than 8 J/cm^2. The size and velocity distributions of the ultra-fine aerosol particles were analysed using the original realization of the particle tracking velocimetry technique.

Keywords: fuel activation; waste-derived fuel; coal-water slurry; laser pulse; syngas; aerosol

---

1. Introduction

An ignition delay time is one of the most important parameters of industrial fuels which makes direct influence onto the design of the furnaces. Numerous investigations are targeted on the optimization of the ignition of industrial fuels [1,2]. Decreasing the ignition delay time of the popular fuels, one can make a step to the smaller heaters or to an essential simplification of the ignition process [3]. This, in turn, will decrease the price of heaters and, therefore, the final price of the heat.

There are some different approaches for ignition optimization: switching to more appropriate temperatures by the usage of the furnace pre-heating and/or fuel activation by different ways (using the finer fuel atomization or introduction of special additions which simplifies an ignition, etc). All mentioned techniques have own benefits and weak points and can be used together for the best effect.

However, most of the ignition-related problems become much sharper when used the cheap waste-derived fuels [4]. Such fuels typically have the very low level of the reactivity at the low temperatures [5]. This automatically requires higher furnace temperatures for a stable ignition. However, even in this case, the too long ignition delay time can make them non-appropriate for most of the existing boilers. There are two real ways for fuel adaptation: decreasing of the fuel particle size during atomization and addition of any ignition stimulating chemicals. Both of these ways have limitations (rheological or commercial) for traditional approaches [6,7]. Therefore, the simple method allowing the fuel activation with minimal changes of the furnace design will be always of interest. It is clear too, that the most effective way should include complex influence on the fuel. Methods combining the finer atomization together with pre-heating, intensification of the volatiles release or fuel gasification are always of interest.

In this paper we suggest the new approach to the pre-ignition processing of the fuel droplets by the high-power laser pulses [8–10]. This leads to two main effects *combined altogether*: from the one hand, laser pulse initiates an additional fine atomization of the fuel droplet. From the other hand, it leads to partial gasification [11–13] of the fuel, producing the cloud of combustible gas around the initially injected fuel droplet. Both these effects transform certain part of the fuel matter into the highly reactive state (micrometer scale particles and volatiles have much lower ignition delays times). The numerous investigations about ignition of different fuels [14–16] make a proof that increasing the volatiles release rate together with increase of the atomization quality always stimulate an ignition. In principle, the proposed method can work with different types of non-transparent liquid or solid fuels. An activation occurs due to mechanical and thermal effect of light pulses and, therefore, it is independent enough of its wavelength.

The proposed approach can be applied immediately inside the furnace and it means that the fuel activation will occur before ignition. That is the fuel can be stored for a long term in natural low reactivity state satisfying the fire safety requirements. The light-induced fuel activation does not require any modification of the chemical content of the fuel as well as expensive modifications of the fuel injection systems. All what is needed is a small side hole for laser beam input.

2. Materials and Methods

We have prepared the fuel composition with low reactivity whose ignition parameters are more or less clear at this moment from the previous investigations [1,17]. The waste-derived coal-water slurry (CWS) was made of the fiter cake of fiery coal (particle size ~140 μ) [17]. The water content of the slurry was ~40 wt %. The coal powder shows strong *hydrophilic* behaviour. The static contact angle of water droplet on the surface of the pressed dry filter cake is ~130° (measured by the sessile droplet method [18,19]). The powder mixed with water form a slurry which looks like a black viscous substance (density 1.5–1.54 kg/m^3, viscosity 10–15 mPa· s) which can absorb most part of an incident light of visible range.

The surface structure of the coal particles (in the dry state pores contain up to 30 wt % of bound volatiles) as well as a lot of flocculants on the particle surface (introduced during the coal enrichment by floatation) determine the very good adhesion between the solid and liquid components of the slurry. Thus, there is not essential natural separation of the liquid and solid components before the fuel atomization. We have analysed the effect of laser pulses using the single fuel droplet fixed by the special holder under the laser irradiation.

Each laser pulse ($\lambda = 533$ nm, pulse duration ~10 ns, light intensity inside the spot $E_{spot} = 12$ J/cm^2, pulse repetition rate 2.5 Hz) causes the micro-explosions on the surface of slurry droplet (d~2 mm). Thus, the solid slurry matter partially ablates [8,9] as well as some of water evaporates after the pulse came. Additionally, some of the matter fly out of the sample surface in the form of ejected micro droplets. The experimental setup is shown in Figure 1.

The CW-laser (power is 300 mW at 533 nm) was used for illumination of the flow of micro explosion products by the astigmatic beam having ~ 10° angle with the propagation direction of the pulsed beam. The illuminated particles, flying in the plane of the CW-beam, are observed by the high-speed video camera Phantom V411 (800 × 600 px. and 10,000 fps) with macro optics. This realization of the particle tracking velocimetry technique [20,21] allows easy gathering of the particle size and velocity distributions as well as visualization of the trajectory of gas/aerosol cloud. The overall spatial resolution of the PTV system was at the level of 7 μm.

The corresponding changes in the sample weight are measured by the analytic balances Vibra AF 225DRCE. The sample droplet of the slurry was placed on the plate of the balances specially for monitoring of the gasification efficiency. The chemical composition of the atmosphere inside the reaction volume was analysed by the gas analyser Boner Test-1 adapted for measurements of the low concentrations of gases.

Usually we recorded fifty videos showing interaction of light pulses with a fuel to accumulate the statistics for size and velocity distributions of the aerosol particles. All measurements (including the control of weights of atomized matter and syngas chemical content) were repeated three times to avoid the influence of any random factors on the results. The statistical distributions, typical travel distances for aerosol cloud, syngas weights and concentrations show the variation of key values for less than 5%.

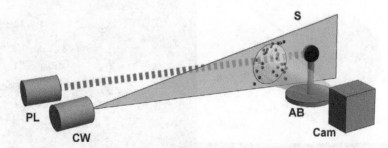

Figure 1. Scheme of the experimental setup including the nanosecond pulsed laser (PL), CW-laser (CW), analytic balances platform (AB), high-speed camera (Cam), sample of the fuel (S).

3. Interaction of The Laser Pulse with Slurry Fuel

An interaction of the focused light with matter can occur by two principally different ways. The continuous wave light absorbed by the sample surface trivially heats the sample gradually increasing its temperature with time. The high-power laser pulses interact with a matter by another way. In this work, we sent the 10 ns pulses with 100 mJ energy toward the sample surface (an illuminated area with radius like 0.5 mm). Each pulse causes the micro explosion on the fuel surface producing the ejection of micro particles (both water droplets and wet coal). The cloud of these particles propagates along the pulse path in backward direction with a certain transverse dispersion. The video of this process is shown in Supplementary Materials.

As one can see there are two types of objects: the sharp fast spots are mentioned ejected particles and the slow diffuse spots are the clouds of gasified volatiles. The high-power nanosecond pulse increases the temperature on the surface of coal particle up to 2000–2500 K [22,23]. Of course, such extreme state has a duration comparable with a pulse duration with further fast cooling. The depth of an extremal heating is like some micrometers. Therefore, the small part of the matter is ablated producing the mixture of the water steam and gaseous oxides of solid part of the fuel. The chemical content of the used filter cake (as well as for most of coals) allows effective production of CO, SO_2, different nitrogen oxides and a little of methane. The CO_2 usually is absent when coal particles are wet as it is shown in our previous investigations [11,12]. The massive release of water steam from the fuel sample simultaneously with carbon ablation leads to decrease of the partial pressure of oxygen near the place of carbon evaporation. This, in turn, suppresses the production of the CO_2. However, the contribution of water is much wider. When the temperature on the surface of coal particle overcomes the certain value (typically it have to be approx. 1000 K or higher) water starts the partial oxidation of carbon according to reaction: $H_2O + C \rightarrow H_2 + CO$.

It worth to note, that typical CWS contains less than 50 wt % of water (due to the rheological requirements [4,15]) and, thus, this reaction goes in our case with an evident deficit of water. It explains the fact that production of hydrogen is one order magnitude less than production of CO (which has an additional channel of production due to the carbon oxidation by the oxygen from air).

The Figure 2b–d shows some video frames with propagating gas-aerosol cloud which correspond to some moments after pulse-fuel interaction. As one can see, the quasi-straightforward propagation of the cloud occurs during approx. 4–4.5 ms. After this time the center of mass of an ejected matter cloud almost stops at approximately 1 cm distance from the initial fuel portion. Further evolution of gas-aerosol cloud is a trivial diffusion in all around direction with slow decrease of volume concentration. Of course, this experiment is done in a steady state atmosphere whereas the realistic case is a fuel droplet flying in hot furnace. However, this simple case demonstrates the processes going around the fuel droplet under the effect of the laser pulses.

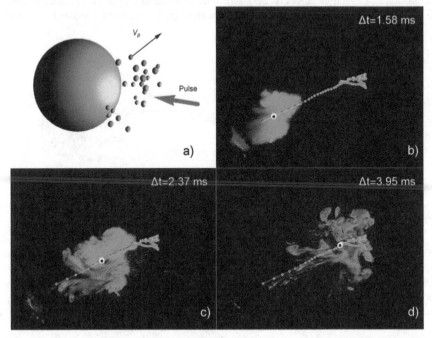

Figure 2. An interaction of the light pulse with an initial fuel droplet (**a**). The sequence of video frames showing the propagation of the gas-aerosol cloud (**b–d**). The green diffuse spot is a gas cloud, small green spots are ejected particles. The trajectory of the center of mass of cloud is shown by yellow line with red points (its position at moment Δ*t* is marked by the white ring).

The repetition rate of laser pulses was low enough to observe the effect of the single pulse. Typically the cloud was observable during less than 20 ms which is much less than inter-pulse time gap. However, the long series of pulses lead to an increase of the residual concentration of the finely atomized fuel around the initial droplet. As result, the cloud decay time increases up to 50 ms after absorption of 20–25 pulses in a row.

An estimation of the weight of the fuel which is typically atomized by the pulsed laser processing was done through monitoring of the weight of initial fuel droplet under the laser irradiation. The difference between initial and current weight of the droplet is an equal to the weight of finely atomized matter. This includes both the syngas and fuel aerosol. The production of an atomized matter with absorption of the laser pulses shown in Figure 3a. As one can see, the amount of ejected matter grows almost linearly with number of absorbed pulses. It achieves approximately 5% of initial fuel weight after hundred of the pulses when pulse intensity exceeds 8 J/cm². An evident gap between the curves corresponding to 7.9 J/cm² and higher intensity corresponds to contribution of the gasification. The Figure 3b shows the changes of the CO production with increase of the laser pulse intensity. The 8 J/cm² looks as an evident threshold value when an effective gasification starts. The CO is a main component of the syngas which appears under the effect of the high-power light flow [11,12].

Therefore, we can estimate that the ratio between the syngas and fuel aerosol is not so far from 1:2. This means that approximately 2 wt % of an initial fuel droplet transforms to the syngas after absorption of hundred of laser pulses as well as 4 wt % becomes the aerosol particles.

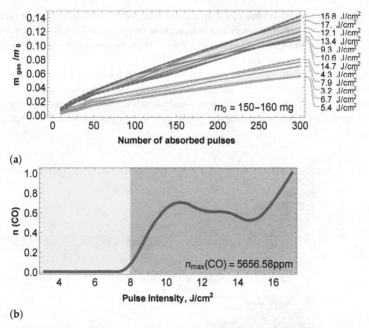

(a)

(b)

Figure 3. The dynamics of production of the finely atomized fuel under the effect of the laser pulses for different intensities (**a**). The dependence of the CO (main combustible component of the syngas) production efficiency (**b**) on the pulse intensity. The background in (**b**) shows the light intensity ranges below the threshold of an effective CO production (yellow) and above it (red).

Thus, we can conclude that an accumulation of the finely dispersed fuel (together with produced syngas) occurs around the initial fuel droplet. Changing the pulse repetition rate, we can variate the density of the finely atomized fuel. Taking in account the maximal distance of the cloud propagation, we can say that laser pulse processing effectively transforms each initial fuel droplet to a sphere (in steady state atmosphere) with 1 cm radius filled by a mixture of syngas, microscopic particles of coal and air. It is evident, that such objects are much better for a low-temperature ignition than an initial CWS droplets.

Parameters of the Obtained Aerosol

The dependence of ignition delay time on the droplet size for different types of the waste-derived coal-water slurry was analysed by Valiullin in [24,25]. It was shown that an ignition delay almost linearly decreases with fuel droplet size. The typical injection systems for industrial heaters allow fuel atomization to the average size like 1–5 mm. It corresponds to ignition delay times in range 3–15 s (at $T_{furnace}$ ~950 K) that is much longer than for dry coal powder. Such delays make the CWS almost useless for heaters whose design is developed for coal dust which has much shorter ignition delays.

The laser pulse makes ultra-fine atomization of the fuel in comparison with effect of usual nozzles for highly viscose fuels. The particle tracking velocimetry approach [26,27] allows easy analysis of the dynamic properties of an ejected matter. The size and velocity distributions of the produced aerosol particles are shown in Figure 4. One can see, that most-probable size of the finely atomized fuel particles is less than 35 μ. The most part of the produced aerosol particles have a sizes in range below 100 μ. The maximal sizes observed during the experiments are up to 280 μ. Such particle sizes are approximately one order of magnitude smaller than typically tested for ignition in numerous investigation. They potentially allow the ignition delay times in sub-second range for earlier mentioned furnace temperatures (~950 K).

The Figure 4b shows the velocity distribution detected in aerosol flow going from the place of pulse impact. The most probable velocity is in narrow range above 0.1 m/s. The average velocity is approx. 1.1 m/s and the maximal one is up to 3 m/s. Thus, the aerosol cloud gets totally stopped the linear propagation in average after 10 ms.

Therefore, the pulse repetition rate can be increased for up to 100 Hz to speed up the fuel processing within the general approach developed in this paper. Such a way allows creation a very dense layer of the volatiles around the initial fuel particles. These volatiles together with ultra-fine aerosol potentially paves the way to ignition delays fully comparable with those that present for dry coal dust.

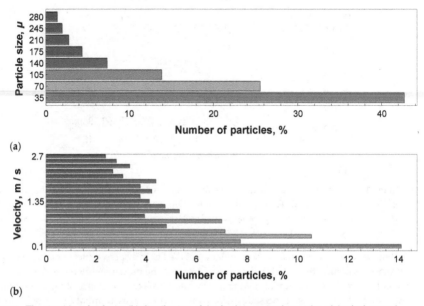

(a)

(b)

Figure 4. The size (**a**) and velocity (**b**) distribution of the finely atomized particles of the fuel aerosol.

Typically, the heterogeneous ignition of coal starts from the ignition of volatiles which are released during the coal heating. Our approach allows the fast conversion of 5–7 wt % of the CWS into the finely atomized form. Most of this are finely dispersed coal particles whose ignition delay time will be comparable with such parameter of volatiles. Increasing the repetition rate of the laser pulses we can increase the aerosol concentration for up to the order of magnitude using the commercially available laser modules. Additionally, the laser processing of the fuel inside the industrial systems can be realized as a propagation of the fuel particles through the sequence of parallel beams like in Figure 5. Such sequence can be cheap enough realized using the fiber-optics beam-splitters with one high-power laser module. Such a way gives a multiplication of the fuel atomization effect proportionally to the number of laser beams. Of course, different design of the laser processing stage can be successfully realized for different purposes. An application of the light pulses for an in-furnace fuel processing looks a convenient way for optimization of ignition and combustion of the low-reactivity fuels.

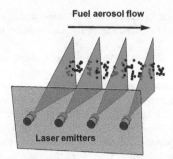

Figure 5. The sequential application of the laser processing for a flow of fuel aerosol from a usual injection nozzle.

4. Conclusions

We have proposed the simple technique that allows an effective increase of the amount of highly flammable fraction in the injected fuel flow inside a furnace. The high-power laser pulses interacting with injected CWS droplets with sizes that typically present in furnaces after injection nozzles allows formation of the ultra-fine fuel aerosol around initial droplets. Partial gasification of the coal by the light pulse makes an additional syngas shell which is fully similar to an addition of combustible volatiles. Typically, up to 5% of the droplet weight can be transformed to finely atomized state by nanosecond pulses with intensity like 8 J/cm^2 or higher.

Introduction of the aerosol and syngas clouds inside the furnace volume potentially allows essential acceleration of the fuel ignition. Physically it can work with different types of the solid or liquid fuels which optical density allows an effective absorption of the laser light.

The proposed approach can serve for in-furnace activation of the fuels which in general have the very low reactivity. Therefore, it can be the very practical way for adaptation of waste-derived fuels for the existing industrial furnaces.

In future we plan the detailed investigation of the physical basics of the observed processes. An influence of different factors like the fuel humidity onto the atomization properties as well as on syngas/aerosol ratio will be investigated.

Supplementary Materials: Supplementary materials are available online at http://www.mdpi.com/1996-1073/11/11/3167/s1.

Author Contributions: Conceptualization, R.I.E.; methodology, R.I.E. and A.S.Z.; formal analysis, E.A.S.; investigation, A.S.Z.; writing—original draft preparation, R.I.E. and E.A.S.; writing—review and editing, E.A.S.

Funding: This research was performed within the framework of the strategic plan for the development of National Research Tomsk Polytechnic University as one of the world-leading universities (project VIU-ISHFVP-299/2018).

Conflicts of Interest: The authors declare no conflict of interest.

References

1. Bhuiyan, A.A.; Naser, J. CFD modelling of co-firing of biomass with coal under oxy-fuel combustion in a large scale power plant. *Fuel* **2015**, *159*, 150–168. [CrossRef]

2. Wang, X.; Niu, B.; Deng, S.; Liu, Y.; Tan, H. Optimization study on air distribution of an actual agriculture up-draft biomass gasification stove. *Energy Procedia* **2014**, *61*, 2335–2338. [CrossRef]

3. Tan, H.; Niu, Y.; Wang, X.; Xu, T.; Hui, S. Study of optimal pulverized coal concentration in a four-wall tangentially fired furnace. *Appl. Energy* **2011**, *88*, 1164–1168. [CrossRef]

4. Wang, Z.; Hong, C.; Xing, Y.; Li, Y.; Feng, L.; Jia, M. Combustion behaviors and kinetics of sewage sludge blended with pulverized coal: With and without catalysts. *Waste Manag.* **2018**, *74*, 288–296. [CrossRef] [PubMed]

5. Di Sarli, V.; Cammarota, F.; Salzano, E. Explosion parameters of wood chip-derived syngas in air. *J. Loss Prev. Process Ind.* **2014**, *32*, 399–403. [CrossRef]

6. He, T.; Chen, Z.; Zhu, L.; Zhang, Q. The influence of alcohol additives and EGR on the combustion and emission characteristics of diesel engine under high-load condition. *Appl. Therm. Eng.* **2018**, *140*, 363–372. [CrossRef]

7. Belal, H.; Han, C.W.; Gunduz, I.E.; Ortalan, V.; Son, S.F. Ignition and combustion behavior of mechanically activated Al–Mg particles in composite solid propellants. *Combust. Flame* **2018**, *194*, 410–418. [CrossRef]

8. Chichkov, B.N.; Momma, C.; Nolte, S.; Von Alvensleben, F.; Tünnermann, A. Femtosecond, picosecond and nanosecond laser ablation of solids. *Appl. Phys. A Mater. Sci. Proc.* **1996**, *63*, 109–155. [CrossRef]

9. Stuart, B.C.; Feit, M.D.; Herman, S.; Rubenchik, A.M.; Shore, B.W.; Perry, M.D. Optical ablation by high-power short-pulse lasers. *JOSA B Opt. Phys.* **1996**, *13*, 459–468. [CrossRef]

10. Stancalie, A.; Ciobanu, S.S.; Sporea, D. Investigation of the effect of laser parameters on the target, plume and plasma behavior during and after laser-solid interaction. *Appl. Surf. Sci.* **2017**, *417*, 124–129. [CrossRef]

11. Egorov, R.I.; Strizhak, P.A. The light-induced gasification of waste-derived fuel. *Fuel* **2017**, *197*, 28–30. [CrossRef]

12. Zaitsev, A.S.; Egorov, R.I.; Strizhak, P.A. Light-induced gasification of the coal-processing waste: Possible products and regimes. *Fuel* **2018**, *212*, 347–352. [CrossRef]

13. Slyusarskiy, K.V.; Larionov, K.B.; Osipov, V.I.; Yankovsky, S.A.; Gubin, V.E.; Gromov, A.A. Non-isothermal kinetic study of bituminous coal and lignite conversion in air and in argon/air mixtures. *Fuel* **2017**, *191*, 383–392. [CrossRef]

14. Bhargava, S.K.; Tardio, J.; Prasad, J.; Föger, K.; Akolekar, D.B.; Grocott, S.C. Wet oxidation and catalytic wet oxidation. *Ind. Eng. Chem. Res.* **2017**, *45*, 1221–1258. [CrossRef]

15. Sazhin, S.S. Modelling of fuel droplet heating and evaporation: Recent results and unsolved problems. *Fuel* **2017**, *196*, 69–101. [CrossRef]

16. Lishtvan, I.I.; Falyushin, P.L.; Smolyachkova, E.A.; Kovrik, S.I. Fuel suspensions based on fuel oil, peat, waste wood, and charcoal. *Solid Fuel Chem.* **2009**, *43*, 1–4. [CrossRef]

17. Iegorov, R.I.; Strizhak, P.A.; Chernetskiy, M.Y. The review of ignition and combustion processes for water-coal fuels. In Proceedings of the European Physical Journal Conferences, Tomsk, Russia, 13–15 October 2015. [CrossRef]

18. Kuznetsov, G.V.; Feoktistov, D.V.; Orlova, E.G. Regimes of Spreading of a Water Droplet Over Substrates with Varying Wettability. *J. Eng. Phys. Thermophys.* **2016**, *89*, 317–322. [CrossRef]

19. Zahiri, B.; Sow, P.K.; Kung, C.H.; Mérida, W. Understanding the wettability of rough surfaces using simultaneous optical and electrochemical analysis of sessile droplets. *J. Colloid Interface Sci.* **2017**, *501*, 34–44. [CrossRef] [PubMed]

20. Damiani, D.; Meillot, E.; Tarlet, D. A Particle-tracking-velocimetry (PTV) investigation of liquid injection in a dc plasma jet. *J. Therm. Spray Technol.* **2014**, *23*, 340–353. [CrossRef]

21. Hadad, T.; Gurka, R. Effects of particle size, concentration and surface coating on turbulent flow properties obtained using PIV/PTV. *Exp. Therm. Fluid Sci.* **2013**, *45*, 203–212. [CrossRef]

22. Murastov, G.; Tsipilev, V.; Ovchinnikov, V.; Yakovlev, A. About laser heat absorbing impurities in the transparence matrix of pentaerythritol tetranitrate. *J. Phys. Conf. Ser.* **2017**, *830*. [CrossRef]

23. Burkina, R.S.; Morozova, E.Y.; Tsipilev, V.P. Initiation of a reactive material by a radiation beam absorbed by optical heterogeneities of the material. *Combust. Explos. Shock Waves* **2011**, *47*, 581–590. [CrossRef]

24. Valiullin, T.R.; Strizhak, P.A.; Shevyrev S.A. Low temperature combustion of organic coal-water fuel droplets containing petrochemicals while soaring in a combustion chamber model. *Therm. Sci.* **2017**, *21*, 1057–1066. [CrossRef]

25. Valiullin, T.R.; Strizhak P.A. Influence of the shape of soaring particle based on coal-water slurry containing petrochemicals on ignition characteristics. *Therm. Sci.* **2017**, *21*, 1399–1408. [CrossRef]

26. Volkov, R.S.; Kuznetsov, G.V.; Strizhak, P.A. Influence of droplet concentration on evaporation in a high-temperature gas. *Int. J. Heat Mass Transf.* **2016**, *96*, 20–28. [CrossRef]

27. Young, C.N.; Johnson, D.A.; Weckman, E.J. A model-based validation framework for PIV and PTV. *Exp. Fluids* **2004**, *36*, 23–35. [CrossRef]

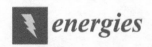

Article

Impact of Holder Materials on the Heating and Explosive Breakup of Two-Component Droplets

Dmitry Antonov [1,*], Jérôme Bellettre [2], Dominique Tarlet [2], Patrizio Massoli [3], Olga Vysokomornaya [1] and Maxim Piskunov [1]

[1] National Research Tomsk Polytechnic University, Tomsk 634050, Russia; vysokomornaja@tpu.ru (O.V.); piskunovmv@tpu.ru (M.P.)
[2] Université de Nantes, Rue Christian Pauc, BP 50609, 44306 Nantes CEDEX 3, France; Jerome.bellettre@univ-nantes.fr (J.B.); dominique.tarlet@univ-nantes.fr (D.T.)
[3] Istituto Motori–Consiglio Nazionale delle Ricerche, Via Marconi 8, 80125 Napoli, Italy; p.massoli@im.cnr.it
* Correspondence: antonovdv132@gmail.com; Tel.: +7-(913)-879-43-88

Received: 25 October 2018; Accepted: 23 November 2018; Published: 27 November 2018

Abstract: The heating of two-component droplets and the following explosive breakup of those droplets have been extensively studied over the most recent years. These processes are of high interest, since they can significantly improve the performance of many technologies in fuel ignition, thermal and flame liquid treatment, heat carriers based on flue gases, vapors and water droplets, etc. Research throughout the world involves various schemes of droplet heating and supply (or, less frequently, injection) to heating chambers. The most popular scheme features the introduction of a two-component or multi-component droplet onto a holder into the heating chamber. In this research, we study how holder materials affect the conditions and integral characteristics of droplet heating and explosive breakup: heating time until boiling temperature; minimum temperature sufficient for droplet breakup; number and size of fragments in the resulting droplet aerosol, etc. Experiments involve droplets that are produced from flammable (oil) and non-flammable (water) components with significantly different thermophysical and optical properties, as well as boiling temperature and heat of vaporization. The most popular elements with the scientific community, such as ceramic, steel, aluminum, copper, and phosphorus rods, as well as a nichrome wire, serve as holders. We establish the roles of energy inflow from a holder to a droplet, and energy outflow in the opposite direction. We compare the holder results with a supporting thermocouple, recording the drop temperature under a heat transfer provided at 350°C. Finally, we forecast the conditions that are required for a significant improvement in the performance of thermal and flame water treatment through the explosive breakup of two-component droplets.

Keywords: two-component droplet; heating; evaporation; explosive breakup; disintegration; droplet holder material

---

1. Introduction

1.1. Motivation

To improve the thermal treatment of sewage and service water (in particular, in the form of an atomized flow) and to develop new, more effective technologies for it, we need to explore the physics of droplets of water solutions, slurries, and emulsions traveling through high-temperature gases. Their temperature exceeds 500 °C, and the most frequently used gases are hot air, fuel combustion products, and their mixtures. Unfortunately, there is still no theory of interconnected heat and mass transfer and phase transformations for such conditions. However, over the recent years, researchers have obtained experimental results (e.g., [1–3]) that can become the premises for such a theory. No research findings on these processes have been published so far, because mathematical modeling becomes difficult for a large number of interfaces with highly nonlinear boundary conditions of rapid vaporization. Sazhin [4] outlined these difficulties in his review paper analyzing the reasons behind the slow development of models simulating the rapid heating and evaporation of droplets of fuels and emulsions based on them.

Experimental results [2] established that the leading droplets in a flow through hot gases significantly affect the heat exchange of the following droplets with the surrounding medium. Volkov et al. hypothesize [3] that, due to rapid vaporization, the first droplets considerably reduce the gas temperature in the front of all the following droplets. So, a thermal insulation of sorts is created for the following droplets in the form of a vapor curtain with a lower temperature, as compared to that of the gas medium in front of the first droplets. Until now, there have been no experimental or theoretical research findings on the thermal insulation of rapidly evaporating liquid droplets. It is important to obtain reliable experimental data and use them to develop adequate physical and mathematical models of heat and mass transfer. When analyzing the overview by Sazhin [4], we concluded that most likely, it is only possible to solve the formulated problem using optic techniques. Reliable information is necessary on temperature distributions in droplets of water and water-based solutions, slurries, and emulsions when rapidly heated. At the same time, we can infer from the findings by Sazhin [4] that the unsteady heating of a droplet has a significant impact on its lifetime. Under such conditions, the assumption of a constant temperature field of an evaporating and shrinking droplet cannot really be considered valid.

Snegirev [5] made attempts at analyzing the temperature gradient of an evaporating droplet to develop simplified mathematical models of phase transformations. He formulated dimensionless criteria to estimate the temperature gradient within a droplet, and its impact on liquid evaporation rate. However, no experimental data to support the reliability of such estimates have been published so far. The task also becomes more complex, because the research needs to be done at relatively high temperatures of the gas medium (over 500 °C). Vysokomornaya et al. [6] show that traditional evaporation models also known as kinetic and diffusion models based on the assumed dominating process [7–9] provide a good agreement between the theoretical research and experimental data only at moderate gas medium temperatures (under 500 °C).

The evaporation of liquids also remains understudied because its intensity depends on the surface temperature of the phase transition and the concentration of liquid vapors in the small-size area next to the interface region. Diffusion and heat transfer in this area are the main drivers of evaporation. Experimental data on the main characteristics of heat and mass transfer near the surface of evaporating droplets are not yet published.

In the considered research area, an unsolved problem is the need to provide the controlled conditions for the crushing droplets due to overheating and micro-explosions. The use of the controlled effects of explosive breakup will solve a number of problems in the areas of unmixed and mixed fuels: combustion stabilization throughout the combustion chamber, reducing heating and ignition costs, increasing calorific value, reducing anthropogenic emissions, improving rheological properties, etc. [10]. These impact on the research in this area.

Explosive breakup of water emulsion and slurry droplets in a high-temperature gas environment was studied experimentally [11–13]. For the explosive disintegration of heterogeneous droplets to happen, the temperature at the interface must reach that of water boiling. A non-contact method, planar laser-induced fluorescence (PLIF), made it possible to establish that the temperature near this interface reached 100–120 °C before disintegration. The authors determined the threshold temperatures at the onset of this effect for a group of solid and liquid organic additives (slurry and emulsion components). As a result of the explosive fragmentation of multi-component droplets, the evaporation surface area increases up to 15 times. It is important to expand the experimental database with the evaporation characteristics of typical sewage and service water compositions to improve their treatment.

The most popular approach to the experimental research into the breakup of boiling liquid, solution, emulsion, and slurry droplets is placing them on a holder into a heated gas flow (e.g., [11–16]). Some setups do not include different holders [14–16] or use substrates [17]. Each of these recording schemes has its own strengths and weaknesses [10]. In terms of the costs and difficulty of the experiment, as well as the reliability of the recording procedure, a holder seems to be the most rational option. However, the choice of the holder material for the fragmentation of boiling droplets of liquids, solutions, emulsions, and slurries is yet to be studied. It is of interest to study how this factor affects the heating and disintegration of typical two-component droplets using a large group of popular materials. A solution to this problem is of principal importance for the development of high-potential gas-vapor-droplet technologies considered in [18–27]. Vershinina et al. [10] established the impact of the holder material on the ignition of fuel slurry droplets. They show that there are two temperature ranges. Above 600 °C, the impact of the holder material is negligible, while below 600 °C, the properties of the said material have a significant influence on heat transfer. It is important to make such estimates for a group of promising two-component droplets.

Another approach [28,29] consists of suspending the emulsion drop onto a thermocouple junction, enabling to measure its temperature during its evaporation under heating. Its results are compared to the present study, since the experimental conditions are similar concerning both emulsion properties and heat source temperature.

The purpose of this work is to study experimentally how the holder material affects the heating, evaporation, and explosive breakup of two-component droplets.

1.2. Review of Time Ranges of Droplet Breakup through Microexplosion

This subsection presents a review of the time ranges of droplet breakup through microexplosion. We considered the experimental results of microexplosion times published in studies [30–33]. Table 1 contains the main suitable data from these papers.

Table 1. Review of time ranges of droplet microexplosion established in the experiments accomplished by using different experimental techniques.

Article	Components	Material of Holder	Range of Two-Component Droplet Breakup Times	Experimental Setup
[30]	Water + n-dodecane Water + n-tetradecane Droplet size V_d = 5–15 μm	Quartz fiber D = 0.25 mm	On the holder (0.22–0.85 s) During fall (0.25–0.95 s)	A droplet is placed on the holder inside the combustion chamber with a temperature of 30 °C. After that, the droplet ignites by an electrically heated wire. The temperature of the droplet is measured by the Pt–PtRh thermocouple. A video camera records the microexplosion process. The fall process lies in the simultaneous motion of the chamber and the droplet during ~ 1 s.
[31]	Pure bio-oils D_0 = 1.12 mm Pure bio-oils D_0 = 1.08 mm	A droplet is fixed on a thermocouple junction (K-type)	t~7s (T_a = 300 °C) t~4s (T_a = 500 °C)	A droplet is fixed on a thermocouple. By using a linear module, it is introduced into the space between two plates heated by electricity.
[32]	Ethanol + Jet A-1; D_0 = 2 mcl	Quartz holder D = 0.2 mm	(1.5–2.3 s)	By using a dispenser, a droplet is placed on a holder. The droplet ignites by using a nichrome wire. The process under study is recorded by a high-speed video camera.
[33]	Heptane C7H16 + Hexadecane C16H34	Without holder	(170–205 ms)	A device is applied to collide two droplets of the required size, and to form a two-component droplet. The droplet moves through the combustion chamber heated up to 1050 °C. High-speed video recording allows the determination of droplet lifetimes and their breakup times. In addition, as a comparison, the experiments are performed with the preliminary formed two-component droplets.
[28,29]	Sunflower oil, distilled water, non-ionic surfactant SPAN 83	K-type thermocouple (Nickel–Chromium, Nickel–Alumel)	(0.9–1.3 s)	A bare K-type thermocouple (wire diameter 76.2 μm) is heated from below by the means of a highly resistive coil with its asymptotic temperature of 350 °C. The emulsion drop is maintained on the thermocouple junction by interfacial tension. The thermocouple signal is acquired by an oscilloscope, and the shadowgraph frames are visualized using a high speed camera (10,000 fps).

An analysis of data presented in Table 1 enables the conclusion that when using the different holders examined in the study, the explosive breakup times of the droplets are the upper estimates of the actual values in the practical applications. Therefore, the values presented in the research can be used to predict the maximum possible times of heating until droplet breakup.

In addition, the challenging task is to determine the times of the heating until breakup, the complete volatilization of impurities, and the ignition of different heterogeneous droplets at free fall, i.e., without the holder [33]. Nowadays, such experiments are labor-intensive and expensive. Thus, the various holders, including those used in the study, will be employed for a rather long time. In such a situation, a utilization of the research results can help to predict differences between droplet-heating characteristics for a large group of experimental studies performed or planned to be carried out by using various holders.

2. Experimental Setup and Procedure

2.1. Components of Two-Component Droplets and their Production Procedure

The experimental research featured two components: water (with a specialized dye–fluorophore Rhodamine B) and transformer oil. The main properties of components are presented in Table 2. The component concentrations were varied over a wide range as per recommendations. The fusion of these liquids resulted in a two-component droplet. The Rhodamine B dye was used to control water temperature in a two-component droplet, similar to the methods used in [16,17]. It was important to provide the same conditions as those used in experiments [16,17] in order to extrapolate the experimental results to various schemes of energy supply to a two-component droplet. Unsteady and inhomogeneous temperature fields of droplets obtained experimentally were in good agreement with the results from [16,17]. Therefore, further analysis will focus on the impact of holder materials on heating.

Table 2. Main properties of the liquids under study.

Component	Thermal Physical Properties	Kinematic Viscosity, m^2/s	Surface Tension, N/m	Boiling Temperature, °C	Heat of Vaporization, MJ/kg
Transformer Oil	$\rho = 877$ kg/m^3, $\lambda = 0.12$ W/(m·°C), $C = 1670$ J/(kg·°C), $a = 8 \cdot 10^{-8}$ m^2/s	$22 \cdot 10^{-6}$ m^2/s at 20 °C, $0.295 \cdot 10^{-6}$ m^2/s at 100 °C	$26.15 \cdot 10^{-3}$	320	0.209
Water	$\rho = 1000$ kg/m^3, $\lambda = 0.6$ W/(m·°C), $C = 4200$ J/(kg·°C), $a = 14 \cdot 10^{-8}$ m^2/s	$1.006 \cdot 10^{-6}$ m^2/s at 20 °C, $2.56 \cdot 10^{-6}$ m^2/s at 100 °C	$72.86 \cdot 10^{-3}$	100	2.258
Sunflower Oil	$\rho = 865$ kg/m^3, $\lambda = 0.165$ W/(m·°C), $C = 2500$ J/(kg·°C)	$6.03 \cdot 10^{-5}$ m^2/s at 25 °C	$33.7 \cdot 10^{-3}$	225	0.21

2.2. Holder Materials

Copper, aluminum, ceramics, steel, nichrome, and phosphorus were the main materials used to produce the holders for two-component droplets under study, since these materials have a wide range of values of thermal and physical characteristics (Table 3).

Table 3. Thermal and physical characteristics of holder materials (average values for the temperature range of 200–500 °C in line with the experiments).

Material	λ, W/(m·°C)	C, J/(kg·°C)	ρ, kg/m^3	$a \cdot 10^6$, m^2/s
Copper	376.86	416.12	8770.31	103.4
Aluminum	229.56	1044.76	2642.526	83.62
Ceramic	1.4	770	2355	0.772
Steel	42.8	561.8	7723	9.912
Nichrome	22.5	460	8660	5.648
Phosphorus	0.236	23.82	1820	5.444

Figure 1 shows the images of the holders used in the experiments: 1—ceramic; 2—steel tube; 3—aluminum; 4—copper; 5—nichrome; 6—phosphorus; 7—steel. When using each of the holders, we measured its contact area with the droplet. The largest holder/droplet contact area was found in the experiments with ceramic and aluminum rods, and the smallest one, with a nichrome wire. The contact surface area of the droplet and holder surface (S_h) mainly depended on the droplet radius (R_d) and the holder size (d_h), considering that the radius of an evaporating droplet decreases nonlinearly. The contact area was calculated using the formula from Figure 1.

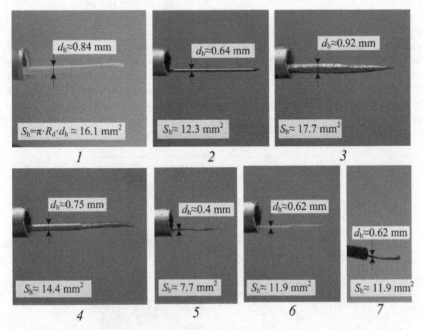

Figure 1. Appearance of holders used in the experiments (their size and contact area with a droplet are specified) 1—ceramic; 2—steel tube; 3—aluminum; 4—copper; 5—nichrome; 6—phosphorus; 7—steel.

The holders were chosen to provide similar schemes of droplet fixation and contact surface. This made it possible to record in the experiments quite a similar geometry for the contact line (interface) between the liquid component and the holder. Under such conditions, the heating or cooling rates of a droplet mostly depended on the thermal and physical properties of the holder material (Table 3). The results of this analysis are presented further.

2.3. Methods for Studying the Disintegration of Boiling Droplets

Figure 2a shows the two-component droplet heating scheme at convective heating, as well as Figure 2b illustrates the actual photo of the two-component droplet during experiments. The Leister CH 6060 hot air blower (air velocity 0.5–5 m/s) (LEISTER Technologies AG, Switzerland) and a Leister LE 5000 HT air heater (temperature range 20–1000 °C) were used as a heating system, generating the necessary parameters of the flow of high-temperature gases (flow rate U_a and temperature T_a). The flow of high-temperature gases was formed in a hollow transparent cylindrical channel (internal diameter 0.1 m, wall thickness 2 mm). A two-component droplet was placed on the holders under study (Figure 1), which were introduced into the flow of high-temperature gases using a motorized coordinate device (motorized manipulator).

We recorded the heating, boiling, and disintegration of two-component droplets by a high-speed video camera. The recordings were processed using the Tema Automotive and ActualFlow software packages for the continuous tracking of moving objects. In the course of processing, we determined the initial droplet radius R_d and the total liquid evaporation surface area, S. The video recordings were processed in two stages. At first, we tracked how the frontal cross-sectional area S_m of an evaporating and deforming droplet changed until it finally broke up (Figure 2c). Using the Airbag and Advanced Airbag tracking algorithms, we observed the changes in the shape of an evaporating droplet. After that, the frontal cross-sectional area of a droplet was calculated, and the curves $S_m(t)$ were plotted. The droplet was assumed to be spherical and its frontal cross-sectional area to be a circle. Using the formula $R_d = (S_m/\pi)^{0.5}$, we calculated the average droplet radius R_d. The errors of the R_d calculation did not exceed 2.5%. After that, the total area of the droplet evaporation surface was calculated using the formula $S = 4\pi R_d^2$.

Video recordings following the explosive breakup of the heterogeneous droplet into separate smaller fragments were analyzed at the second stage. A polydispersed aerosol was usually formed. The shadow image was analyzed using the Actual Flow software to determine the location, boundaries, and dimensions of separate droplets. Median, Low Pass, and Average software filters were used to screen off the noises, and Laplace Edge Detection was used to determine the boundaries of droplet surface. When determining the droplet dimensions, we applied the Bubble Identification algorithm. The error of the R_d calculation using this approach was under 3%.

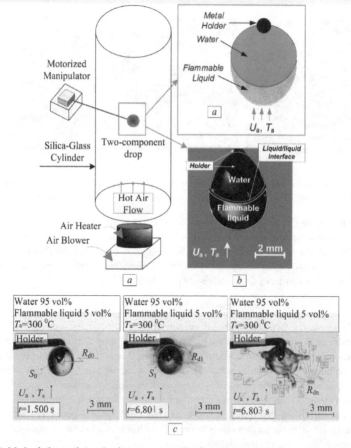

Figure 2. Methods for studying the disintegration of boiling droplets: (**a**): The scheme of registration of the heating process of two-component droplet; (**b**): Photo of the formed two-component droplet; (**c**): Scheme of recording heated droplet breakup and aerosol generation.

To obtain the droplet size distributions, all of the droplets were classified into m groups. We then determined the number of droplets n and the average droplet size R_{dn} in each group. The liquid evaporation surface for the droplets in each group was calculated using the formula $S_n = n \cdot 4\pi R_{dn}^2$. As the final step, we calculated the overall evaporation surface area: $S = S_{n(1)} + S_{n(2)} + \ldots + S_{n(m)}$.

2.4. Main Registered Parameters and Tolerances

Table 4 presents the parameters recorded in the experiments, and the systematic errors of the measurement tools. The next section outlines the random errors calculated as part of statistical analysis of the results in a series of experiments.

Table 4. Main registered parameters and tolerances.

Physical Magnitude	Droplet Volume (V_d)	Droplet Radius (R_d)	Temperature Inside the Droplet (T_d)	Two-Component Droplet Breakup Times (τ) and Lifetimes (τ_h)	Air Temperature (T_a)	Air flow Velocity (U_a)
Measurement Tool/Technique	Finnpipette Novus dispensers	High-speed cameras Phantom Miro M310 and Photron Fastcam SA1, Tema Automotive software	Planar Laser Induced Fluorescence (PLIF)	High-speed cameras Phantom Miro M310, Photron Fastcam SA1, and Phantom V 411, Tema Automotive software	Temperature meter (IT-8)	Particle Image Velocimetry (PIV)
Systematic Errors	$\pm 0.05\ \mu L$	$\leq 4\%$	± 1.5–$2\ ^\circ C$	$\leq 4\%$	$\pm(0.2 + 0.001T)\ ^\circ C$	$\pm 2\%$

3. Results and Discussion

3.1. Droplet Disintegration Regimes

In this section, we present typical video frames showing the heating and disintegration of two-component droplets on holders made of different materials (Figure 3). The smallest droplets were formed in the cases when holders were made of materials with low thermal diffusivity (ceramics and phosphorus).

On a copper holder, the droplets did not reach the temperature sufficient for an explosive breakup, but gradually evaporated in a wide range of the main parameters: air temperature and component concentration. Explosive breakup was only observed at 450 °C, and with a 50/50 concentration of the flammable and non-flammable components. The heating time before disintegration was 25–30 s, most likely because copper is good at removing heat from the droplet. Copper holders have the highest thermal diffusivity (around $117 \cdot 10^{-6}\ m^2/s$). Droplets do not reach the conditions of micro-explosion (even boiling is not observed). The video frames of the experiments only showed monotonous evaporation.

Aluminum, however, with its thermal diffusivity of $90 \cdot 10^{-6}\ m^2/s$ (close to that of copper), provided quite a stable explosive breakup. The heating time until explosive breakup was longer than with the other holders, except for copper. The experiments established two factors prolonging the time of two-component droplet disintegration: heat removal from a droplet due to relatively high thermal diffusivity, formation of thermal stresses, nucleation, and growth of bubbles with high pressure in a droplet, which disintegrates to form smog or mist.

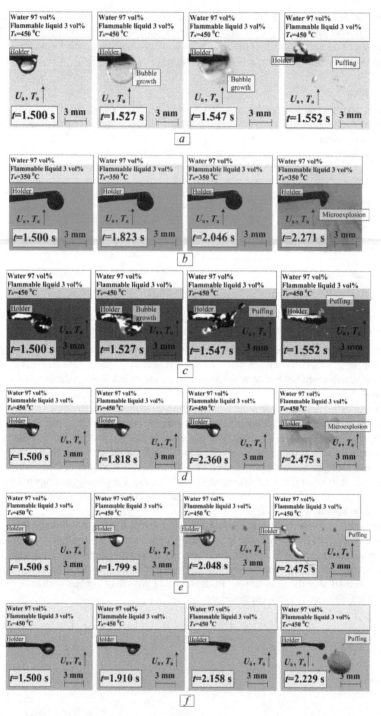

Figure 3. Snapshots of breakup or fragmentation on various holders: (**a**): ceramics; (**b**): steel; (**c**): aluminum; (**d**): steel tube; (**e**): nichrome; (**f**): phosphorus.

3.2. Impact of Key Factors

Figure 4 shows the times of two-component droplet disintegration vs gas medium temperature. The longest disintegration times were observed when using an aluminum rod as a holder for a two-component droplet. This results from the above-described possible rapid heat removal from a droplet due to high thermal diffusivity of aluminum. This plot also shows that the times of the two-component droplet disintegration decreased rapidly with an increase in the temperature. Such a pattern is typical of each holder material under study.

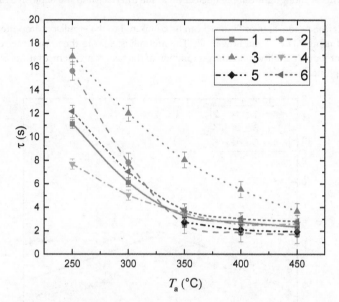

Figure 4. Two-component droplet heating times until explosive breakup vs gas medium temperature with various holders: 1—ceramic; 2—steel; 3—aluminum; 4—steel tube; 5—nichrome; 6—phosphorus. The value $T_a \approx 250\ °C$ is the threshold for explosive breakup.

The temperatures at which the main experimental studies were performed to observe explosive breakup, ranged from 250 °C to 450 °C. The optimal temperature that steadily provided an explosive droplet breakup within a short time was 350 °C. Below that, the disintegration times increased non-linearly, and above that, they remained practically the same. Further increase in the T_a is redundant and unpractical for water treatment and other energy-consuming applications.

Apart from temperature functions, the plots of two-component droplet disintegration times vs flammable component concentration were among the key ones. The resulting functions are highly non-linear, which suggests a significant impact of the flammable liquid concentration on a group of interconnected processes promoting the breakup of the initial two-component droplets (Figure 5).

These results can be compared to the recorded lifetime of the water-sunflower oil emulsion drop, supported by a K-type wire thermocouple [28,29] under the asymptotic temperature of 350 °C. It ranges from 0.9 to 1.3 s, close to the same order of magnitude. This result confirms that reproducibility of lifetimes can be obtained under the same conditions of heat transfer that are mainly determined by the temperature of the heat source. The emulsion temperature acquired by the thermocouple steadily increased until the boiling point of oil, which is less than 250°C (see Table 1).

Figure 5 shows the maximum droplet heating times until breakup with equal relative fractions of the flammable and non-flammable components. This stems from a set of factors and processes that are opposite in terms of their impact. Due to the higher heat capacity and the vaporization heat, water heats up rather slowly as compared to oil, but the thermal conductivity and thermal diffusivity of the latter are several times lower than those of water. Therefore, under identical heating and equal component concentrations, these factors counterbalance each other. As a result, a two-component droplet is heated more slowly until it reaches the conditions of explosive breakup. Moreover, with a low proportion of water in a droplet, the film of the flammable component is thick, and it is heated faster than water. Thus, the heating times of the initial (parent) droplet until breakup are minimum. With the highest possible fraction of water and lowest fraction of the flammable component, the trend changes. A thin film of the flammable component is heated fast and locally overheats the near-surface water layer. This is enough for bubble nucleation at the interface and the explosive breakup of the initial droplet.

Moreover, the results shown Figure 5 can be compared to the emulsion drop lifetime of 0.9 to 1.3 s obtained using 70 vol % sunflower oil. The lifetime is close to the same order of magnitude. It confirms that reproducibility is not only a function of the heat source temperature, but also that of the emulsion properties.

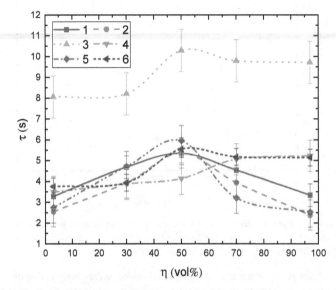

Figure 5. Droplet breakup times vs. relative mass fraction of the flammable component with various holders: 1—ceramic; 2—steel; 3—aluminum; 4—steel tube; 5—nichrome; 6—phosphorus.

3.3. Droplet Disintegration Outcomes

By analyzing the outcomes of the explosive breakup (Figure 6) of a two-component droplet, we have established that hollow steel and phosphorus tubes as well as a nichrome wire used as a holder yield droplet aerosols with a maximum quantity of small fragments. The liquid evaporation surface area increased more than 40 times under such conditions.

In the experiments with an aluminum holder, the evaporation surface area increased massively with the growing concentration of the flammable component. This results from the longer droplet heating time (Figure 5). The longer the period of droplet heating until breakup, the greater the volume of the two liquids that is heated to high temperatures. The droplet broke up into a greater number of fragments, which boiled and disintegrated in the process, into even smaller droplets. Presumably, the chain-like breakup of droplet aerosols may potentially intensify.

For a phosphorus holder, on the contrary, a low concentration of flammable liquid provides the largest evaporation surface area of droplets, most likely due to water boiling that is in contact with the holder surface. Since a phosphorus rod removes very little energy from the droplet, almost all the energy that is supplied is spent on heating the liquid components. The explosive breakup occurred when the liquid–liquid interface was heated to water boiling temperature. A thin flammable component film quickly reached high temperatures, and so did a thin water layer at the interface, which was recorded in the experiments.

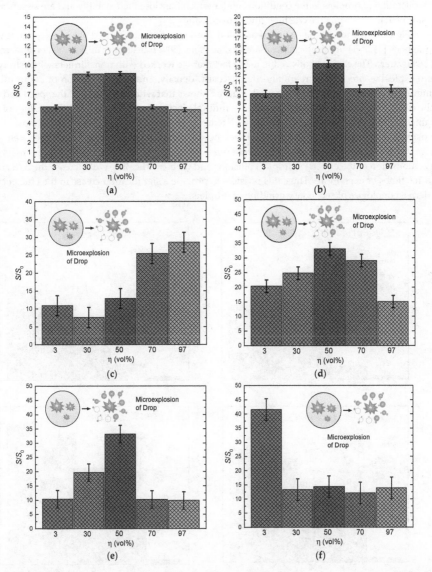

Figure 6. Ratios of surface areas of small droplets formed after the breakup of two-component droplets to their initial areas depending on the concentration of flammable liquid (oil) on various holders: (**a**): ceramics; (**b**):steel; (**c**): aluminum; (**d**): steel tube; (**e**): nichrome; (**f**): phosphorus.

For the other holder materials, the emerging droplets had the largest evaporation surface areas with 50/50 component concentrations. With this concentration of the flammable liquid, the breakup times were the longest. Therefore, a droplet has more time to form the temperature stresses and nucleation sites of vapor bubbles, i.e., to reach the temperatures sufficient for rapid vaporization near the inner water–flammable liquid interface.

A literature analysis shows that it is possible to significantly reduce the size of the droplets of various liquids (respectively, to increase S/S_0), due to several mechanisms. The most common are the following: the impact of the droplets between themselves; the interaction of droplets with an obstacle; the acceleration of droplets to the conditions under which they lose their stability and are significantly transformed; micro-explosive crushing due to overheating.

In Figure 7, we added the results of additional test experiments (carried out in accordance with the methods [34,35]) with droplets of oil–water emulsions (50% transformer oil, 50% water; 50% castor oil, 50% water). The choice of oils is due to the fact that we worked with transformers when studying micro-explosive effects (Figure 6), and the viscosity, density, and surface tension of castor oil are significantly different from transformer oil. Figure 7 shows that with an increase of the speed and size of the colliding drops, it is possible to ensure a multiple increase of the ratio S/S_0. The values of this parameter grow especially on a large-scale with temperature increasing, since the surface tension and viscosity of liquids decrease. At the same time, the scale of growth of this ratio correlates well with Figure 6 at micro-explosive decay. These results are the basis for the formulation of the hypothesis that the combination of the effects of droplet collisions and their overheating will increase the S/S_0 ratio by a 100 times or even more. Thus, it is possible to provide a significant increase in the efficiency of modern technologies of secondary grinding of droplets.

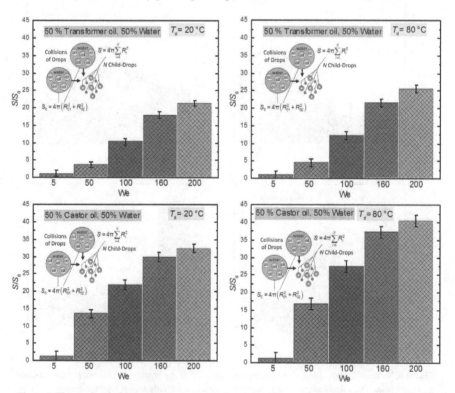

Figure 7. The results of additional test experiments (carried out in accordance with the methods [34,35]) with droplets of oil–water emulsions.

3.4. Generalization of Research Findings

The experimental results made it possible to determine how the holder material affects the heating of multi-component droplets, and to discover that in some cases, additional heating of such droplets due to their contact with the holder intensifies their explosive breakup. A major role belongs to the direct contact of the non-flammable component (water) with the heated holder surface. Moreover, the experiments have established that the heating of the two-component droplets is more rapid if the core of the droplet is made of water, and the envelope, of oil. This happens because oil has a high absorption ability, and less energy is spent on its evaporation. Therefore, oil is heated faster than water, although water has a higher thermal conductivity and diffusivity than oil.

To demonstrate the heating conditions of multi-component droplets on a holder, we have developed a simplified one-dimensional mathematical model of heat transfer in the holder–two-component droplet system (Figure 8) similar to the one in [36]. This model determined the temperature variation trends for the holder, core and envelope of a droplet (Figure 9). Signature domains are divided by different colors in Figure 8. Similarly to the model in [36], we took into account the droplet heating through both thermal conductivity and radiation absorption, according to the Beer–Lambert–Bouguer law. At a first approximation, we used a one-dimensional statement to evaluate the variation of the vertical temperature profile, as shown in Figure 8. From the analysis of the temperature fields established in experiments [16,17], we can conclude that highly unsteady and inhomogeneous temperature profiles, which further determine the intensity of droplet breakup, are formed in such sections.

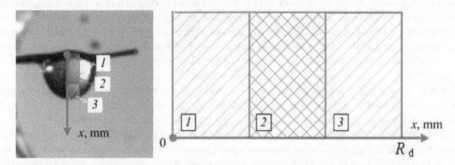

Figure 8. Schematic representation of the solution domain for the problem of two-component droplet heating on a holder: 1—holder, 2—droplet core (water as the first component), 3—droplet envelope (oil as the second component).

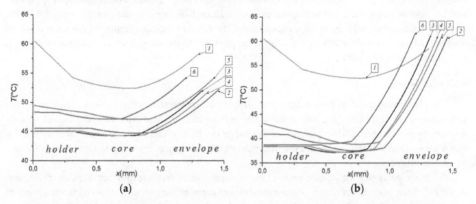

Figure 9. Temperature distributions (along x in a holder–two-component droplet system, see Figure 8) at $T_a = 300$ °C for 10 s (**a**) and $T_a = 400$ °C for 5 s (**b**) using various holders: 1—phosphorus, 2—aluminum, 3—steel, 4—copper, 5—ceramics, 6—nichrome; the holder and droplet dimensions were chosen in line with the conditions of the experiments; the boundaries of areas showing the holder, core, and envelope of a droplet are not marked, since they were slightly different for each of the holders used (see Figure 1).

Theoretical plots (Figure 9) show the following: with aluminum, steel, copper, ceramic, and nichrome holders (2-6), the heat inflow to the droplet comes mostly from the flammable liquid (oil), and in the case of the phosphorus holder, from the holder (1). From this, we can conclude that the longest period of droplet heating will be provided by using a phosphorus holder, since the heat inflow comes from both the holder and the flammable liquid. Moreover, these curves (Figure 9) show that smallest heat inflow from the holder will be provided by using steel (3) and copper (4) holder materials.

In Figure 9, a significant increase in the temperature of a two-component droplet outpaces the increase in the temperature of most holders used. Before the breakup, the water–flammable liquid interface exceeded the water boiling temperature (100–120 °C). It was difficult to show such trends in one measurement system in Figure 9, since for various holders, droplets are heated and they disintegrate at various typical rates (the simulation results for water droplets are considered in research [29]). Therefore, to demonstrate the highly inhomogeneous temperature profile, Figure 9 presents the calculations for temperatures, at which a droplet remains in one piece, i.e., before explosive breakup.

In terms of practical importance, the experimental research proved that it is possible to provide adequate high-temperature liquid treatments by the explosive breakup of droplets containing various components in various proportions. In chambers used for high-temperature evaporation and the burnout of impurities, multi-component droplets swirl through high-temperature turbulent and pulsating gas media. Therefore, droplets are heated almost uniformly throughout their surface until they reach the conditions sufficient for heat removal (droplet cooling), e.g., when a droplet is fixed on a holder [15,16] or its substrates [17]. The fixation scheme of a two-component droplet on a holder, chosen in this research, is fully in line with such conditions. This scheme provides adequate evaluation of the main parameters of high-temperature liquid treatment from any impurities promoting a rapid (explosive) breakup of multi-component droplets (e.g., slurries, emulsions, and solutions).

Experiments with droplets fixed on the holders used in this study (especially the phosphorus one) make it possible to reproduce the conditions of liquid heating in high-temperature chambers. Droplets move in such chambers at almost the same velocities as the carrier medium-heated gas. The carrier medium velocities are as low as several meters per second. Therefore, liquid droplets are mostly heated by the radiative heat flux. Droplet fixation on a holder with a very low thermal diffusivity leads to a slight increase in the convective component of the flux as compared to the real-life evaporation and burnout of impurities. However, our estimates show that these deviations do not exceed 10%, and they decrease with the growing temperature of the carrier medium (Figure 9). Also, the research results can be used for development of effective approaches to the secondary atomization of droplets in fuel technologies [37–39].

4. Conclusions

(i) The breakup of a two-component droplet is connected with the overheating of the *water–flammable liquid* interface above the water boiling temperature (100–120 °C). Liquid surface tension forces suppress the free release of the vapor bubbles formed near the interface. When the vapor pressure in a droplet exceeded the threshold value, the droplet broke up to form a mist, aerosol, or several droplets.

(ii) When analyzing the heating times of the two-component droplets until breakup, we discovered that the disintegration times of two-component droplets are minimum when holders with a low thermal diffusivity are used ($a < 10$ m^2/s), and maximum when thermal diffusivity is high ($a > 80$ m^2/s).

(iii) In comparison with the results obtained onto a suspending thermocouple junction, under similar conditions of heat source and emulsion properties, the lifetime of the drop is close to the same order of magnitude (2–6 s).

Author Contributions: D.A. and M.P. conceived and designed the experiments; D.A. and O.V. performed the experiments; D.A., J.B., D.T., P.M., O.V. and M.P. analyzed the data and wrote the paper.

Funding: This research received no external funding.

Acknowledgments: Research was supported by the Russian Science Foundation (project 18–71–10002).

Conflicts of Interest: The authors declare no conflict of interest.

Nomenclature and Units

a	thermal diffusivity, m^2/s
C	specific heat capacity, $J/(kg \cdot {}^\circ C)$
d_h	holder diameter, mm
m	number of groups
n	number of droplets in each group
R_d	droplet radius, mm
R_{d0}	initial two-component droplet radius, mm
R_{d1}	droplet radius before breakup, mm
R_{dn}	mean radius of droplets in a group, mm
S	total area of droplet evaporation surface after breakup, mm^2
S_0	initial droplet surface area, mm^2
S_1	droplet surface area before breakup, mm^2
S_h	contact surface area of a droplet and holder surface, m^2
S_m	frontal cross-sectional area of droplet, mm^2
S_n	evaporation surface area in each droplet group, mm^2
T	temperature, $^\circ C$
T_a	gas flow temperature, $^\circ C$
T_d	temperature in a droplet, $^\circ C$
t	time, s
U_a	high-temperature gas flow velocity, m/s
V_d	drop volume, μL
We	Weber number
x	coordinate in a one-dimension model, mm
η	flammable liquid concentration, vol%
λ	thermal conductivity, $W/(m \cdot {}^\circ C)$
ρ	density, kg/m^3
τ	two-component droplet breakup times, s
τ_h	two-component droplet lifetimes, s

References

1. Kuznetsov, G.V.; Strizhak, P.A.; Volkov, R.S. The influence of initial sizes and velocities of water droplets on transfer characteristics at high temperature gas flow. *Int. J. Heat Mass Transf.* **2014**, *79*, 838–845. [CrossRef]

2. Kuznetsov, G.V.; Strizhak, P.A.; Volkov, R.S. Experimental investigation of mixtures and foreign inclusions in water droplets influence on integral characteristics of their evaporation during motion through high-temperature gas area. *Int. J. Therm. Sci.* **2015**, *88*, 193–200. [CrossRef]

3. Volkov, R.S.; Kuznetsov, G.V.; Legros, J.C.; Strizhak, P.A. Experimental investigation of consecutive water droplets falling down through high-temperature gas zone. *Int. J. Heat Mass Transf.* **2016**, *95*, 184–197. [CrossRef]

4. Sazhin, S.S. Modelling of fuel droplet heating and evaporation: Recent results and unsolved problems. *Fuel* **2010**, *196*, 69–101. [CrossRef]

5. Snegirev, A.Y. Transient temperature gradient in a single-component vaporizing droplet. *Int. J. Heat Mass Transf.* **2013**, *65*, 80–94. [CrossRef]

6. Vysokomornaya, O.V.; Kuznetsov, G.V.; Strizhak, P.A. Evaporation of water droplets in a high-temperature gaseous medium. *J. Eng. Phys. Thermophys.* **2016**, *89*, 141–151. [CrossRef]

7. Spalding, D.B. *Some Fundamentals of Combustion*; Butterworth's: London, UK, 1955.

8.   Fuchs, N.A. *Evaporation and Droplet Growth in Gaseous Media*; Pergamon Press: London, UK, 1959.

9.   Ranz, W.E.; Marshall, W.R. Evaporation from drops—I. *Chem. Eng. Prog.* **1952**, *48*, 141–146.

10. Vershinina, K.Y.; Egorov, R.I.; Strizhak, P.A. The ignition parameters of the coal-water slurry droplets at the different methods of injection into the hot oxidant flow. *Appl. Therm. Eng.* **2016**, *107*, 10–20. [CrossRef]

11. Watanabe, H.; Harada, T.; Matsushita, Y.; Aoki, H.; Miura, T. The characteristics of puffing of the carbonated emulsified fuel. *Int. J. Heat Mass Transf.* **2009**, *52*, 3676–3684. [CrossRef]

12. Suzuki, Y.; Harada, T.; Watanabe, H.; Shoji, M.; Matsushita, Y.; Aoki, H.; Miura, T. Visualization of aggregation process of dispersed water droplets and the effect of aggregation on secondary atomization of emulsified fuel droplets. *Proc. Combust. Inst.* **2011**, *33*, 2063–2070. [CrossRef]

13. Tarlet, D.; Josset, C.; Bellettre, J. Comparison between unique and coalesced water drops in micro-explosions scanned by differential calorimetry. *Int. J. Heat Mass Transf.* **2016**, *95*, 689–692. [CrossRef]

14. Strizhak, P.A.; Piskunov, M.V.; Volkov, R.S.; Legros, J.C. Evaporation, boiling and explosive breakup of oil-water emulsion drops under intense radiant heating. *Chem. Eng. Res. Des.* **2017**, *127*, 72–80. [CrossRef]

15. Vysokomornaya, O.V.; Piskunov, M.V.; Strizhak, P.A. Breakup of heterogeneous water drop immersed in high-temperature air. *Appl. Therm. Eng.* **2017**, *127*, 1340–1345. [CrossRef]

16. Piskunov, M.V.; Strizhak, P.A. Using Planar Laser Induced Fluorescence to explain the mechanism of heterogeneous water droplet boiling and explosive breakup. *Exp. Therm. Fluid Sci.* **2018**, *91*, 103–116. [CrossRef]

17. Kuznetsov, G.V.; Piskunov, M.V.; Volkov, R.S.; Strizhak, P.A. Unsteady temperature fields of evaporating water droplets exposed to conductive, convective and radiative heating. *Appl. Therm. Eng.* **2018**, *131*, 340–355. [CrossRef]

18. Misyura, S.Y. Evaporation of a sessile water drop and a drop of aqueous salt solution. *Nat. Sci. Rep.* **2017**, *7*, 14759. [CrossRef] [PubMed]

19. Misyura, S.Y. Evaporation and heat and mass transfer of a sessile drop of aqueous salt solution on heated wall. *Int. J. Heat Mass Transf.* **2018**, *116*, 667–674. [CrossRef]

20. Misyura, S.Y. Non-isothermal evaporation in a sessile droplet of water-salt solution. *Int. J. Therm. Sci.* **2018**, *124*, 76–84. [CrossRef]

21. Korobeinichev, O.P.; Shmakov, A.G.; Shvartsberg, V.M.; Chernov, A.A.; Yakimov, S.A.; Koutsenogii, K.P.; Makarov, V.I. Fire suppression by low-volatile chemically active fire suppressants using aerosol technology. *Fire Saf. J.* **2012**, *51*, 102–109. [CrossRef]

22. Varaksin, A.Y. Fluid dynamics and thermal physics of two-phase flows: Problems and achievements. *High Temp.* **2013**, *51*, 377–407. [CrossRef]

23. Tarlet, D.; Allouis, C.; Bellettre, J. The balance between surface and kinetic energies within an optimal micro-explosion. *Int. J. Therm. Sci.* **2016**, *107*, 179–183. [CrossRef]

24. Kichatov, B.; Korshunov, A.; Kiverin, A.; Son, E. Experimental study of foamed emulsion combustion: Influence of solid microparticles, glycerol and surfactant. *Fuel Process. Technol.* **2017**, *166*, 77–85. [CrossRef]

25. Salgansky, E.A.; Zaichenko, A.Y.; Podlesniy, D.N.; Salganskaya, M.V.; Toledo, M. Coal dust gasification in the filtration combustion mode with syngas production. *Int. J. Hydrogen Energy.* **2017**, *42*, 11017–11022. [CrossRef]

26. Zhukov, V.E.; Pavlenko, A.N.; Moiseev, M.I.; Kuznetsov, D.V. Dynamics of interphase surface of self-sustaining evaporation front in liquid with additives of nanosized particles. *High Temp.* **2017**, *55*, 79–86. [CrossRef]

27. Sazhin, S.S.; Shchepakina, E.; Sobolev, V. Order reduction in models of spray ignition and combustion. *Combust. Flame* **2018**, *187*, 122–128. [CrossRef]

28. Mura, E.; Calabria, R.; Califano, V.; Massoli, P.; Bellettre, J. Emulsion droplet micro-explosion: Analysis of two experimental approaches. *Exp. Therm. Fluid Sci.* **2014**, *56*, 69–74. [CrossRef]

29. Strizhak, P.A.; Volkov, R.S.; Castanet, G.; Lemoine, F.; Rybdylova, O.; Sazhin, S.S. Heating and evaporation of suspended water droplets: Experimental studies and modelling. *Int. J. Heat Mass Transf.* **2018**, *127*, 92–106. [CrossRef]

30. Tsue, M.; Yamasaki, H.; Kadota, T.; Segawa, D.; Kono, M. Effect of gravity on onset of microexplosion for an oil-in-water emulsion droplet. *Symp. Comb.* **1998**, *27*, 2587–2593. [CrossRef]

31. Hou, S.-S.; Rizal, F.M.; Lin, T.-H.; Yang, T.-Y.; Wan, H.-P. Microexplosion and ignition of droplets of fuel oil/bio-oil (derived from lauan wood) blends. *Fuel* **2013**, *113*, 31–42. [CrossRef]

32. Chaitanya KumarRao, D.; Syam, S.; Karmakar, S.; Joarder, R. Experimental investigations on nucleation, bubble growth, and micro-explosion characteristics during the combustion of ethanol/Jet A-1 fuel droplets. *Exp. Therm. Fluid Sci.* **2017**, *89*, 284–294. [CrossRef]
33. Wang, C.H.; Hung, W.G.; Fu, S.Y.; Huang, W.C.; Law, C.K. On the burning and microexplosion of collision-generated two-component droplets: Miscible fuels. *Combust. Flame* **2003**, *134*, 289–300. [CrossRef]
34. Antonov, D.V.; Volkov, R.S.; Kuznetsov, G.V.; Strizhak, P.A. Experimental Study of the Effects of Collision of Water Droplets in a Flow of High-Temperature Gases. *J. Eng. Phys. Thermophys.* **2016**, *89*, 100–111. [CrossRef]
35. Volkov, R.S.; Kuznetsov, G.V.; Strizhak, P.A. Statistical analysis of consequences of collisions between two water droplets upon their motion in a high-temperature gas flow. *Tech. Phys. Lett.* **2015**, *41*, 840–843. [CrossRef]
36. Antonov, D.V.; Kuznetsov, G.V.; Strizhak, P.A. Determination of Temperature and Concentration of a Vapor–Gas Mixture in a Wake of Water Droplets Moving through Combustion Products. *J. Eng. Therm.* **2016**, *25*, 337–351. [CrossRef]
37. Mura, E.; Massoli, P.; Josset, C.; Loubar, K.; Bellettre, J. Study of the micro-explosion temperature of water in oil emulsion droplets during the Leidenfrost effect. *Exp. Therm. Fluid Sci.* **2012**, *43*, 63–70. [CrossRef]
38. Tarlet, D.; Mura, E.; Josset, C.; Bellettre, J.; Allouis, C.; Massoli, P. Distribution of thermal energy of child-droplets issued from an optimal micro-explosion. *Int. J. Heat Mass Transf.* **2014**, *77*, 1043–1054. [CrossRef]
39. Moussa, O.; Tarlet, D.; Massoli, P.; Bellettre, J. Parametric study of the micro-explosion occurrence of W/O emulsions. *Int. J. Therm. Sci.* **2018**, *133*, 90–97. [CrossRef]

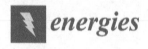

Article

Retention of Hydraulic Fracturing Water in Shale: The Influence of Anionic Surfactant

Hesham Abdulelah [1,*], Syed M. Mahmood [1,*], Sameer Al-Hajri [2], Mohammad Hail Hakimi [3] and Eswaran Padmanabhan [1]

1   Shale Gas Research Group (SGRG), Institute of Hydrocarbon Recovery, Faculty of Petroleum & Geoscience, Universiti Teknologi PETRONAS, Seri Iskandar 32610, Perak, Malaysia; eswaran_padmanabhan@utp.edu.my
2   Department of Petroleum Engineering, Universiti Teknologi PETRONAS, Seri Iskandar 32610, Perak, Malaysia; ensamyo87@gmail.com
3   Geology Department, Faculty of Applied Science, Taiz University, 6803 Taiz, Yemen; ibnalhakimi@yahoo.com
*   Correspondence: heshamsaif09@gmail.com (H.A.); mohammad.mahmood@utp.edu.my (S.M.M.); Tel.: +60-1139628761 (H.A.); +60-5368-7103 (S.M.M.)

Received: 16 October 2018; Accepted: 26 November 2018; Published: 30 November 2018

Abstract: A tremendous amount of water-based fracturing fluid with ancillary chemicals is injected into the shale reservoirs for hydraulic fracturing, nearly half of which is retained within the shale matrix. The fate of the retained fracturing fluid is raising some environmental and technical concerns. Mitigating these issues requires a knowledge of all the factors possibly contributing to the retention process. Many previous studies have discussed the role of shale properties such as mineralogy and capillarity on fracturing fluid retention. However, the role of some surface active agents like surfactants that are added in the hydraulic fracturing mixture in this issue needs to be understood. In this study, the influence of Internal Olefin Sulfate (IOS), which is an anionic surfactant often added in the fracturing fluid cocktail on this problem was investigated. The effect on water retention of treating two shales "BG-2 and KH-2" with IOS was experimentally examined. These shales were characterized for their mineralogy, total organic carbon (TOC) and surface functional groups. The volume of retained water due to IOS treatment increases by 131% in KH-2 and 87% in BG-2 shale. The difference in the volume of retained uptakes in both shales correlates with the difference in their TOC and mineralogy. It was also inferred that the IOS treatment of these shales reduces methane (CH_4) adsorption by 50% in KH-2 and 30% in BG-2. These findings show that the presence of IOS in the composition of fracturing fluid could intensify water retention in shale.

Keywords: hydraulic fracturing; water retention in shale; anionic surfactant; shale gas

1. Introduction

Shale gas reservoirs are known to have ultra-porosity and permeability, thus exploiting them through conventional production methods is not economically feasible [1,2]. Hydraulic fracturing combined with horizontal drilling has been implemented to enhance gas production from shale, and they were proven to be commercial and effective approaches [3–5]. The aim of hydraulic fracturing in shale is to promote its permeability by opening the existing natural fractures and generating new fractures. It is accomplished by injecting a large volume of water-based fluid down a well at a suitable rate and pressure. The resulting fracture networks within shale are typically kept open with proppants to encourage the gas flow from shale to the producing well thus improving gas recovery [6,7]. The fracking fluid is generally composed of water (~99.5%), proppants and a mixture of chemical additives that vary depending on the characteristics of shale reservoir [8,9].

One of the significant issues associated with fracking in shale is that massive amount of fracturing fluid (~5–50%) is retained in the formation after the fracking process [10,11]. For example, Ge [9] and Penny et al. [12] reported that only around 5% of the water is recovered the fracturing processes in shale while Nicot et al. [13] found less than 20%. Yang et al. [4], Makhnov et al. [14] and Reagan et al. [15] disclosed lower than 30% of fracking fluid in some other shale plays to flow-back whereas the other 70% of the injected fluid is believed to be retained by the shale reservoir. In some areas of the Barnet and Marcellus shales, the recovered water after fracking was nearly 50% [9,16].

The fracking water retention issue in shale has raised environmental [17–20] and technical concerns [21,22]. The role of retained water in contaminating the drinking water aquifers is a topic of debate [17]. Vidic et al. [20] stated that the induced fractures outside the target formation could provide pathways for fracking fluid to migrate through. In the town of Pavillion, WY, the U.S. Environmental Protection Agency (EPA) observed water contamination in two shallow monitoring wells [18]. Elevated levels of pH, specific conductance and traces of gas were confirmed in shallow groundwater possibly due to retained fracking water [23]. In Garfield County, the salinity of groundwater was reported to increase with fracking activities in the nearby wells [24]. The rise in salinity with increasing the number of oil and gas wells could trigger the claim that migration from oil/gas wells nearby took place thus contaminating the shallow groundwater [17]. Similarly, an official report by EPA [19] proposed that local water well in West Virginia was found contaminated with gel; conceivably due to leakage of fracturing fluid from an adjacent vertically fractured well. Birdsell et al. [25] concluded based on a two-dimensional conceptual model that the risk of aquifer contamination is reduced ten times by the combined influence of production well and capillary imbibition. Myers [26] estimated the risk of groundwater contamination by fracking water by applying groundwater transport model to a Marcellus shale utilizing the pressure data from a gas well. He found that fracking fluid might reach groundwater aquifers in less than ten years. In Europe, investigating the possible in-situ contamination risk of hydraulic fracturing operation has gained considerable attention. The European Union (EU) has recently sponsored a project called "FracRisk" to explore the likely risks of the fracking operation. Under the "FracRisk" project, some generic and modeling studies were carried out to assess the potential impact of hydraulic fracturing on groundwater aquifers [27,28]. The amount of water used for hydraulic fracturing in shale is massive, the considerable proportion of retained fracturing fluid will necessitate using even more water which will adversely affect the water resources in some shale gas areas, which endure water scarcity [29]. Besides its possible environmental issues, the retained fracking water can significantly impair the production of shale gas. Ge et al. [19], Gallegos et al. [29] and Sharma et al. [30] explained that retention of fracturing fluid was found to develop the water saturation near surfaces of the created fractures, which can prominently impact the gas relative permeability and productivity. Gas production will be significantly reduced as the water saturation reaches 40–50% [31,32].

Mitigating the issue of fracturing fluid retention in shale requires knowledge about the factors controlling this phenomenon. In the literature, more focus was given to the role of shale mineralogy [33] as a significant factor governing water retention during hydraulic fracturing. Many of the published studies [21,34,35] reports that clays, which are one of the primary minerals in shale have the affinity to imbibe water molecules due to their hydrophilic nature. However, the effect of some surface acting agents' that are added into the fracking fluid mixture on water retention in shale was not given enough focus. Common chemical additives in fracturing fluid are often surfactants [7,34,36–38]. Generally, they are intended to increase the viscosity of fracturing fluid to allow it to propagate within the target formation. An anionic surfactant that is added in fracturing fluid mixture is Internal Olefin Sulfate (IOS) [34].

Shales have a mixed-charged surface due to the coexistence of negative surface-charged and positive surface-charged minerals. Anionic surfactants have a negatively charged headgroup and positively charged weak tail. Figure 1 presents a depiction of the interaction between an anionic surfactant and a shale surface. Once anionic surfactant solution comes into contact with shale surface, either its strong headgroup will be attracted to the positively charged site (Figure 1a), or its weak tail will be attracted to the negatively charged sites in shale (Figure 1b). These interactions between the anionic surfactant and shale can alter its wettability and thus causing its water imbibition behavior to increase or decrease [34,39].

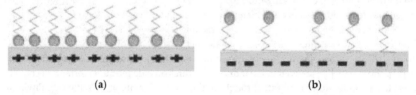

(a)   (b)

Figure 1. Schematic diagram of the electrostatic interaction between (**a**) surfactant headgroup and positive-charged sites in shale, and (**b**) surfactant weak tail and negative-charged sites in shale. Modified after Zhou et al. [39].

In this study, the effect of IOS (anionic surfactant) on water retention in two Malaysian shales was investigated. The shales were characterized for their total organic carbon (TOC), mineralogy, topology, and pore system. Water retention was then examined in two ways; measurements of water uptakes [40] and by utilizing the U.S Bureau of Mine Method (USBM) [34] adsorption/desorption method.

2. Materials and Methods

2.1. Shales

Table 1 lists the properties of the shales used in this study. The two shales differ in their mineralogy and the amount of organic carbon. The two shale shales were collected from two different Paleozoic black shale formations in Peninsular Malaysia. One shale "BG-2" was taken from Batu Gajah formation in Perak district. Batu Gajah formation was described by Baioumy et al. [41] to be a Carboniferous black shale outcrop formation composed of grey and black flaggy shales. The other shale "KH-2" was obtained from Kroh formation in Kedah district, which comes under the Ordovician-Devonian age. The Kroh formation is composed of a sequence of black carbonaceous shale and mudstone.

Table 1. Properties of the BG-2 and KH-2 shales.

Sample ID	Color	Geological Age	Thermal Maturity	Formation	Country
BG-2	Grey	Carboniferous *	Over-matured *	Batu Gajah	Malaysia
KH-2	Black	Ordovician-Devonian *	Over-matured *	Kroh	Malaysia

* Baioumy et al. [41].

2.2. Surfactant

IOS was used to treat the two shales in this study. Table 2 shows the available information about this surfactant. Surfactants are added into the hydraulic fracturing fluid to control its viscosity. In Bakken shale, a surfactant formulation including IOS was used to understand its imbibition behavior [42]. IOS is also suitable to be used in hydraulic fracturing processes [43].

Table 2. Properties of IOS obtained from supplier and literature.

Commercial Name	Type	Key Properties	Supplier
ENORDET O332	Anionic surfactant	Appearance: Colorless. Liquid at room temperature; pH: 9–12 Density: 0.7 g/cm^3 density at 23 °C Active matter (%): 28.03 Carbon atoms numbers: 15–18 Critical micelle concentration: 0.05% *	SHELL

* Abdulelah et al. [34].

The chemical structure of IOS is shown in Figure 2. It has two alkyl groups (R) with 15 to 18 carbon atoms on the tail [44]. To achieve the highest change in wettability, IOS was used at a concentration of 1 wt.%, which is well above its critical micelle concentration (CMC).

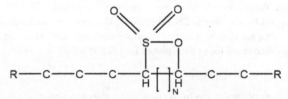

Figure 2. Chemical structure of Internal Olefin Sulfate (IOS).

2.3. Mineralogy and Topology

The mineralogy of the two shales was studied using x-ray diffractometer (Model: XPert3, PANalytical, Seri Iskandar, Malaysia). Powder forms of BG-2 and KH-2 shales were scanned from 5° to 65° with a step size of 0.026°. The basic principle of this technique is that electrons are produced from an X-ray tube and then accelerated towards the sample (shale in this study). Typical x-ray spectra are generated once electrons collide with the sample. The sample is continuously rotated by a motor, and the intensity of diffracted X-rays at an angle (2theta) is plotted. The interatomic spacing (d) is computed using Bragg's law. Each D spacing value is a signature of some minerals that are then identified by comparing these values with the database.

The elements that constitute the mineralogy of both shales were investigated utilizing energy dispersive spectrometry (EDS) by a microscope (ZEISS, Seri Iskandar, Malaysia) at an accelerating voltage of 20 kV. Visualization of the mineralogy and pores in BG-2 and KH-2 shales was acquired exploiting Field Emission Scanning Electron Microscopy (FE-SEM, Model: Zeiss Supra 55VP). The fundamental of EDS and Images is that electrons beam is focused onto the sample surface hence producing secondary electrons, backscattered electrons, and characteristics X-ray. Both secondary and backscattered electrons are used for imaging. Characteristics X-ray is used for EDS. FE-SEM follow the same working principle but produces higher resolution images than SEM [45].

2.4. TOC

Total carbon (TC) analyzer (Model: Multi N/C 3100, Analytik Jena, Seri Iskandar, Malaysia) was utilized to measure the percentage of the TOC in BG-2 and KH-2 shales. Before the measurements, the two shales were treated with Hydrochloric acid (HCL) of 37% concentration to remove the inorganic carbon. The TC analyzer utilizes the combustion approach to determine the TOC. After the sample is loaded into a ceramic boat, the amount of carbon is then determined by combustion in an oxygen environment at 1200 °C. The resulted carbon dioxide is then measured by a detector, and then carbon % can be calculated.

2.5. Fourier-Transform Infrared Spectroscopy (FTIR)

The FTIR spectra of the two shales "BG-2 and KH-2" were obtained using a Perkin Elmer (Seri Iskandar, Malaysia) spectrometer. The measurements were carried out to obtain the abundant structures in both shales. The procedures for FTIR spectroscopy involve emitting a photon to a molecule hence exciting it to higher energy level. The molecular bonds at the higher energy state vibrate at varying wavenumber. Each wavenumber corresponds to a particular functional group (e.g., C=O) [46]. FTIR spectroscopy has been utilized by many researchers to decipher the existing functional groups in many materials including but not limited to rocks [46,47] and chemicals [48]. In this study, FTIR was carried out to unravel the surface functional groups in the two shales to support the mineralogy and TOC results. Before the measurements, the two shales were dried for 12 h.

2.6. Wettability Measurement

To assess the affinity of the two shales towards water and surfactant solution, contact angles between the polished shale surfaces and water/surfactant solution were measured. The measurements were obtained using Vinci's interfacial tension meter (IFT, model IFT 700, Vinci Technology, Seri Iskandar, Malaysia) by the Sessile Drop Method. The baseline wettability of the two shales was assessed using a droplet of pure water on their polished surfaces. Their wettability for anionic surfactants was then determined using a droplet of 1 wt.% IOS solution.

2.7. Direct Measurement of Water Retention

The water retention phenomenon in the two shales was evaluated utilizing the conventional natural stone method [49] at ambient condition under two cases; baseline and with IOS solution. The procedure includes immersing 100 g of each shale in pure water/surfactant solution in a desiccator under continuous vacuuming to remove the trapped air noted. When equilibrium was achieved, the retained water for the two shales was then calculated using mass balance.

2.8. Indirect Measurement of Water Retention

Figure 3 displays the schematic of the adsorption column used in this study. It is well known from the literature that water retention in shale impairs gas flow [19,30,31,50]. The measurements were carried out to investigate the change in gas adsorption in "BG-2 and KH-2" shales due to treatment by IOS solution. To achieve that, the pressure across the adsorption column was monitored during CH_4 adsorption.

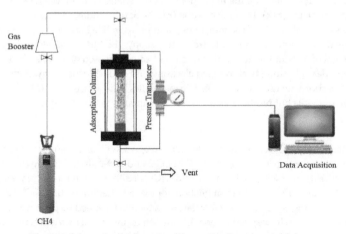

Figure 3. Schematic of Adsorption/Desorption Experimental Set-up.

The readings were then plotted versus time to unravel the adsorption behavior under the effect of water retention. The investigation was carried out in "BG-2 and KH-2" shales before and after treatment with 1wt.% of IOS solution. The U.S Bureau of Mines (USBM) adsorption procedures were followed [34].

3. Results and Discussion

3.1. Mineralogy of Shales

The mineralogy results from X-ray Powder Diffraction (XRD) of the two shales "BG-2 and KH-2" used in this study are presented in Table 3. It can be seen that BG-2 shale contains a higher amount of clay and a lower amount of non-clay minerals as compared to KH-2. The difference in mineralogy in both shales will help better explain the water retention results.

Table 3. Quantitative mineralogy of BG-2 and KH-2 shales from XRD measurement.

Sample ID	Non-Clay Minerals (wt.%)				Clay Minerals (wt.%)							
	Quartz	Kpsar	Clacite	Total (%)	Kaolinite	Smectite	Illite	Chlorite	Muscovite	Biotite	Dickite	Total (%)
BG-2	29.3	12.8	1.3	43.4	6.3	12.7	23.1	5	9.3	0.1	NA	56.6
KH-2	66	7.2	0.8	74.5	2.8	9.7	9.7	1.9	NA	0.9	0.5	25.5

The FE-SEM images of the two shales are shown in Figure 4. The platy/flaky structure of grains indicates the presence of clay. It can be seen that BG-2 (a) shale has more clay as compared to the KH-2 shale (b), which supports the XRD results.

(a)   (b)

Figure 4. FE-SEM images for (a) BG-2 and (b) KH-2 shales.

The Energy-Dispersive X-Ray (EDX) spectra for BG-2 (a) and KH-2 (b) shales is shown in Figure 5 with a corresponding miniature image of Figure 4 embedded in the graph for ease of comparison.

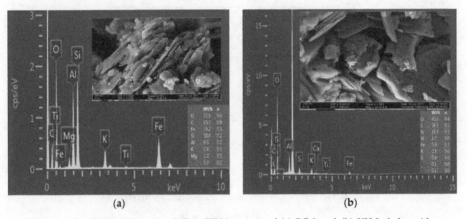

(a)   (b)

Figure 5. The Energy-Dispersive X-Ray (EDX) spectra of (a) BG-2 and (b) KH-2 shales with a corresponding miniature FE-SEM images.

The BG-2 shale spectrum shows high values of counts per second per electron-volt (cps/eV) at around 1.5 and 1.7 KeV identify Al and Si minerals respectively, thus indicating the presence of iron-rich platy crystals of clays such as kaolinite (Si, Al) and/or illite (Al, Si, K, Fe). While comparing the elements of KH-2 with BG-2, it is found that KH-2 shale is richer in carbon (C) and silicon (Si). However, it is lower in aluminum (Al), potassium (K), and iron (Fe). Furthermore, it contains a trace of calcium as opposed to magnesium (Mg) that was found in BG-2. Thus, the KH-2 shale is richer in silica and organic matters and has a lower clay content.

3.2. TOC

Figure 6 shows that KH-2 has a TOC of 12.1%, while BG-2 has only 2.1%. TOC is associated with the presence of organic matter in the shale. From the perspective of wettability, the organic matter contributes towards the hydrophobicity of shale [34]. As such, KH-2 is presumably more hydrophobic than BG-2.

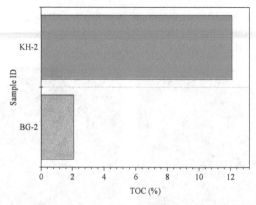

Figure 6. The total organic carbon (TOC) percentage in BG-2 and KH-2 shales.

3.3. Fourier-Transform Infrared Spectroscopy (FTIR)

Figure 7 presents the FTIR spectra of BG-2 and KH-2 shales. The broadband existing between 3404 and 3627 cm^{-1} probably corresponds to O-H stretching of hydroxyl groups which could be attributed to the existence of clay minerals.

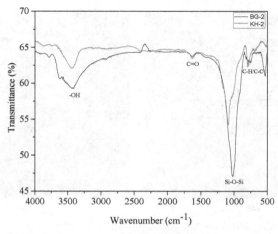

Figure 7. Fourier-Transform Infrared Spectroscopy (FTIR) Spectrum of BG-2 and KH-2 shales.

The presence of clay minerals and quartz is also inferred from the stretching of Si–O–Si band at about 1021 cm^{-1} [51]. The intensity of these bands in BG-2 is greater than KH-2 inferring that BG-2 is richer in clay than KH-2, which confirms XRD results in Table 3. The presence of organic matter is evidenced by the peaks at 1635 cm^{-1} and 676 cm^{-1}. These peaks are attributed to C=O vibration of carboxylates and deformation of CH group; respectively. These two functional groups come from organic matter [51–53].

3.4. Wettability

Figure 8 display the wettability results by contact angle method of BG-2 (a) and KH-2 (b) shales with pure water and IOS solution. The presented contact angles are the average of the contact angles taken throughout three minutes to ensure the stability of the recorded contact angles. The lower contact angle of a solution on a surface indicates that it is more wetting than pure water. In BG-2 shale, the contact angle recorded with IOS solution was nearly 3.5°, which is lower than the contact angle with pure water that was about 22°. Therefore, it can be deduced that IOS solution was more wetting than pure water for BG-2 shale. Similar but less obvious behavior was seen in KH-2 shale, where a contact angle of approximately 19.5° was measured with IOS solution compared to 37° contact angle obtained with pure water.

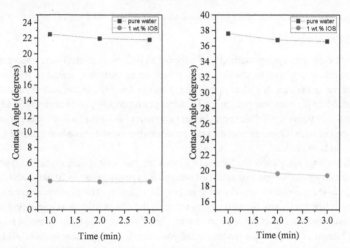

Figure 8. Measured contact angles at (**a**) BG-2 shale surface with pure water and 1 wt.% IOS solution (**b**) KH-2 shale surface with pure water and 1wt.% IOS solution.

3.5. Water Uptake

Figure 9 displays the water uptake volume in BG-2 and KH-2 shales before and after treatment with 1 wt.% of IOS solution. The amount of retained water in both shales increased significantly after treating the shales with the IOS solution. When BG-2 shale was immersed in pure distilled water, the volume of water uptake was noted to be 11.2 mL. It increased to 21 mL (an 87% increase) when immersed in a 1 wt.% IOS solution. Similar but less significant behavior was seen in KH-2 shale in which the amount of water uptake increased drastically from 8 mL in pure distilled water to 18.5 mL (a 131% increase) in 1 wt.% IOS solution. It is noteworthy to mention that the upsurge in water uptakes is possibly due to wettability alteration of shales by IOS towards more water wet. In a previous study [34] in two shales from the same formations, it was noticed that IOS changed the wettability of shales into more water- wet. These findings correlate well with the wettability results in Figure 8. IOS was more wetting than pure water in both shales hence imbibing more than pure water.

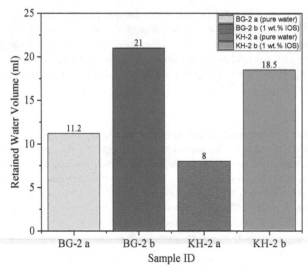

Figure 9. Water uptakes volume in BG-2 and KH-2 shales.

3.6. End Cycle Pressure vs. Time

The USBM's gas adsorption method was adopted in this study to study the adsorption behavior of shales. According to previous studies, water retention in shale was found to impair gas flow in shale. The decline in the quantity of adsorbed CH_4 was used to explain the water retention behaviors in both shales. More pressure drop during CH_4 adsorption process presumably indicates higher gas adsorption and vice versa. CH_4 adsorption in the two shales "BG-2 and KH-2" was performed before and after treatment with IOS surfactant and the pressure drop versus time was utilized to explain the water uptakes in both shales.

A crushed shale specimen (100 g) was placed in the adsorption column which was then closed/sealed and tested for gas leakage. The column was pressurized to 20 bar with CH_4, and then the connection to the gas source was closed to isolate the chamber. The pressure was noted at the end of the 2 h adsorption cycle, and the valve connecting to the CH_4 source was re-opened momentarily, and the column was re-pressurized to 20 bar again. Ten of such cycles were repeated for each shale.

Figure 10 displays the pressure reading at the end of each cycle vs. time for BG-2 and KH-2 shales. The black curve is for pure distilled water treated while the red curve is for the IOS treated shale. In the case of pure distilled water treated shale, the pressure continued decreasing till the end of the 5th adsorption cycles (10 h) indicating the continued CH_4 adsorption. The pressure at the end of the 6th cycle was similar to the 5th cycle suggesting that the rate of adsorption is leveling off. The lowest pressure observed during the test was 14 bar (a 30% change in pressure), and it occurred after about 12 h. The pressure change at the end of each subsequent cycle became gradually smaller until the last two cycles in which there was no change in pressure indicating that CH_4 adsorption is completed. When BG-2 shale was treated with IOS, the end of cycle pressure profile drastically changed. The lowest pressure observed during the test was 18 bar (only 10% change in pressure), and it occurred after just the 1st cycle (2 h). The next two cycles saw a slow reduction of the end of cycle pressure, but there was no change in pressure reduction after the 5th cycle. When comparing the behavior of the two curves, it becomes evident that IOS treatment has reduced the adsorption capacity of the shale. For the pure distilled water treated KH-2 shale, the cycle end pressure continued decreasing till the end of the 4th adsorption cycles indicating the continued CH_4 adsorption. The pressures at the end of the 5th–7th cycles were similar to the 4th cycle suggesting that the rate of adsorption was leveling off.

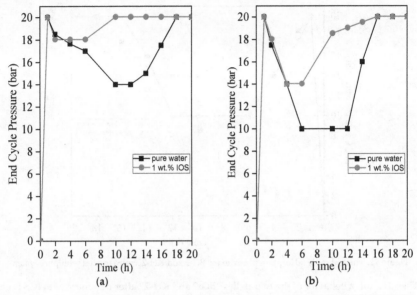

Figure 10. Column pressure at the end of adsorption cycle vs. time for (**a**) BG-2 and (**b**) KH-2 shales.

The lowest pressure observed during the test was 10 bar (a 50% change in pressure), and it occurred after about 6 h (3rd cycle). The pressure changes at the end of each subsequent cycle became gradually smaller until the last three cycles in which there was no change in pressure indicating that complete CH_4 adsorption was achieved. In the case of KH-2 that was treated with IOS, the end of cycle pressure profile significantly changed. The lowest pressure observed during the test was 14 bar (a 30% change in pressure), and it occurred after only the third cycle. The pressures at the end of 4th cycles were similar to the 3rd cycle suggesting no change in the rate of adsorption. The next four cycles saw a gradual reduction of the end of cycle pressure, but there was no change in pressure reduction after the 7th cycle. While comparing the behavior of the two curves, it becomes evident that IOS treatment has reduced the adsorption capacity of KH-2 shale.

Figure 11 shows a comparison of the two shales "BG-2 and KH-2" after treatment with pure distilled water only. The pressure at the end of each adsorption cycle was plotted vs. time. The solid curve is for KH-2 water-treated shale, while the dotted curve is for BG-2 water-treated shale. In the first adsorption cycle, a similar cycle-end pressure was noted in both shales. Later, KH-2 shale showed a more drastic change in cycle-end pressure till the end of the 3rd cycle (a 50% reduction in pressure). A less pronounced decrease in pressure was seen in BG-2 shale, where only a 30% reduction in pressure was noted at the end of the 5th cycle. When comparing the cycle-end pressure of the two shales, it becomes clear that the amount adsorbed CH_4 was more in KH-2 shale. It is most likely attributed to the fact that KH-2 is richer than BG-2 in organic matter.

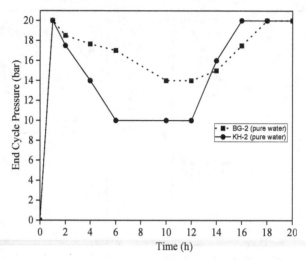

Figure 11. The end cycle adsorption pressure for BG-2 and KH-2 shales treated with pure water.

The adsorption behavior of the two shales "BG-2 and KH-2" after treatment with IOS is shown in Figure 12, where the end-cycle pressure is plotted vs. time. A reduction in pressure was noted in KH-2 shale (solid curve) up to the end of the 3rd cycle (a 30% reduction in pressure). However, it was not as significant as with the purely distilled water case cycle (was a 50% reduction in pressure). Less significant reduction in end-cycle pressure was seen in BG-2 shale, where only a 10% reduction in pressure was noted at the end of the 2nd cycle. When comparing the end cycle pressure reading in both shales, it is found that KH-2 showed higher CH_4 adsorption than BG-2, as was the case with pure water treatment.

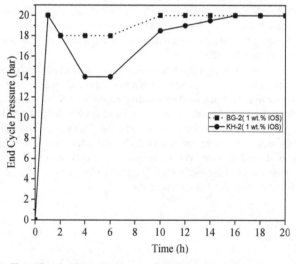

Figure 12. The end cycle adsorption pressure for BG-2 and KH-2 shales treated with IOS

It is evident from Figures 11 and 12 that IOS treatment has reduced the CH_4 adsorption in both shales as compared to pure water treatment. It should be noted that the IOS had increased the water retention in both shales as compared to pure water treatment. Presumably, the higher water retention in both shales occupied some of the available adsorption sites, thus resulting in lower gas adsorption.

4. Conclusions

In this study, the influence of anionic surfactant on water retention in shales was investigated. Two well-characterized Malaysian shales "BG-2 and KH-2" were treated with 1 wt.% IOS solution, and the changes in water uptake were noted. The water retention phenomenon was inferred from the results of the water imbibition and gas adsorption tests. When BG-2 and KH-2 shales were treated with 1 wt.% IOS solution, their water retention and CH_4 adsorption characteristics changed compared to the case when they were immersed in pure water. The water uptakes dramatically increased by 131% in KH-2 and 87% in BG-2, while CH_4 adsorption was reduced by 50% in KH-2 and 30% in BG-2. It is presumed that the higher water retention in both shales occupied some of the available adsorption sites, thus resulting in lower gas adsorption. The mineralogical analysis of the two shales showed that higher water retention and consequent lower gas adsorption was observed in BG-2 shale which had a higher clay content of 57% and low TOC of 2.1% as compared to the KH-2. The difference in the amount of retained water in both shales was found to correlate with their TOC and mineralogy. The higher affinity of BG-2 to retain a significant amount of water is possibly attributed to its high clay content and poor organic material. The relatively lower water uptake in KH-2 is presumably attributed to its high TOC of 12.1% and low clay content of 26%. As opposed to clay, organic matter is hydrophobic and thus hindering water imbibition. These results also suggest that the addition of anionic surfactant into the fracking fluid cocktail for hydraulic fracturing of shales could increase the water retention issue.

Author Contributions: Conceptualization, H.A.; methodology, H.A. and S.M.M.; software, S.A.-H.; validation, H.A., S.M.M. and M.H.H.; formal analysis, H.A.; investigation, H.A.; resources, S.M.M.; data curation, H.A and M.H.H.; writing—original draft preparation, H.A.; writing—review and editing, S.A.-H.; S.M.M. and H.A.; visualization, H.A. and S.M.M.; supervision, S.M.M. project administration, E.P.; funding acquisition, E.P.

Funding: This research was funded by Shale Gas Research Group (SGRG) and PRF-Research Grant (Cost Center 0153AB-A33).

Acknowledgments: We acknowledge the Shale Gas Research Group (SGRG) in UTP and Shale PRF project (cost center # 0153AB-A33) for the financial support. We also thank SHELL for providing the surfactant.

Conflicts of Interest: The authors declare no conflict of interest.

References

1. Mokhatab, S.; Poe, W.A. *Handbook of Natural Gas Transmission and Processing*; Gulf Professional Publishing: Boston, MA, USA, 2012.
2. Speight, J.G. *The Chemistry and Technology of Petroleum*; CRC Press: Boca Raton, FL, USA, 2014.
3. Novlesky, A.; Kumar, A.; Merkle, S. Shale Gas Modeling Workflow: From Microseismic to Simulation–A Horn River Case Study. In Proceedings of the 2011 Canadian Unconventional Resources Conference, Calgary, AB, Canada, 15–17 November 2011; p. 24.
4. Yang, L.; Ge, H.; Shi, X.; Cheng, Y.; Zhang, K.; Chen, H.; Shen, Y.; Zhang, J.; Qu, X. The effect of microstructure and rock mineralogy on water imbibition characteristics in tight reservoirs. *J. Nat. Gas Sci. Eng.* **2016**, *34*, 1461–1471. [CrossRef]
5. Rivard, C.; Lavoie, D.; Lefebvre, R.; Séjourné, S.; Lamontagne, C.; Duchesne, M. An overview of Canadian shale gas production and environmental concerns. *Int. J. Coal Geol.* **2014**, *126*, 64–76. [CrossRef]
6. Sydansk, R.D. Hydraulic Fracturing Process. U.S. Patent 5,711,376, 27 January 1998.
7. Atherton, F.; Bradfield, M.; Christmas, K.; Dalton, S.; Dusseault, M.; Gagnon, G.; Hayes, B.; MacIntosh, C.; Mauro, I.; Ritcey, R.; et al. Report of the Nova Scotia Independent Panel on Hydraulic Fracturing. Available online: https://energy.novascotia.ca/sites/default/files/Report%20of%20the%20Nova%20Scotia%20Independent%20Panel%20on%20Hydraulic%20Fracturing.pdf (accessed on 30 November 2018).
8. Cherry, J.; Ben-Eli, M.; Bharadwaj, L.; Chalaturnyk, R.; Dusseault, M.B.; Goldstein, B.; Lacoursière, J.-P.; Matthews, R.; Mayer, B.; Molson, J. *Environmental Impacts of Shale Gas Extraction in Canada*; Council of Canadian Academies: Ottawa, ON, Canada, 2014.

9. King, G.E. Hydraulic fracturing 101: What every representative, environmentalist, regulator, reporter, investor, university researcher, neighbor, and engineer should know about hydraulic fracturing risk. *J. Petrol. Technol.* **2012**, *64*, 34–42. [CrossRef]

10. Shen, Y.; Ge, H.; Meng, M.; Jiang, Z.; Yang, X. Effect of water imbibition on shale permeability and its influence on gas production. *Energy Fuels* **2017**, *31*, 4973–4980. [CrossRef]

11. Engelder, T.; Cathles, L.M.; Bryndzia, L.T. The fate of residual treatment water in gas shale. *J. Unconv. Oil Gas Resour.* **2014**, *7*, 33–48. [CrossRef]

12. Penny, G.S.; Dobkins, T.A.; Pursley, J.T. Field Study of Completion Fluids To Enhance Gas Production in the Barnett Shale. In Proceedings of the 2006 SPE Gas Technology Symposium, Calgary, AB, Canada, 15–18 May 2006; p. 10.

13. Nicot, J.-P.; Scanlon, B.R. Water use for shale-gas production in Texas, US. *Environ. Sci. Technol.* **2012**, *46*, 3580–3586. [CrossRef] [PubMed]

14. Makhanov, K.; Habibi, A.; Dehghanpour, H.; Kuru, E. Liquid uptake of gas shales: A workflow to estimate water loss during shut-in periods after fracturing operations. *J. Unconv. Oil Gas Resour.* **2014**, *7*, 22–32. [CrossRef]

15. Reagan, M.T.; Moridis, G.J.; Keen, N.D.; Johnson, J.N. Numerical simulation of the environmental impact of hydraulic fracturing of tight/shale gas reservoirs on near-surface groundwater: Background, base cases, shallow reservoirs, short-term gas, and water transport. *Water Resour. Res.* **2015**, *51*, 2543–2573. [CrossRef] [PubMed]

16. Sun, Y.; Bai, B.; Wei, M. Microfracture and surfactant impact on linear cocurrent brine imbibition in gas-saturated shale. *Energy Fuels* **2015**, *29*, 1438–1446. [CrossRef]

17. Vengosh, A.; Jackson, R.B.; Warner, N.; Darrah, T.H.; Kondash, A. A critical review of the risks to water resources from unconventional shale gas development and hydraulic fracturing in the United States. *Environ. Sci. Technol.* **2014**, *48*, 8334–8348. [CrossRef] [PubMed]

18. DiGiulio, D.C.; Wilkin, R.T.; Miller, C.; Oberley, G. Investigation of Ground Water Contamination Near Pavillion. Presented at the Wyoming Workgroup Meeting, Pavillion, WY, USA, 30 November 2011.

19. Ge, H.-K.; Yang, L.; Shen, Y.-H.; Ren, K.; Meng, F.-B.; Ji, W.-M.; Wu, S. Experimental investigation of shale imbibition capacity and the factors influencing loss of hydraulic fracturing fluids. *Pet. Sci.* **2015**, *12*, 636–650. [CrossRef]

20. Vidic, R.D.; Brantley, S.L.; Vandenbossche, J.M.; Yoxtheimer, D.; Abad, J.D. Impact of shale gas development on regional water quality. *Science* **2013**, *340*, 1235009. [CrossRef] [PubMed]

21. Cheng, Y. Impact of water dynamics in fractures on the performance of hydraulically fractured wells in gas-shale reservoirs. *J. Can. Pet. Technol.* **2012**, *51*, 143–151. [CrossRef]

22. Yan, Q.; Lemanski, C.; Karpyn, Z.T.; Ayala, L. Experimental investigation of shale gas production impairment due to fracturing fluid migration during shut-in time. *J. Nat. Gas Sci. Eng.* **2015**, *24*, 99–105. [CrossRef]

23. Wright, P.R.; McMahon, P.B.; Mueller, D.K.; Clark, M.L. *Groundwater-Quality and Quality-Control Data for Two Monitoring wells Near Pavillion, Wyoming, April and May 2012*; 2327-638X; US Geological Survey: Reston, VA, USA, 2012.

24. Thyne, G. Review of Phase II Hydrogeologic Study. Available online: https://www.garfield-county.com/oil-gas/documents/Thyne%20FINAL%20Report%2012[1].20.08.pdf (accessed on 30 November 2018).

25. Birdsell, D.T.; Rajaram, H.; Dempsey, D.; Viswanathan, H.S. Hydraulic fracturing fluid migration in the subsurface: A review and expanded modeling results. *Water Resour. Res.* **2015**, *51*, 7159–7188. [CrossRef]

26. Myers, T. Potential contaminant pathways from hydraulically fractured shale to aquifers. *Groundwater* **2012**, *50*, 872–882. [CrossRef] [PubMed]

27. Taherdangkoo, R.; Tatomir, A.; Taylor, R.; Sauter, M. Numerical investigations of upward migration of fracking fluid along a fault zone during and after stimulation. *Energy Procedia* **2017**, *125*, 126–135. [CrossRef]

28. Tatomir, A.; McDermott, C.; Bensabat, J.; Class, H.; Edlmann, K.; Taherdangkoo, R.; Sauter, M. Conceptual model development using a generic Features, Events, and Processes (FEP) database for assessing the potential impact of hydraulic fracturing on groundwater aquifers. *Adv. Geosci.* **2018**, *45*, 185–192. [CrossRef]

29. Gallegos, T.J.; Varela, B.A.; Haines, S.S.; Engle, M.A. Hydraulic fracturing water use variability in the United States and potential environmental implications. *Water Resour. Res.* **2015**, *51*, 5839–5845. [CrossRef] [PubMed]

30. Sharma, M.; Agrawal, S. Impact of Liquid Loading in Hydraulic Fractures on Well Productivity. In Proceedings of the 2013 SPE Hydraulic Fracturing Technology Conference, The Woodlands, TX, USA, 4–6 February 2013; p. 16.

31. Shanley, K.W.; Cluff, R.M.; Robinson, J.W. Factors controlling prolific gas production from low-permeability sandstone reservoirs: Implications for resource assessment, prospect development, and risk analysis. *AAPG Bull.* **2004**, *88*, 1083–1121. [CrossRef]

32. Hematpour, H.; Mahmood, S.M.; Nasr, N.H.; Elraies, K.A. Foam flow in porous media: Concepts, models and challenges. *J. Nat. Gas Sci. Eng.* **2018**. [CrossRef]

33. Dehghanpour, H.; Lan, Q.; Saeed, Y.; Fei, H.; Qi, Z. Spontaneous imbibition of brine and oil in gas shales: Effect of water adsorption and resulting microfractures. *Energy Fuels* **2013**, *27*, 3039–3049. [CrossRef]

34. Abdulelah, H.; Mahmood, S.M.; Al-Mutarreb, A. The Effect of Anionic Surfactant on the Wettability of Shale and its Implication on Gas Adsorption/Desorption Behavior. *Energy Fuels* **2018**. [CrossRef]

35. Settari, A.; Sullivan, R.B.; Bachman, R.C. The Modeling of the Effect of Water Blockage and Geomechanics in Waterfracs. In Proceedings of the 2002 SPE Annual Technical Conference and Exhibition, San Antonio, TX, USA, 29 September–2 October 2002; p. 7.

36. Fisher, M.K.; Warpinski, N.R. Hydraulic-fracture-height growth: Real data. *SPE. Prod. Oper.* **2012**, *27*, 8–19. [CrossRef]

37. Huynh, U.T. *Surfactant Characterization to Improve Water Recovery in Shale Gas Reservoirs*; The University of Texas at Austin: Austin, TX, USA, 2013.

38. Patel, P.S.; Robart, C.J.; Ruegamer, M.; Yang, A. Analysis of US Hydraulic Fracturing Fluid System and Proppant Trends. In Proceedings of the 2014 SPE Hydraulic Fracturing Technology Conference, The Woodlands, TX, USA, 4–6 February 2014; p. 20.

39. Zhou, L.; Das, S.; Ellis, B.R. Effect of surfactant adsorption on the wettability alteration of gas-bearing shales. *Environ. Eng. Sci.* **2016**, *33*, 766–777. [CrossRef]

40. Al-Mutarreb, A.; Jufar, S.R.; Abdulelah, H.; Padmanabhan, E. Influence of Water Immersion on Pore System and Methane Desorption of Shales: A Case Study of Batu Gajah and Kroh Shale Formations in Malaysia. *Energies* **2018**, *11*, 1511. [CrossRef]

41. Baioumy, H.; Ulfa, Y.; Nawawi, M.; Padmanabhan, E.; Anuar, M.N.A. Mineralogy and geochemistry of Palaeozoic black shales from Peninsular Malaysia: Implications for their origin and maturation. *Int. J. Coal Geol.* **2016**, *165*, 90–105. [CrossRef]

42. Wang, D.; Butler, R.; Zhang, J.; Seright, R. Wettability survey in Bakken shale with surfactant-formulation imbibition. *SPE Reservoir Eval. Eng.* **2012**, *15*, 695–705. [CrossRef]

43. Semple, T.C.; Reznik, C.; Barnes, J.R.; Buechele, J.L.; Dubey, S.T.; King, T.E. Use of Long Chain Internal Olefin Sulfonates. U.S. Patent 2014/0353250 A1, 4 December 2014.

44. Nasr, N.H.; Mahmood, S.M.; Hematpur, H. A rigorous approach to analyze bulk and coreflood foam screening tests. *J. Pet. Explor. Prod. Technol.* **2018**, 1–14.

45. Mutalib, M.A.; Rahman, M.; Othman, M.; Ismail, A.; Jaafar, J. Scanning Electron Microscopy (SEM) and Energy-Dispersive X-Ray (EDX) Spectroscopy. In *Membrane Characterization*; Elsevier: Amsterdam, The Netherlands, 2017; pp. 161–179.

46. Chen, Y.; Zou, C.; Mastalerz, M.; Hu, S.; Gasaway, C.; Tao, X. Applications of micro-fourier transform infrared spectroscopy (FTIR) in the geological sciences—A review. *Int. J. Mol. Sci.* **2015**, *16*, 30223–30250. [CrossRef] [PubMed]

47. Cronauer, D.; Snyder, R.; Painter, P. Characterization of Oil Shale by FTIR Spectroscopy. Available online: https://web.anl.gov/PCS/acsfuel/preprint%20archive/Files/27_2_LAS%20VEGAS_03-82_0122.pdf (accessed on 28 November 2018).

48. Akbari, S.; Mahmood, S.M.; Tan, I.M.; Ghaedi, H.; Ling, O.L. Assessment of Polyacrylamide Based Co-Polymers Enhanced by Functional Group Modifications with Regards to Salinity and Hardness. *Polymers* **2017**, *9*, 647. [CrossRef]

49. En, U. *Natural Stone Test Methods-Determination of Water Absorption at Atmospheric Pressure*; British Standards Institution: London, UK, 2008.

50. Ford, W.G.; Penny, G.S.; Briscoe, J.E. Enhanced water recovery improves stimulation results. *SPE Prod. Eng.* **1988**, *3*, 515–521. [CrossRef]

51. Khraisheh, M.A.; Al-degs, Y.S.; Mcminn, W.A. Remediation of wastewater containing heavy metals using raw and modified diatomite. *Chem. Eng. J.* **2004**, *99*, 177–184. [CrossRef]
52. Van der Marel, H.W.; Beutelspacher, H. *Atlas of Infrared Spectroscopy of Clay Minerals and Their Admixtures*; Elsevier Publishing Company: Amsterdam, The Netherlands, 1976.
53. Haberhauer, G.; Rafferty, B.; Strebl, F.; Gerzabek, M. Comparison of the composition of forest soil litter derived from three different sites at various decompositional stages using FTIR spectroscopy. *Geoderma* **1998**, *83*, 331–342. [CrossRef]

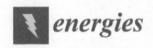

Article

Closed-Form Solution of Radial Transport of Tracers in Porous Media Influenced by Linear Drift

Lateef T. Akanji [1],* and Gabriel K. Falade [2]

[1] Division of Petroleum Engineering, School of Engineering, University of Aberdeen, Aberdeen AB24 3FX, UK
[2] Department of Petroleum Engineering, University of Ibadan, Ibadan 23402, Nigeria; faladegk@gmail.com
* Correspondence: l.akanji@abdn.ac.uk; Tel.: +44-1-224-27-2793

Received: 3 November 2018; Accepted: 17 December 2018; Published: 22 December 2018

Abstract: A new closed-form analytical solution to the radial transport of tracers in porous media under the influence of linear drift is presented. Specifically, the transport of tracers under convection–diffusion-dominated flow is considered. First, the radial transport equation was cast in the form of the Whittaker equation by defining a set of transformation relations. Then, linear drift was incorporated by considering a coordinate-independent scalar velocity field within the porous medium. A special case of low-intensity tracer injection where molecular diffusion controls tracer propagation but convection with linear velocity drift plays a significant role was presented and solved in Laplace space. Furthermore, a weak-form numerical solution of the nonlinear problem was obtained and used to analyse tracer concentration behaviour in a porous medium, where drift effects predominate and influence the flow pattern. Application in enhanced oil recovery (EOR) processes where linear drift may interfere with the flow path was also evaluated within the solution to obtain concentration profiles for different injection models. The results of the analyses indicated that the effect of linear drift on the tracer concentration profile is dependent on system heterogeneity and progressively becomes more pronounced at later times. This new solution demonstrates the necessity to consider the impact of drift on the transport of tracers, as arrival times may be significantly influenced by drift intensity.

Keywords: transport of tracers; linear drift effect; convection–diffusion equation; enhanced oil recovery; closed-form analytical solution

1. Introduction

The study of the transport of tracers has become an essential technique for porous media characterisation, particularly in enhanced oil recovery (EOR) in hydrocarbon reservoirs (e.g., Baldwin [1]), hydrology (e.g., Rubin and James [2]), nuclear (e.g., Moreno et al. [3] and Herbert et al. [4]), drug transport in blood vessels (e.g., Mabuza et al. [5]) and geothermal engineering (e.g., Vetter and Zinnow [6]). Multiple processes and mechanisms are usually involved in the chemical interaction of the constituent components when the tracer is being transported through a porous medium. Two major processes involved in the transport phenomenon include convection and hydrodynamic dispersion. The convection process involves bulk movement of fluids, while hydrodynamic dispersion describes the dual actions of molecular dispersion and shear or mechanical mixing process. These complementary transport processes are adequately captured by the well-known convection–dispersion–diffusion equations with or without chemical reactions (e.g., Tomich et al. [7], Bear [8], Zhou and Zhan [9]).

Surfactant or biosurfactant partitioning and transport in the oil phase during enhanced oil recovery processes is usually neglected, because, it requires the solution of a system of nonlinear coupled partial differential equations whose solution is numerically challenging [10]. These diffusion equations are based on linear or one-dimensional geometry due to the relative ease with which such equations can be solved analytically. Recently, several authors have studied hydrodynamic transport in porous media using the random walk method (see a review paper by Noetinger et al. [11] and references cited therein). Approximate solutions have also been presented in modelling of radial geometry under conditions of shear mixing, albeit approximate in nature [12]. Exact analytical solutions have been obtained in cases where convective velocity and hydrodynamic dispersion functions were assumed constant (e.g., Carslaw and Jaeger [13]) and in porous media where tracer adsorption, non-uniform convection and variable dispersion manifest (e.g., Falade and Brigham [14]).

Attinger and Abdulle [15] studied the effective drift of transport problems in heterogeneous compressible flows. They discussed the impact of a mean drift and showed that static compressible flow with mean drift can produce a heterogeneity-driven large-scale drift or ballistic transport. A similar study was carried out by Vergassola and Avellaneda [16], where it was demonstrated that for static compressible flow without mean drift, there is no impact on the large-scale drift. The calculation of the effective ballistic velocity V_b was reduced to the solution of one auxiliary equation. They derived an analytic expression for V_b for some special instances where flow depends on a single coordinate, random with short correlation times and slightly compressible cellular flow. Transport will be depleted due of the trapping for arbitrary time-independent potential flow and for time-dependent potential flow or generic compressible flow, transport will be enhanced or depleted depending on the velocity field. Vergassola and Avellaneda [16] also discovered that trapping due to flow compressibility may enhance particle spreading, leading to ballistic transport that is very efficient.

In field applications, particularly during EOR involving chemical injections such as surfactants, alkali or polymer, fluid migration in an active or partially active aquifer formation may lead to displacement of the injected chemical during the shut-in period. Linear drift may also occur as a result of interference by the production/injection well, which is hydraulically connected to the formation of interest. Investigation conducted by Tomich et al. [7] indicated that in a single-well test involving the injection of ethyl acetate, fluid migration in the formation due to a reservoir water drive might lead to displacement of the tracer bank during the shut-in period. The injection of the ethyl acetate was followed by the injection of a water bank, allowing the system to hydrolyse during the chemical reaction to form ethanol, a secondary tracer. The difference in magnitude of the velocity of arrival of the two tracers was used in estimating the residual oil saturation.

The occurrence of linear drift may lead to the flow path being rerouted, leading to inaccurate and inconclusive tests with ultimate financial implications. Moench and Ogata [17] applied Laplace transform as described by Stehfest [18] to solve the dispersion in a radial flow in a porous medium. The resulting Airy function was computed using the series representation for $|z| > 1$ [19]. Mashayekhizadeh et al. [20] applied Fourier series methods to numerically solve the Laplace transform of a pressure distribution equation for radial flow in porous media. Other authors (e.g., De-Hoog et al. [21], Dubner and Abate [22], Zakian [23] and Schapery [24] and Brzeziński and Ostalczyk [25]) have proposed improved techniques for numerical inversion of Laplace transforms, typically by accelerating the convergence of the Fourier series.

The occurrence of advection–dispersion with the influence of drift is vast and may occur in petroleum reservoirs with underlying aquifer, CO_2–EOR processes and in contaminant hydrology. Understanding flow and transport behaviour in porous media where drift may occur is important in radioactive waste management due to the possible longevity of radionuclide materials and the possibility of being rerouted to the surface environment during transport processes. Despite the extensive research in this field, particularly in solving the advection–dispersion equation (ADE) both analytically and numerically, there is yet to be a consideration for the closed-form solution of the ADE in systems where the effect of linear drift may predominate.

A Fickian solution (Fick [26]) to the the convection–diffusion equation can be easily obtained for the simple cases where velocity and hydrodynamic dispersion are constant and the reaction term is either zero or first order in concentration (e.g., Falade and Brigham [14] and Skellam [27]). However, in cases where hydrodynamic dispersion is radially distributed and linear drift predominates, an exact analytical solution to the transport equation has not been reported in the literature. In this work, a closed-form solution of the transport of tracers in porous media under the influence of linear drift is presented. First, the radial transport equation is cast in the form of the Whittaker equation [28] by defining a set of transformation relations and a change of variables. Linear drift is incorporated by considering a coordinate-independent scalar velocity field within the porous medium. A special case of low intensity tracer injection where molecular diffusion controls tracer propagation but convection with linear velocity drift plays a significant role is presented and solved in Laplace space. Second, the concentration distribution around the source of tracer injection is solved analytically in radial coordinate and the obtained result transformed to the equivalent Cartesian coordinates system. A weak-form numerical solution is then obtained and used to analyse tracer concentration behaviour in enhanced oil recovery (EOR) processes where linear drift effect may interfere with the fluid flow path.

2. Radial Diffusion Models with Drift

Figure 1 shows a schematic representation of a chemical tracer injection in a single-well test, indicating (a) the injection of a chemical, (b) the reaction between the injected chemical and the injected water bank and (c) the production stage without the influence of drift. Figure 1d–f shows the same process as highlighted in Figure 1a–c, but, with underlying aquifer causing a noticeable drift effect during the production stage (f) (e.g., Tomich et al. [7]).

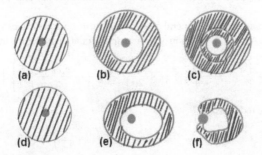

Figure 1. Schematic representation of chemical tracer method in a single well test involving injection of tracer (a–c) without drift and (d–f) with drift.

The transport of tracers in a constant flow of carrier fluid flowing in a porous medium governed by the convection–diffusion equation expressed in terms of resident concentration in radial coordinates can be written as (see, for instance, Falade and Brigham [14]) :

$$\frac{1}{r}\frac{\partial}{\partial r}\left(r\phi D\frac{\partial C}{\partial r}\right) - \frac{1}{r}\frac{\partial}{\partial r}\left(r\phi v C\right) - \gamma\left(\kappa_r + s\right)C = \frac{\partial \phi C}{\partial t}, \tag{1}$$

$$v = \left(v_x \cos\theta + \frac{q_i}{2\pi r\phi h}\right) \tag{2}$$

$$= \left(v_d + \frac{\alpha}{r}\right), \tag{3}$$

where the composite velocity v (m/s) consists of the linear flow velocity v_d (m/s) superimposed on a radial flow of the injected tracer of strength q_i (m^3/s). When the flow of the tracer is influenced by linear drift, the linear flow velocity will consist of both radial and tangential velocity components:

$$v_d = v_r + v_t \tag{4}$$

$$= \frac{dr}{dt}\hat{e}_r + r\frac{d\theta}{dt}\hat{e}_\theta. \tag{5}$$

For low intensity tracer injection, the tangential velocity is negligibly small compared to the radial velocity. Other variables α, γ and D are defined as follows:

$$\alpha = \frac{q_i}{2\pi r \phi h}, \tag{6}$$

$$\gamma = \theta \left(\frac{S_{mp} + \kappa_l \left(1 - S_{mp} \right)}{S_{mp}} \right), \tag{7}$$

and D is the flow hydrodynamic dispersion (m^2/s), s is Laplace parameter, S_m is the mobile fluid phase saturation, q_i is the tracer injection rate (m^3/s), r is radial distance (m), ϕ is porosity and h is porous media thickness (m). Hydrodynamic dispersion D is generally believed to be made up of two components—molecular diffusion and shear mixing—which can be expressed as:

$$D = D_m + D_o|v_{mr_D}|. \tag{8}$$

In Equation (8), D_m is the molecular diffusion constant (m^2/s) and D_o is the shear mixing constant (m).

The dimensionless form of the general convection–diffusion equation can be written in Laplace space as (see Appendix A for the derivation of the transport equation under the influence of linear drift):

$$\frac{d^2\Psi}{dr_D^2} - \left\{ \frac{\alpha^2 (\omega r_D + 1)^2 - \kappa^2}{4(\kappa r_D + \beta)^2} + \frac{\alpha\omega}{2(\kappa r_D + \beta_w)} + \frac{\phi(\kappa_r + s)r_D}{(\kappa r_D + \beta)} \right\} \Psi = 0, \tag{9}$$

where the variables are redefined as:

$$D(r_{Dk}) = D_m + v_m D_o \left(\omega + \frac{1}{r_D} \right), \tag{10}$$

$$= \kappa + \frac{\beta}{r_D}, \tag{11}$$

and

$$\kappa = D_m + \beta\omega \tag{12}$$

$$v_w = \frac{\alpha}{r_w}, \tag{13}$$

$$\beta = \frac{\beta_w(\alpha D_o)}{r_w}, \tag{14}$$

$$= v_w D_o, \tag{15}$$

$$\gamma = \theta \left(\frac{S_m + \kappa_l \left(1 - S_m \right)}{S_m} \right) r_w^2. \tag{16}$$

3. Mathematical Formulation of the Radial Transport Equation with Linear Drift

In order to establish the radial transport equation where linear drift effect can be incorporated, the following transformation relations are defined:

$$\eta = (\kappa r_D + \beta), \tag{17}$$

$$r_D = (\eta - \beta)\frac{1}{\kappa}, \tag{18}$$

$$\frac{d\eta}{dr_D} = \kappa, \tag{19}$$

$$\frac{d^2\Psi}{dr_D^2} = \kappa^2 \frac{d^2\Psi}{d\eta^2}, \tag{20}$$

and applying the transformation to Equation (9):

$$\kappa^2 \frac{d^2\Psi}{d\eta^2} - \left\{ \overset{\underset{i}{\downarrow}}{\frac{\alpha^2 \left[\frac{\omega}{\kappa}(\eta - \beta) + 1\right]^2 - \kappa^2}{4\eta^2}} + \frac{\alpha\omega}{2\eta} + \overset{\underset{ii}{\downarrow}}{\frac{\phi(\kappa_r + s)\frac{1}{\kappa}(\eta - \beta)}{\eta}} \right\} \Psi = 0. \tag{21}$$

Note:

$$\left[\frac{\omega}{\kappa}(\eta - \beta) + 1\right]^2 = \frac{\omega^2}{\kappa^2}(\eta - \beta)^2 + \frac{2\omega}{\kappa}(\eta - \beta) + 1$$

$$= \frac{\omega^2}{\kappa^2}\eta^2 - \frac{2\omega}{\kappa}\left(\frac{\omega\beta}{\kappa} - 1\right)\eta + \left(\frac{\omega\beta}{\kappa} - 1\right)^2. \tag{22}$$

Therefore, the term highlighted as (i) in Equation (21) can be written out by considering the relational expression (Equation (22)), thus:

$$\frac{\alpha^2 \left[\frac{\omega}{\kappa}(\eta - \beta) + 1\right]^2 - \kappa^2}{4\eta^2} = \left[\frac{\alpha^2\omega^2}{\kappa^2}\eta^2 - \frac{2\alpha^2\omega}{\kappa}\left(\frac{\omega\beta}{\kappa} - 1\right)\eta + \alpha^2\left(\frac{\omega\beta}{\kappa} - 1\right)^2 - \kappa^2\right]\frac{1}{4\eta^2}$$

$$= \frac{\alpha^2\omega^2}{4\kappa^2} - \frac{\alpha^2\omega}{2\kappa}\left(\frac{\omega\beta}{\kappa} - 1\right)\frac{1}{\eta} + \frac{1}{4\eta^2}\left[\alpha^2\left(\frac{\omega\beta}{\kappa} - 1\right)^2 - \kappa^2\right]. \tag{23}$$

Similarly, the term highlighted as (ii) in Equation (21) can be rewritten thus:

$$\frac{\phi(\kappa_r + s)(\eta - \beta)\frac{1}{\kappa}}{\eta} = \frac{\phi(\kappa_r + s)}{\kappa} - \frac{\beta\phi(\kappa_r + s)}{\kappa\eta}. \tag{24}$$

Hence, Equation (21) can now be written as:

$$\kappa^2 \frac{d^2\Psi}{d\eta^2} = 0 - \left\{ \overset{\underset{i}{\downarrow}}{\frac{1}{4\eta^2}\left[\alpha^2\left(\frac{\omega\beta}{\kappa} - 1\right)^2 - \kappa^2\right]} - \frac{\omega}{2}\left[\frac{\alpha^2}{\kappa}\left(\frac{\omega\beta}{\kappa} - 1\right) - \alpha + \overset{\underset{ii}{\downarrow}}{\frac{2\beta\phi(\kappa_r + s)}{\omega\kappa}}\right]\frac{1}{\eta} \right\} \Psi$$

$$+ \left\{\frac{\alpha^2\omega^2}{4\kappa^2} + \frac{\phi(\kappa_r + s)}{\kappa}\right\}\Psi = 0. \tag{25}$$

Expanding the terms highlighted as i and ii in Equation (25) and rearranging:

$$\frac{d^2\Psi}{d\eta^2} - \left[\frac{\frac{\alpha^2}{\kappa^2}\left(\frac{\omega\beta}{\kappa}-1\right)^2 - 1}{4\eta^2} - \frac{\omega}{2\kappa^2}\left[\frac{\alpha^2}{\kappa}\left(\frac{\omega\beta}{\kappa}-1\right) - \alpha + \frac{2\beta\phi(\kappa_r+s)}{\omega\kappa}\right]\frac{1}{\eta} \right]\Psi$$

$$+ \left[\left(\frac{\alpha^2\omega^2}{4\kappa^2} + \frac{\phi(\kappa_r+s)}{\kappa} \right) \frac{1}{\kappa^2} \right]\Psi = 0. \quad (26)$$

Simplifying, Equation (26):

$$\frac{d^2\Psi}{d\eta^2} + \left[\frac{1-\frac{\alpha^2}{\kappa^2}\left(\frac{\omega\beta}{\kappa}\right)^2 - 1}{4\eta^2} + \frac{\omega}{2\kappa}\left[\frac{\alpha^2}{\kappa^2}\left(\frac{\omega\beta}{\kappa}-1\right) - \frac{\alpha}{\kappa} + \frac{2\beta\phi(\kappa_r+s)}{\omega\kappa^2}\right]\frac{1}{\eta} \right]\Psi$$

$$- \left[\left(\frac{\alpha^2\omega^2}{4\kappa^2} + \frac{\phi(\kappa_r+s)}{\kappa} \right) \frac{1}{\kappa^2} \right]\Psi = 0. \quad (27)$$

Equation (27) can be cast in the form of the Whittaker equation if a change of variable is defined thus:

$$\xi = 2\eta\sqrt{a}. \quad (28)$$

where:

$$a = \frac{1}{\kappa^2}\left(\frac{\alpha^2\omega^2}{4\kappa^2} + \frac{\phi(\kappa_r+s)}{\kappa} \right) \quad (29)$$

$$= f(s)^2. \quad (30)$$

Then:

$$\frac{d\Psi}{d\eta} = \frac{d\Psi}{d\xi}\frac{d\xi}{d\eta} \quad (31)$$

$$= 2f(s)\frac{d\Psi}{d\eta}, \quad (32)$$

and

$$\frac{d^2\Psi}{d\eta^2} = 4f(s)^2\frac{d^2\Psi}{d\xi^2}. \quad (33)$$

Using Equations (32) and (33) in Equation (27) gives:

$$\frac{d^2\Psi}{d\xi^2} + \left[\frac{1-\frac{\alpha^2}{\kappa^2}\left(\frac{\omega\beta}{\kappa}-1\right)^2}{4\xi^2} + \frac{\omega}{4\sqrt{a\kappa}}\left[\frac{\alpha^2}{\kappa^2}\left(\frac{\omega\beta}{\kappa}-1\right) - \frac{\alpha}{\kappa} + \frac{2\beta\phi(\kappa_r+s)}{\omega\kappa^2}\right]\frac{1}{\xi} - \frac{1}{4} \right]\Psi = 0 \quad (34)$$

$$\frac{d^2\Psi}{d\xi^2} + \left[\frac{1-\frac{\alpha^2}{\kappa^2}\left(\frac{\omega\beta}{\kappa}-1\right)^2}{4\xi^2} + \frac{\omega}{4\kappa f(s)}\left[\frac{\alpha^2}{\kappa^2}\left(\frac{\omega\beta}{\kappa}-1\right) - \frac{\alpha}{\kappa} + \frac{2\beta}{\omega\kappa}\left(\kappa^2 f(s)^2 - \frac{\alpha^2\omega^2}{4\kappa^2}\right)\right]\frac{1}{\xi} - \frac{1}{4} \right]\Psi = 0 \quad (35)$$

$$\frac{d^2\Psi}{d\xi^2} + \left[\frac{1-\frac{\alpha^2}{\kappa^2}\left(\frac{\omega\beta}{\kappa}-1\right)^2}{4\xi^2} + \frac{\frac{\omega\alpha^2}{4\kappa^3}\left(\frac{\omega\beta}{2\kappa}-1\right) - \frac{\alpha\omega}{4\kappa^2}}{f(s)} - \frac{\beta}{2}f(s)\right]\frac{1}{\xi} - \frac{1}{4} \right]\Psi = 0 \quad (36)$$

$$\frac{d^2\Psi}{d\varsigma^2} + \left\{ \frac{1 - \frac{\alpha^2}{\kappa^2}\left(\frac{\omega\beta}{\kappa} - 1\right)^2}{4\varsigma^2} + \left[\frac{\frac{\alpha\omega}{4\kappa^2}\left[\frac{\alpha}{\kappa}\left(\frac{\omega\beta}{\kappa} - 1\right) - 1\right]}{f(s)} - \frac{\beta}{2}f(s) \right]\frac{1}{\varsigma} - \frac{1}{4} \right\}\Psi = 0, \tag{37}$$

which can be expressed in the form of the Whittaker equation [28] as:

$$\frac{d^2\Psi}{dz^2} + \left[-\frac{1}{4} + \frac{K}{z} + \frac{\left(\frac{1}{4} - \mu^2\right)}{z^2} \right]\Psi = 0, \tag{38}$$

where:

$$\mu = \frac{\alpha}{2\kappa}\left(\frac{\omega\beta}{\kappa} - 1\right), \tag{39}$$

and:

$$K = \left[\frac{\frac{\alpha\omega}{4\kappa^2}\left[\frac{\alpha}{\kappa}\left(\frac{\omega\beta}{\kappa} - 1\right) - 1\right]}{f(s)} - \frac{\beta}{2}f(s) \right], \tag{40}$$

$$f(s) = \frac{1}{\kappa}\sqrt{\left(\frac{\alpha^2\omega^2}{4\kappa^2} + \frac{\gamma(\kappa_r + s)}{\kappa}\right)} \tag{41}$$

$$\varsigma = 2(\kappa r_D + \beta)f(s), \tag{42}$$
$$= 2\eta f(s) \tag{43}$$

3.1. Introducing the Linear Drift

Linear drift can be introduced to the convection–diffusion equation by applying it as a scalar velocity field v_d, since it is coordinate-independent, having a magnitude that acts on every point within the porous body. For the purpose of this analysis, linear drift is applied on the x-direction only. Using the general hydrodynamic description of the diffusivity coefficient:

$$D(r) = D_m + D_o|v_d|, \tag{44}$$
$$= D_m + \alpha D_o\left(\overline{\omega} + \frac{1}{r}\right), \tag{45}$$
$$= D_m + v_m D_o\left(\omega + \frac{1}{r_D}\right), \tag{46}$$

and defining the following relational variables:

$$\alpha\overline{\omega} = v_m\omega, \tag{47}$$
$$\frac{\alpha}{v_m} = \frac{\omega}{\overline{\omega}} = r_w, \tag{48}$$

and dimensionless variables:

$$\frac{\alpha}{r} = \frac{v_m}{r_D} \tag{49}$$
$$r_D = \frac{r}{r_w} \tag{50}$$
$$v_d = v - \frac{\alpha}{r}. \tag{51}$$

Equation (46) can be written out after expanding and rearranging thus:

$$D(r) = D_m + \beta_w \left(\varpi + \frac{1}{r}\right) \tag{52}$$

$$= \kappa + \frac{\beta_w}{r}, \tag{53}$$

where:

$$\kappa = D_m + \beta_w \bar{\varpi}, \tag{54}$$

$$\bar{\varpi} = \left(\frac{v_d \cos\theta}{\alpha}\right), \tag{55}$$

$$\beta_w = \alpha D_o, \tag{56}$$

$$\alpha = \left(\frac{Q}{2\pi h\theta}\right). \tag{57}$$

3.2. Analytical Solution

The general solution of the Whittaker equation [28] is given as:

$$\Psi(\eta, s) = A(s) M_{\kappa,\mu}(\xi) + B(s) W_{\kappa,\mu}(\xi). \tag{58}$$

In Equation (58), $A(s)$ and $B(s)$ are arbitrary functions of 's' to be determined by the requirements of the boundary conditions, while $M_{\kappa,\mu}(\xi)$ and $W_{\kappa,\mu}(\xi)$ are the Whittaker function, which can also be defined in terms of Kummer's confluent hypergeometric functions as:

$$M_{\kappa,\mu}(\xi) = e^{-\frac{\xi}{2}} \xi^{\frac{1}{2}+\mu} M\left(\frac{1}{2} + \mu - \kappa, 1 + 2\mu, \xi\right) \tag{59}$$

$$W_{\kappa,\mu}(\xi) = e^{-\frac{\xi}{2}} \xi^{\frac{1}{2}+\mu} U\left(\frac{1}{2} + \mu - \kappa, 1 + 2\mu, \xi\right). \tag{60}$$

Therefore, Equation (58) can be presented in terms of the Kummer's function as:

$$\psi(\xi, s) = \left(A(s) M(a, b, \xi) + B(s) U(a, b, \xi)\right) e^{-\frac{\xi}{2}} \xi^{\frac{1}{2}+\mu} \tag{61}$$

In general, the Kummer's function of the first kind:

$$M(a, b, \xi \to \infty) \to \frac{\Gamma(b)}{\Gamma(a)} \xi^{a-b} e^{\xi}, \tag{62}$$

implies that the $M(a, b, \xi)$ function becomes unbounded when ξ becomes large. Therefore, the arbitrary function coefficient $A(s)$ of Equation (61) must become zero for Equation (62) to satisfy the external boundary condition specified for the system. Equation (61) therefore reduces to:

$$\psi(\xi, s) = e^{-\frac{\xi}{2}} \xi^{\frac{1}{2}+\mu} B(s) U\left(\frac{1}{2} + \mu - \kappa, 1 + 2\mu, \xi\right). \tag{63}$$

The coefficient $B(s)$ is obtained from the application of the inner boundary condition. The function $U(a, b, \xi)$, variously referred to as the Kummer's function of the second kind or the Tricomi function, decreases exponentially as ξ increases and vanishes as ξ becomes infinitely large as required by the inner boundary condition of this problem.

The tracer concentration $C(\eta, s)$ can be now be written as:

$$C(\eta(r), s) = \Psi e^{-\left\{\frac{1}{2}\int \left(\frac{\kappa-\alpha}{\kappa r + \beta} + \frac{\alpha \omega r}{\kappa r + \beta}\right) dr\right\}}.$$

(64)

Detailed mathematical transformation in dimensionless form and inverse Laplace transfom of the general solution are available in Appendix B. The solution to Equations (34)–(37) can therefore be expressed as:

$$\psi(\xi) = e^{-\frac{\xi}{2}} \xi^{\frac{1}{2}+\mu} U\left(\frac{1}{2} + \mu - \kappa, 1 + 2\mu, \xi\right).$$

(65)

However, the concentration C is defined in terms of ψ as:

$$C = \psi e^{\left\{-\frac{1}{2}\int \left(\frac{D'-v}{D} + \frac{1}{r_D}\right) dr_D\right\}},$$

(66)

or:

$$C = \psi e^{\left\{-\frac{1}{2}\int \left(\frac{\kappa-\alpha}{\kappa r_D + \beta} + \frac{\alpha \omega r_D}{\kappa r_D + \beta}\right) dr_D\right\}}$$

(67)

$$C = \psi (\kappa r_D + \beta)^{\left(\frac{\kappa^2 - \alpha\kappa - \alpha\omega\beta}{\kappa^2}\right)} e^{\left\{-\frac{1}{2}\frac{\alpha\omega(\kappa r_D + \beta)}{\kappa^2}\right\}}$$

(68)

$$C = (\kappa r_D + \beta)^{\left(\frac{\kappa^2 - \alpha\kappa - \alpha\omega\beta}{\kappa^2}\right)} e^{\left\{-\frac{1}{2}\frac{\alpha\omega(\kappa r_D + \beta)}{\kappa^2}\right\}} e^{-\frac{\xi}{2}} \xi^{\frac{1}{2}+\mu} U\left(\frac{1}{2} + \mu - \kappa, 1 + 2\mu, \xi\right),$$

(69)

so that:

$$C = (\eta)^{\left(\frac{\kappa^2 - \alpha\kappa - \alpha\omega\beta}{\kappa^2}\right) + \left(\frac{1}{2}+\mu\right)} (2\sqrt{a})^{\frac{1}{2}+\mu} e^{\left\{-\left(\frac{1}{2}\frac{\alpha\omega}{\kappa^2} + \sqrt{a}\right)\eta\right\}} U\left(\frac{1}{2} + \mu - \kappa, 1 + 2\mu, (2\sqrt{a})\eta\right),$$

(70)

or:

$$C = (\xi)^{\left(\frac{\kappa^2 - \alpha\kappa - \alpha\omega\beta}{\kappa^2}\right) + \left(\frac{1}{2}+\mu\right)} (2\sqrt{a})^{-\left(\frac{\kappa^2 - \alpha\kappa - \alpha\omega\beta}{\kappa^2}\right)} Exp\left\{-\left(\frac{1}{2}\frac{\alpha\omega}{(2\sqrt{a})\kappa^2} + 1\right)\xi\right\} U\left(\frac{1}{2} + \mu - \kappa, 1 + 2\mu, \xi\right),$$

(71)

where 'a' is given as:

$$\begin{aligned} a &= \frac{1}{k^2}\left(\frac{a^2\omega^2}{4k^2} + \frac{\phi(k_r + s)}{k}\right) \\ &= \frac{1}{4k^4}(a^2\omega^2 + 4k\phi(k_r + s)), \end{aligned}$$

(72)

and:

$$\begin{aligned} \xi &= (2\sqrt{a})\eta \\ &= (2\sqrt{a})(kr_D + \beta) \\ &= \frac{1}{k^2}\sqrt{(a^2\omega^2 + 4k\phi(k_r + s))}(kr_D + \beta) \end{aligned}$$

(73)

$$\xi_{rD} = \frac{1}{k^2}\sqrt{(a^2\omega^2 + 4k\phi(k_r + s))(k + \beta)}.$$

(74)

3.3. Weak-Form Numerical Solution of the Tracer Transport Equation

The weak-form solution of Equation (71) is now presented by considering (i) a *'pot'* diffusion case where tracer flow is controlled by molecular diffusion with no hydrodynamic dispersion but with velocity drift and (ii) cases where convection dominated the molecular diffusion effect. In order to achieve this, the separation of variables is adopted with X-parameter (X_p) and Y-parameter (Y_p), defined thus:

$$C(x,y,s) = X_p(x)Y_p(y,s). \tag{75}$$

The X-parameter (X_p) can be expressed as:

$$X_p(x) = \tau U\left(\frac{1}{2} + \mu - \kappa, 1 + 2\mu, \sigma\xi_{X_p}\right)e^{-\frac{1}{2}\left(\frac{1}{\sigma}-1\right)\xi_{X_p}}, \tag{76}$$

where the components and arguments of the Tricomi Kummer function $U(a, b, x)$ are defined thus:

$$\mu = -\frac{1}{2}, \tag{77}$$

$$\xi_{X_p} = \left[\frac{\beta v_0}{D_0 d} + \frac{v_0}{D_0}x\right], \tag{78}$$

$$\sigma = \sqrt{1 - \frac{4\omega^2 D_0}{v_0^2 d}}, \tag{79}$$

$$\tau = \frac{\Gamma(\frac{i}{k})}{\Gamma(j)}, \tag{80}$$

$$h = \frac{\omega^2 \beta}{v_0 d^2}, \tag{81}$$

$$k = \sqrt{1 - j}, \tag{82}$$

and Y-parameter (Y_p):

$$Y_p(y,s) = \left(\frac{1}{3}\xi_{Y_p}^{\frac{1}{2}+\mu}\right)e^{-\frac{v_0 y}{2D_0}}\left[I_{-\frac{1}{3}}\left(\frac{2}{3}L\xi_{Y_p}^{\frac{3}{2}+\mu}\right) + I_{+\frac{1}{3}}\left(\frac{2}{3}L\xi_{Y_p}^{\frac{3}{2}+\mu}\right)\right], \tag{83}$$

$$\mu = 0, \tag{84}$$

$$\xi_{Y_p} = \left[\frac{1}{4}\left(\frac{v_0}{D_0}\right)^2 + \frac{Rs + R\kappa + \omega^2}{D_0\lambda}y\right], \tag{85}$$

$$L = \frac{D_0\lambda}{Rs + R\kappa + \omega^2}. \tag{86}$$

I is modified Bessel functions of the first kind and decays to zero rapidly with the concentration distribution of the $Y(y,s)$ component in the negative half, mirroring the positive half.

3.3.1. Separation Constant (ω^2)

The constant of separation (ω^2) is obtained by rewriting the general advection–dispersion equation (ADE) for the flow of reactive tracers under the influence of linear drift thus:

$$D_0\left(\frac{\beta}{x}+d\right)C_{xx} - v_0\left(\frac{\beta}{x}+d\right)C_x + D_0\frac{\lambda}{y}C_{yy} - v_0\frac{\lambda}{y}C_y - R\kappa C = RC_t, \tag{87}$$

where the linear drift ratio is written as $d = \frac{v_d}{v_0}$. The linear drift ratio, d, is coordinate-independent; therefore, its magnitude can be applied equally to the y-axis or, in the case of a 3D system, the z-axis.

Substituting Equation (75) into Equation (87), dividing through by $X(x)Y(y,t)$ and rearranging (see Appendix C) gives:

$$D_0\left(\frac{\beta}{x}+d\right)X'' - v_0\left(\frac{\beta}{x}+d\right)X' + \omega^2 X(x) = 0, \tag{88}$$

$$D_0\frac{\lambda}{y}Y'' - v_0\frac{\lambda}{y}Y' - (Rs + R\kappa + \omega^2)Y(y,s) = 0, \tag{89}$$

where the component Equation (88) is time-independent, while Equation (89) is time-dependent and expressed in Laplace space with Laplace parameter s. Considering the following transformation parameters:

$$\beta = r_w \cos^2\theta, \tag{90}$$

$$\lambda = v_y r_w \sin^2\theta, \tag{91}$$

$$x_w = r_w \cos\theta, \tag{92}$$

$$y_w = r_w \sin\theta, \tag{93}$$

Equations (88) and (89) can be rewritten as:

$$D_0\left(\cos\theta + d\right)X'' - v_0\left(\cos\theta + d\right)X' + \omega^2 X(x_w) = 0, \tag{94}$$

$$\mathcal{L}^{-1}\left[D_0(v_y\sin\theta)Y'' - v_0(v_y\sin\theta)Y' - (Rs + R\kappa + \omega^2)Y(y_w,s)\right] = 0. \tag{95}$$

Equation (94) is time-independent, while Equation (95) is time-dependent and expressed in Laplace space with Laplace parameter s.

3.3.2. Boundary and Initial Conditions

Equation (87) is governed by the following boundary conditions:

$$C(x,y,t=0) = C_i(x,y) \qquad \text{for } x = y = \mathbb{R} \tag{96}$$

$$C(x=\pm\infty, y, t) = 0 \qquad \text{for } y = \mathbb{R}, t > 0 \tag{97}$$

$$C(x, y=\pm\infty, t) = 0 \qquad \text{for } x = \mathbb{R}, t > 0 \tag{98}$$

$$C(x_w, y_w, t) = C_0 \qquad \text{for } t > 0 \tag{99}$$

where, the transformation from the polar (radial) coordinate to Cartesian coordinate is given as $x_w = r_w \cos\theta$ and $y_w = r_w \sin\theta$. The concentration $C(x_w, y_w, t)$ of the tracer is known within the wellbore during a tracer test; thus, the solution for $t > 0$ can be obtained by solving Equation (87) within the porous formation outside the wellbore.

The Y-parameter Y_p is in Laplace space and requires a numerical inversion scheme, such as the Gaver–Stehfest algorithm [18,29], Talbot inversion algorithm [30] or Euler inversion [31,32] algorithm. The scripts for these algorithms are open source and are readily available for download from the Mathworks website [33]. Out of the two algorithms attempted for the numerical inverse Laplace operation, Gaver–Stehfest and Euler inversion, only the Euler inversion algorithm produced a stable result (see also Avdis and Whitt [34]). Hence, Euler's Inversion Algorithm was used for the numerical inverse Laplace operation in the numerical code developed for this work.

Flow convection depends on the velocity of the system and is modelled by considering a heterogeneous system. Anisotropic porosity distribution was generated using the random probability density function (PDF) allowing for non-uniform velocity distribution to be computed. Typical porosity distribution is shown in Figure 2 with a mean value of 0.25 and standard deviation of 0.64. The corresponding computed velocity distribution is shown in Figure 3.

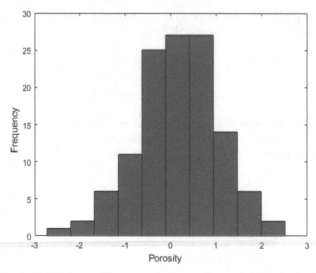

Figure 2. Porosity distribution profile generated using random probability density function. [−] denotes that the variable has no unit.

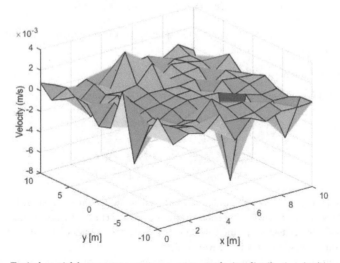

Figure 3. Typical spatial heterogeneous porous system velocity distribution (m/s) used in the numerical computation.

4. Analysis of Results

In this work, the effect of linear drift on the tracer propagation profile in a typical formation of thickness $h = 9$ m and injection well of radius $r_w = 0.127$ m was investigated. Injected particles are considered to be components of surfactants or polymers used in EOR processes, but in this case, the chemical reaction was neglected. The injection rate was fixed at 1.4×10^{-3} m^3/s and dispersion coefficient $D_0 = 1.4 \times 10^{-3}$ m^2/s was applied. The tracer concentration distribution ratio C_i was monitored under three continuous injection periods of ten (10) days, fifty (50) days and seventy (70) days respectively.

The developed solution was first tested by evaluating the error limit associated with the separation of variable parameter (ω^2) for different angle θ. Simulation runs involved one hundred (100) values of

ω^2 as defined by Equation (81)—ranging from $0.01 \times \omega_1^2$ corresponding to a value of $\omega^2 = 2.33 \times 10^{-8}$ to ω_{100}^2 corresponding to a value of $\omega^2 = 2.59 \times 10^{-6}$—with an incremental value of $0.01 \times \omega_{100}^2$. The minimum error corresponds to the value of separation variable $\omega^2 = 2.33 \times 10^{-8}$, as indicated by $X_p, Y_{p,t}, X - Y_{p,t}$ plots (Figures 4–6). The combined solution $X - Y_{p,t}$ indicated that there exist points of singularity at $0°, 180°$ and $360°$ when taking the inverse of the coordinate point y.

A typical result of the concentration distribution as a function of angle θ obtained from the solution of Equation (75) is shown in Figure 6 for five 5 selected ranges of values of ω^2. The concentration profile basically grows with increasing values of ω^2 between $\omega^2 = 2.33 \times 10^{-8}$ and $\omega^2 = 1.06 \times 10^{-6}$ and between $\theta = 150°$ and $\theta = 300°$, after which the impact of drift sets in. With the onset of drift, the concentration profile shortens but with a much wider coverage for $\omega^2 = 1.58 \times 10^{-6}$. The region covered by $\omega^2 = 2.10 \times 10^{-6}$ can be seen to have drifted to $\theta = 240°$ to $-300°$.

In order to further examine the impact of the drift parameter on separation constant ω^2, X_p and angle $\theta°$, a full simulation run at different time intervals of $t = 10$ d, $t = 30$ d, 50 d and 70 d was carried out and results presented in Figures 7–10. The magnitude of error due to ω^2 generally reduces with increasing computational time. In Figure 10, the influence of drift is more pronounced. A typical concentration distribution within the porous media with linear drift after 10 days is shown in Figure 11.

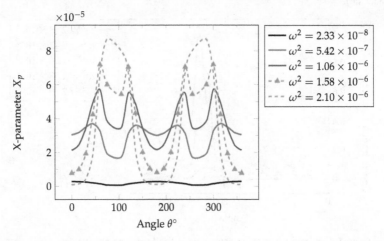

Figure 4. X-parameter (X_p) as a function of angle ($\theta°$) at a fixed time $t = 10$ d and varying separation constant (ω^2).

Linear Drift Effect and Concentration Distribution Profile

In a single-well tracer or surfactant injection systems, a primary tracer bank is first injected into a formation containing oil at residual saturation. The bank is then followed by a bank of tracer-free water. The well is then shut in for a period of time, after which the well is produced and the concentration profiles monitored. Where there is fluid migration in the formation due to the movement of an underlying basal water displacing the injected tracer banks during shut-in, the production profile may be distorted and fluid pathway rerouted. In this situation, the effect of linear drift on fluid flow behaviour will have to be investigated.

Generally, in isotropic and homogeneous systems, where there is no linear drift or natural convection, the tracer propagation profile is expected to follow a cyclic pattern. The concentration distribution will be equal at an equidistant radial position from the injection well. In this work, a system with variable porosity distribution was modelled and the corresponding non-uniform velocity profile used in the computation of the drift ratio. In this case, the tracer propagation profile is expected to follow a natural pattern determined by the interplay of the forces associated with the system variables, such as the tracer injection rate.

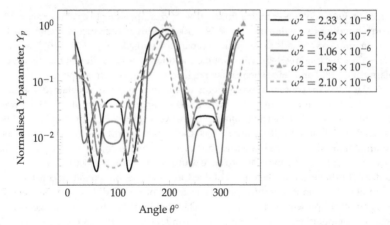

Figure 5. Y-parameter (Y_p) as a function of angle ($\theta°$) at a fixed time $t = 10$ d and varying separation constant (ω^2).

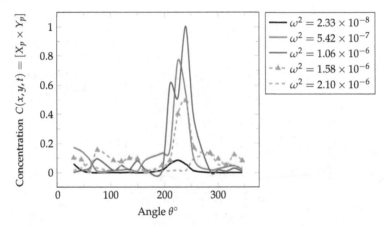

Figure 6. Concentration profile versus angle ($\theta°$) and varying separation constant (ω^2).

In order to investigate the effect of the drift intensity on the tracer concentration profile, three (3) values of drift ratio $d = 0.03, 0.09$ and 0.2 were evaluated for time duration $t = 10$ d, 50 d and 70 d. In the presence of linear drift in a heterogeneous system, however, there exists an unequal distribution of tracers along the principal x-axis and varies in a non-uniform manner along the positive and negative radial distance. Where the angle θ increases in the +ve and −ve y-axis, the system convection will lead to an increase or decrease in the tracer concentration distribution depending on the degree of system heterogeneity. It is important to note that an increase in the tracer concentration distribution will be observed in a homogeneous system.

The results of the tracer tests for different drift ratios at different time intervals are shown in Figure 12. At a fixed drift ratio (e.g., $d = 0.2$), the effect of the linear drift on the tracer concentration profile is progressively more pronounced at later times; with the lowest tracer concentration ratio at later time indicating a high drift effect. Similarly, at any particular point in time (e.g., at time $t = 70$ d), linear drift has a greater effect at a higher drift ratio (e.g., $d = 0.2$).

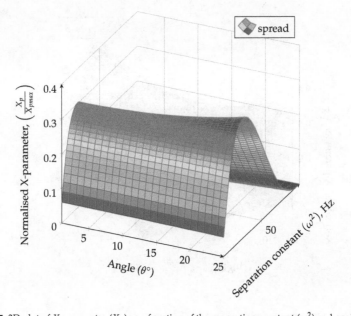

Figure 7. 3D plot of X-parameter (X_p) as a function of the separation constant (ω^2) and angle ($\theta°$) at time $t = 10$ d.

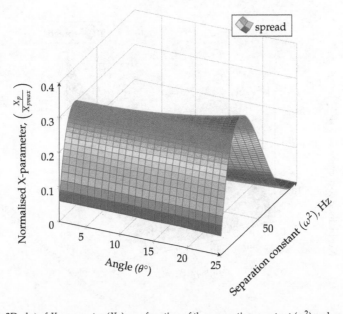

Figure 8. 3D plot of X-parameter (X_p) as a function of the separation constant (ω^2) and angle ($\theta°$) at time $t = 30$ d.

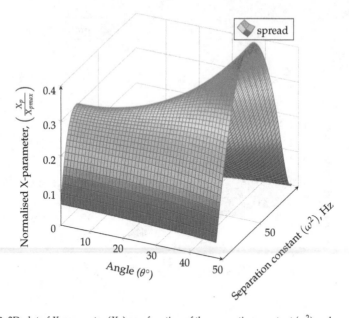

Figure 9. 3D plot of X-parameter (X_p) as a function of the separation constant (ω^2) and angle ($\theta°$) at time $t = 50$ d.

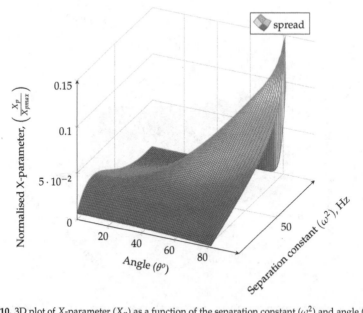

Figure 10. 3D plot of X-parameter (X_p) as a function of the separation constant (ω^2) and angle ($\theta°$) at time $t = 70$ d.

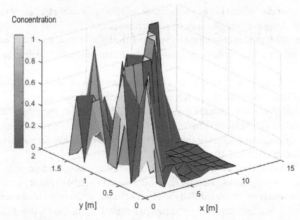

Figure 11. Typical concentration distribution within the porous media with low linear drift ratio $d = 0.006$ after 10 days.

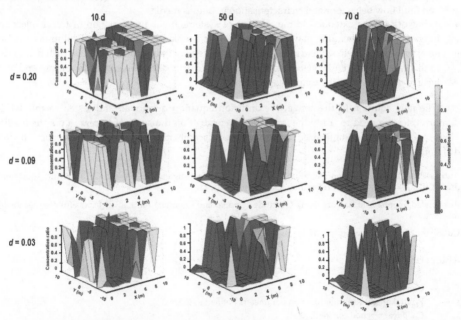

Figure 12. Concentration distribution within the porous media with linear drift ratio $d = 0.03, 0.09$ and 0.2 and time interval $t = 10$ d, 50 d and 70 d.

5. Conclusions

The study of the transport of tracers with linear drift is an important aspect of porous media characterisation. Despite the extensive research in this field, particularly in solving the ADE both analytically and numerically, there is yet to be a consideration for the closed-form solution of the ADE in systems where the effect of linear drift may predominate and an exact analytical solution has not been reported in the literature. The following conclusions can be drawn from this work:

- A new closed-form analytical solution to the radial transport of tracers in porous media under the influence of linear drift and radial convection was developed. The radial transport equation was cast in the form of the Whittaker equation after adopting variable transformation and an exact solution for the tracer concentration derived therefrom.

- The weak-form solution was developed by splitting the transformed equation, adopting a common separation constant and invoking inverse Laplace transformation using the Euler inversion algorithm.
- Variable transformation from a radial to a Cartesian coordinate system was used to analyse the concentration distribution profiles in three-dimensional graphical plots.
- The obtained solutions are generally stable and dependent on the precision with which the separation constant (ω^2) can be determined. This is important because the exponential term in the inversion formula may amplify the numerical error. The maximum error quantified by the separation constant is $\omega^2 = 2.10 \times 10^{-6}$.
- The influence of linear drift on the concentration profiles was evaluated in the x-direction for a system with nonhomogeneous porosity distribution and variable velocity profiles.
- The results of the analyses indicated that the effect of linear drift on the tracer concentration profile is dependent on system heterogeneity and progressively becomes more pronounced at later times.
- Practical application was demonstrated in a typical EOR process involving the injection of chemicals (e.g., surfactants or polymers), but without a chemical reaction. Another possible application is a single-well chemical tracer injection method for measuring residual oil saturation and fluid flow behaviour and characterisation in porous media.
- This work can be extended to the analysis of systems involving variation of tracer injection intensity, where spreading may occur in the r-θ or x-y plane. The developed solution can also be extended to systems where moderate-to-high intensity tracer flow with linear drift manifests. In this case, the tangential velocity component of the drift velocity becomes significant and will have to be included in the solution approach.

The new solution to the convective-diffusion equation developed and tested in this work demonstrates the need to study the impact of linear drift on transport of tracers in porous media. This is particularly important, since the arrival times of tracers may be significantly influenced by the drift intensity.

Author Contributions: Conceptualisation, L.T.A. and G.K.F.; Investigation, L.T.A. and G.K.F.; Writing—original draft, L.T.A.; Writing—review and editing, G.K.F. and L.T.A.

Acknowledgments: The authors thank Olakunle Popoola and Ofomana Emmanuel for useful discussions of this topic.

Conflicts of Interest: The authors declare no conflict of interest.

Abbreviations

$\Gamma(x)$	Gamma function
κ, μ	Whittaker and Kummer function parameters defined in the text
κ_r	Chemical reaction constant, (d^{-1})
$\mathcal{L}$	Laplace operator
ω^2	Absolute value of the separation constant
ϕ	Porosity, dimensionless
θ	Angle, (°)
A	Constant of integration
$Ai(x)$	Airy function of the 1st kind
$Bi(x)$	Airy function of the 2nd kind
C	Tracer concentration
D	Flow hydrodynamic dispersion (m²/s)
d	Drift ratio
D_m	Molecular diffusion constant
D_o	Shear mixing constant

h	Formation thickness, (m)
i	Unit vector along the x-axis
j	Unit vector along the y-axis
k	Unit vector along the z-axis
L	Length of dispersion, (m)
Q	Injection rate, (m^3/s)
r	Radial distance from the well, (m)
r_D	Dimensionless well-bore radius
r_w	Well-bore radius, (m)
s	Laplace parameter
S_m	Saturation of the mobile fluid phase
S_{mp}	Mobile fluid phase saturation, dimensionless
t	Time, (s or d)
u	Velocity along the x-axis without linear drift, (m/s)
$U(a,b,z)$	Tricomi Kummer U-function with parameters (a,b,z)
u_d	Linear drift velocity, (m/s)
u_o	Velocity at the well-bore (m/s)
v_y	Velocity along the y-axis, (m/s)
$W_{\kappa,\mu}(\xi)$	Whittaker function with parameters (κ, μ)
x	x-coordinate variable
y	y-coordinate variable
D	dimensionless
d	drift
m	molecular
p	phase
r	reaction
w	well-bore

Appendix A. Derivation of the Transport Flow Equation with Linear Drift

The flow Equation (1) can be expanded by considering a steady-state condition thus:

$$\frac{1}{r}\left\{ D\frac{dC}{dr} + rD'\frac{dC}{dr} + rD\frac{d^2C}{dr^2} \right\} - \frac{1}{r}\left\{ vC + rv'C + rv\frac{dC}{dr} \right\} - \phi(\kappa_r + s)C = 0, \tag{A1}$$

$$\frac{d^2C}{dr^2} + \left(\frac{D'-v}{D} + \frac{1}{r} \right)\frac{dC}{dr} - \left\{ \frac{rv'+v}{Dr} + \frac{\phi(\kappa_r + s)}{D} \right\}C = 0, \tag{A2}$$

where all variables are as defined in the main text. Let:

$$C = \Psi \exp\left\{ -\frac{1}{2}\int \left(\frac{D'-v}{D} + \frac{1}{r} \right)dr \right\}. \tag{A3}$$

Then:

$$\frac{dC}{dr} = \frac{d\Psi}{dr} \cdot e^{\left[-\frac{1}{2}\left(\frac{D'-v}{D} + \frac{1}{r} \right) \right]} - \left[\frac{1}{2}\left(\frac{D'-v}{D} + \frac{1}{r} \right) \right]\Psi e^{\left[-\frac{1}{2}\left(\frac{D'-v}{D} + \frac{1}{r} \right) \right]}, \tag{A4}$$

and:

$$\frac{d^2C}{dr^2} = \frac{d}{dr}\left\{\frac{d\Psi}{dr}\cdot e^{\left[-\frac{1}{2}\left(\frac{D'-v}{D}+\frac{1}{r}\right)\right]} - \left[\frac{1}{2}\left(\frac{D'-v}{D}+\frac{1}{r}\right)\right]\Psi e^{\left[-\frac{1}{2}\left(\frac{D'-v}{D}+\frac{1}{r}\right)\right]}\right\}$$

$$= \frac{d^2\Psi}{dr^2}\cdot e^{\left[-\frac{1}{2}\left(\frac{D'-v}{D}+\frac{1}{r}\right)\right]} - \left[\frac{1}{2}\left(\frac{D'-v}{D}+\frac{1}{r}\right)\right]\frac{d\Psi}{dr}e^{\left[-\frac{1}{2}\left(\frac{D'-v}{D}+\frac{1}{r}\right)\right]}$$

$$-\left[\frac{1}{2}\left(\frac{D'-v}{D}+\frac{1}{r}\right)\right]\Psi e^{\left[-\frac{1}{2}\left(\frac{D'-v}{D}+\frac{1}{r}\right)\right]} - \left[\frac{1}{2}\left(\frac{D'-v}{D}+\frac{1}{r}\right)\right]\frac{d\Psi}{dr}e^{\left[-\frac{1}{2}\left(\frac{D'-v}{D}+\frac{1}{r}\right)\right]}$$

$$+\left[\frac{1}{4}\left(\frac{D'-v}{D}+\frac{1}{r}\right)^2\right]\Psi e^{\left[-\frac{1}{2}\left(\frac{D'-v}{D}+\frac{1}{r}\right)\right]} \tag{A5}$$

$$\frac{d^2C}{dr^2} = \left\{\frac{d^2\Psi}{dr^2} - \left(\frac{D'-v}{D}+\frac{1}{r}\right)\frac{d\Psi}{dr} - \left[\frac{1}{2}\left(\frac{D'-v}{D}+\frac{1}{r}\right) - \frac{1}{4}\left(\frac{D'-v}{D}+\frac{1}{r}\right)^2\right]\Psi\right\}$$

$$e^{\left[-\frac{1}{2}\left(\frac{D'-v}{D}+\frac{1}{r}\right)\right]}. \tag{A6}$$

Let:

$$\frac{d^2C}{dr^2} + \left(\frac{D'-v}{D}+\frac{1}{r}\right)\frac{dC}{dr} = I, \tag{A7}$$

with the first order differential component $\left(\frac{dC}{dr}\right)$ defined by Equation (A4), then:

$$I = \frac{d^2\Psi}{dr^2}e^{\left[-\frac{1}{2}\left(\frac{D'-v}{D}+\frac{1}{r}\right)\right]}$$

$$-\left(\frac{D'-v}{D}+\frac{1}{r}\right)\frac{d\Psi}{dr}e^{\left[-\frac{1}{2}\left(\frac{D'-v}{D}+\frac{1}{r}\right)\right]}$$

$$-\left[\frac{1}{2}\left(\frac{D'-v}{D}+\frac{1}{r}\right) - \frac{1}{4}\left(\frac{D'-v}{D}+\frac{1}{r}\right)^2\right]\Psi e^{\left[-\frac{1}{2}\left(\frac{D'-v}{D}+\frac{1}{r}\right)\right]}$$

$$+\left(\frac{D'-v}{D}+\frac{1}{r}\right)\left[e^{\left[-\frac{1}{2}\left(\frac{D'-v}{D}+\frac{1}{r}\right)\right]}\frac{d\Psi}{dr} - \frac{1}{2}\left(\frac{D'-v}{D}+\frac{1}{r}\right)\Psi\right]e^{\left[-\frac{1}{2}\left(\frac{D'-v}{D}+\frac{1}{r}\right)\right]}. \tag{A8}$$

Applying Equations (A3) and (A8) in (A1) and rearranging yields:

$$\frac{d^2\Psi}{dr^2} - \left\{\frac{1}{2}\left(\frac{D'-v}{D}+\frac{1}{r}\right) + \frac{1}{4}\left(\frac{D'-v}{D}+\frac{1}{r}\right)^2 + \left(\frac{rv'+v}{Dr} + \frac{\phi(\kappa_r+s)}{D}\right)\right\}\Psi = 0. \tag{A9}$$

Expressing the effective fluid velocity v in dimensionless form:

$$v = v_d + \frac{\alpha}{r}, \tag{A10}$$

$$D'(r) = -\frac{\beta_w}{r^2}, \tag{A11}$$

$$-v' = -\frac{\alpha}{r^2}. \tag{A12}$$

Thus:

$$D'(r) - v = -\frac{\beta_w}{r^2} - \alpha\left(\overline{\omega}+\frac{1}{r}\right) \tag{A13}$$

$$= -\frac{\beta_w}{r^2} - \frac{\alpha(\overline{\omega}r+1)}{r} \tag{A14}$$

$$\frac{D'(r) - v}{D(r)} = -\frac{\beta_w}{r^2}\frac{r}{\kappa r + \beta_w} - \frac{\alpha(\overline{\omega}r + 1)}{\kappa r + \beta_w} \tag{A15}$$

$$= -\frac{\beta_w}{r(\kappa r + \beta_w)} - \frac{\alpha(\overline{\omega}r + 1)}{\kappa r + \beta_w}. \tag{A16}$$

Let:

$$I \equiv \frac{-\beta_w}{r(\kappa r + \beta_w)} \tag{A17}$$

$$\equiv \frac{A}{r} + \frac{C}{\kappa r + \beta_w}, \tag{A18}$$

then:

$$(A\kappa + C)r + A\beta_w = -\beta_w. \tag{A19}$$

Thus, $A = -1$ and $C = \kappa$, so that:

$$\frac{-\beta_w}{r(\kappa r + \beta_w)} = \frac{\kappa}{(\kappa r + \beta_w)} - \frac{1}{r}. \tag{A20}$$

Therefore:

$$\frac{D'(r) - v}{D(r)} = \frac{\kappa}{(\kappa r + \beta_w)} - \frac{1}{r} - \frac{\alpha(\overline{\omega}r + 1)}{(\kappa r + \beta_w)}, \tag{A21}$$

$$\left(\frac{D'(r) - v}{D(r)} + \frac{1}{r}\right) = \frac{\kappa}{(\kappa r + \beta_w)} - \frac{\alpha(\overline{\omega}r + 1)}{(\kappa r + \beta_w)}. \tag{A22}$$

Similarly:

$$\left(\frac{D'(r) - v}{D(r)} + \frac{1}{r}\right)' = \frac{-\kappa^2}{(\kappa r + \beta_w)^2} - \frac{\alpha\overline{\omega}}{(\kappa r + \beta_w)} + \frac{\alpha\kappa(\overline{\omega}r + 1)}{(\kappa r + \beta_w)^2}$$

$$= \frac{\alpha\kappa(\overline{\omega}r - \kappa^2)}{(\kappa r + \beta_w)^2} - \frac{\alpha\overline{\omega}}{(\kappa r + \beta_w)}. \tag{A23}$$

Additionally:

$$rv' = -r \cdot \frac{\alpha}{r^2}$$

$$= -\frac{\alpha}{r}, \tag{A24}$$

and:

$$rv' + v = -\frac{\alpha}{r} + \alpha\left(\overline{\omega} + \frac{1}{r}\right)$$

$$= -\frac{\alpha}{r} + \alpha\left(\frac{\overline{\omega}r + 1}{r}\right)$$

$$= \alpha\overline{\omega}. \tag{A25}$$

Thus:

$$\frac{rv' + v}{D(r)r} = \frac{\alpha\overline{\omega}}{(\kappa r + \beta_w)}. \tag{A26}$$

Therefore, Equation (A9) becomes:

$$\frac{d^2\Psi}{dr^2} - \left\{ \frac{1}{2}\left[\frac{\alpha\kappa(\overline{\omega}r+1)-\kappa^2}{(\kappa r+\beta_w)^2} - \frac{\alpha\overline{\omega}}{(\kappa r+\beta_w)}\right] + \frac{1}{4}\left[\frac{\kappa-\alpha(\overline{\omega}r+1)}{(\kappa r+\beta_w)}\right]^2 + \left[\frac{\alpha\overline{\omega}+\phi(\kappa_r+s)r}{\kappa r+\beta_w}\right] \right\}\Psi = 0, \quad (A27)$$

or:

$$\frac{d^2\Psi}{dr^2} - \left\{ \left[\frac{1}{2}\frac{\alpha\kappa(\overline{\omega}r+1)-\kappa^2}{(\kappa r+\beta_w)^2} - \frac{\alpha\overline{\omega}}{2(\kappa r+\beta_w)}\right] \right\}\Psi$$
$$+ \left\{ \frac{1}{4}\left[\frac{\kappa^2}{(\kappa r+\beta_w)^2} + \frac{\alpha^2(\overline{\omega}r+1)^2}{(\kappa r+\beta_w)^2} - \frac{2\alpha\kappa(\overline{\omega}r+1)}{(\kappa r+\beta_w)^2}\right] \right\}\Psi$$
$$+ \left\{ \left[\frac{\alpha\overline{\omega}+\phi(\kappa_r+s)r}{\kappa r+\beta_w}\right] \right\}\Psi = 0,$$

$$\frac{d^2\Psi}{dr^2} - \left\{ \frac{\alpha\kappa(\overline{\omega}r+1)}{2(\kappa r+\beta_w)^2} - \frac{\kappa^2}{2(\kappa r+\beta_w)^2} - \frac{\alpha\overline{\omega}}{2(\kappa r+\beta_w)} + \frac{\kappa^2}{4(\kappa r+\beta_w)^2} \right\}\Psi$$
$$+ \left\{ \frac{\alpha^2(\overline{\omega}r+1)^2}{4(\kappa r+\beta_w)^2} - \frac{\alpha\kappa(\overline{\omega}r+1)}{2(\kappa r+\beta_w)^2} + \frac{\alpha\overline{\omega}}{\kappa r+\beta_w} + \frac{\phi(\kappa_r+s)r}{\kappa r+\beta_w} \right\}\Psi = 0.$$

An expression of the form:

$$\frac{d^2\Psi}{dr^2} - \left\{ \frac{\alpha^2(\overline{\omega}r+1)^2-\kappa^2}{4(\kappa r+\beta_w)^2} + \frac{\alpha\overline{\omega}}{2(\kappa r+\beta_w)} + \frac{\phi(\kappa_r+s)r}{(\kappa r+\beta_w)} \right\}\Psi = 0, \quad (A28)$$

is therefore obtained for the general convection–diffusion equation in Laplace space. This can be written in dimensionless form as:

$$\frac{d^2\Psi}{dr_D^2} - \left\{ \frac{\alpha^2(\omega r_D+1)^2-\kappa^2}{4(\kappa r_D+\beta)^2} + \frac{\alpha\omega}{2(\kappa r_D+\beta_w)} + \frac{\phi(\kappa_r+s)r_D}{(\kappa r_D+\beta)} \right\}\Psi = 0, \quad (A29)$$

where the variables are redefined as:

$$D(r_{Dk}) = D_m + v_m D_o\left(\omega+\frac{1}{r_D}\right), \quad (A30)$$

$$= \kappa + \frac{\beta}{r_D}, \quad (A31)$$

and:

$$\kappa = D_m + \beta\omega. \quad (A32)$$

Using Equations (23) and (24) in Equation (A29) yields:

$$\kappa^2\frac{d^2\Psi}{d\eta^2} - \left\{ \frac{\alpha^2\omega^2}{4\kappa^2} - \frac{\alpha^2\omega}{2\kappa}\left(\frac{\omega\beta}{\kappa}-1\right)\frac{1}{\eta} + \frac{1}{4\eta^2}\left[\alpha^2\left(\frac{\omega\beta}{\kappa}-1\right)^2-\kappa^2\right] \right\}\Psi$$
$$+ \left\{ \frac{\alpha\omega}{2\eta} + \frac{\phi(\kappa_r+s)}{\kappa} - \frac{\beta\phi(\kappa_r+s)}{\kappa\eta} \right\}\Psi = 0. \quad (A33)$$

Appendix B. Analytical Solution—Dimensionless Representation and Inverse Laplace Transform

In dimensionless radial length r_D, $C(\eta(r), s)$ can be written as:

$$C(\eta(r), s) = \Psi e^{\left\{ -\frac{1}{2} \int \left(\frac{\kappa - \alpha}{\kappa r_D + \beta} + \frac{\alpha \omega r_D}{\kappa r_D + \beta} \right) dr_D \right\}} \tag{A34}$$

$$= \Psi e^{\left\{ -\frac{1}{2\kappa} \int \left(\frac{\kappa - \alpha}{\eta} + \frac{\alpha \omega (\eta - \beta)}{\eta} \right) d\eta \right\}} \tag{A35}$$

$$C(\eta(r), s) = \Psi e^{\left\{ -\frac{1}{2} \int \left(\frac{\kappa - \alpha - \frac{\alpha \omega}{\kappa}}{\eta} + \frac{\alpha \omega}{\kappa} \right) d\eta \right\}} \tag{A36}$$

$$\psi(\xi, s) = e^{-\frac{\xi}{2}} \xi^{\frac{1}{2} + \mu} U\left(\frac{1}{2} + \mu - \kappa, 1 + 2\mu, \xi \right) \tag{A37}$$

$$A(s) = (2)^{\left(\frac{k^2 - \alpha k - \alpha \omega \beta}{2k^2} \right)} \xi^{\frac{k^2 - \alpha k - \alpha \omega \beta}{2k^2} - \mu - \frac{1}{2}} (f(s))^{\left(\frac{k^2 - \alpha k - \alpha \omega \beta}{2k^2} \right)}, \tag{A38}$$

but:

$$B(s) = \frac{C(\xi, s)}{(2)^{\left(\frac{k^2 - \alpha k - \alpha \omega \beta}{2k^2} \right)} \xi^{-\left(\frac{k^2 - \alpha k - \alpha \omega \beta}{2k^2} - \mu - \frac{1}{2} \right)} (f(s))^{\left(\frac{k^2 - \alpha k - \alpha \omega \beta}{2k^2} \right)} Exp\left\{ -\frac{\xi}{2} \left(1 + \frac{\alpha \omega}{2k^2 f(s)} \right) \right\} U\left(\frac{1}{2} + \mu - \kappa, 1 + 2\mu, \xi \right)} \tag{A39}$$

Therefore:

$$\frac{dC(\xi, s)}{d\xi} = \frac{\left((\mu + \frac{1}{2}) \xi^{-1} U - \frac{1}{2} (1 + \frac{\alpha (\frac{\omega \beta}{k})}{2kf(s)}) U - (\frac{1}{2} + \mu - \kappa) \frac{dU(\frac{1}{2} + \mu - \kappa, 1 + 2\mu, \xi)}{d\xi} \right) C(\xi, s) A(s) e^{-\frac{\xi}{2} (1 + \frac{\alpha (\frac{\omega \beta}{k} - 1) - k}{2f(s)})}}{(2)^{\left(\frac{k^2 - \alpha k - \alpha \omega \beta}{2k^2} \right)} \xi^{-\left(\frac{k^2 - \alpha k - \alpha \omega \beta}{2k^2} - \mu - \frac{1}{2} \right)} (f(s))^{\left(\frac{k^2 - \alpha k - \alpha \omega \beta}{2k^2} \right)} Exp\left\{ -\frac{\xi}{2} \left(1 + \frac{\alpha \omega}{2k^2 f(s)} \right) \right\} U(\frac{1}{2} + \mu - \kappa, 1 + 2\mu, \xi)} \tag{A40}$$

$$\frac{dC(\xi, s)}{d\xi} = \frac{\left\{ (\mu + \frac{1}{2}) \xi^{-1} U - \frac{1}{2} (1 + \frac{\alpha (\frac{\omega \beta}{k})}{2kf(s)}) U - (\frac{1}{2} + \mu - \kappa) \frac{dU(\frac{1}{2} + \mu - \kappa, 1 + 2\mu, \xi)}{d\xi} \right\} C(\xi, s)}{U(\frac{1}{2} + \mu - \kappa, 1 + 2\mu, \xi)} \tag{A41}$$

$$\frac{dC(\xi, s)}{d\xi} = \left\{ \frac{1}{\xi} (\mu + \frac{1}{2}) - \frac{1}{2} (1 + \frac{\alpha (\frac{\omega \beta}{k})}{2kf(s)}) - (\frac{1}{2} + \mu - K) \left[\frac{dU'(\frac{1}{2} + \mu - \kappa, 1 + 2\mu, \xi)}{dU(\frac{1}{2} + \mu - \kappa, 1 + 2\mu, \xi)} \right] \right\} C(\xi, s). \tag{A42}$$

Therefore:

$$\frac{dC(\xi, s)}{dr_D} = 2kf(s) \left\{ \frac{1}{\xi} (\mu + \frac{1}{2}) - \frac{1}{2} (1 + \frac{\alpha (\frac{\omega \beta}{k})}{2kf(s)}) - (\frac{1}{2} + \mu - \kappa) \left[\frac{dU'(\frac{1}{2} + \mu - \kappa, 1 + 2\mu, \xi)}{dU(\frac{1}{2} + \mu - \kappa, 1 + 2\mu, \xi)} \right] \right\} C(\xi, s) \tag{A43}$$

but:

$$\left[\frac{U'(\frac{1}{2} + \mu - \kappa, 1 + 2\mu, \xi)}{U(\frac{1}{2} + \mu - \kappa, 1 + 2\mu, \xi)} \right] = \frac{1}{\xi}$$

$$= \frac{1}{2\eta f(s)} \tag{A44}$$

$$\frac{dC(\xi, s)}{d\eta} = 2kf(s) \left\{ \frac{1}{\xi} (\mu + \frac{1}{2}) - \frac{1}{2} (1 + \frac{\alpha (\frac{\omega \beta}{k})}{2kf(s)}) - (\frac{1}{2} + \mu - \kappa) \frac{1}{\xi} \right\} C(\xi, s) \tag{A45}$$

$$\frac{1}{f(s)}\frac{dC(\xi,s)}{dr_D} = 2\kappa\left\{\frac{K}{\xi} - \frac{1}{2}\left(1 + \frac{\alpha\left(\frac{\omega\beta}{\kappa}\right)}{2\kappa f(s)}\right)\right\}C(\xi,s)$$

$$= \kappa\left\{\frac{K}{\eta f(s)} - 1 - \frac{\alpha\left(\frac{\omega\beta}{\kappa}\right)}{2\kappa f(s)}\right\}C(\xi,s). \tag{A46}$$

However:

$$K = \left[\frac{\frac{\omega\alpha}{4\kappa^2}\left(\frac{\alpha}{\kappa}\left(\frac{\omega\beta}{2\kappa} - 1\right) - 1\right)}{f(s)} - \frac{\beta}{2}f(s)\right] \tag{A47}$$

$$\frac{1}{f(s)}\frac{dC(\xi,s)}{d\eta} = \left\{\frac{\frac{\omega\alpha}{2\kappa}\left(\frac{\alpha}{\kappa}\left(\frac{\omega\beta}{2\kappa} - 1\right) - 1\right)}{f(s)} - \frac{k\beta}{2}f(s)}{\eta f(s)} - k - \frac{\alpha\left(\frac{\omega\beta}{\kappa}\right)}{2f(s)}\right\}C(\xi,s) \tag{A48}$$

$$\frac{1}{f(s)}\frac{dC(\xi,s)}{d\eta} = \left\{\frac{\frac{\omega\alpha}{2\kappa}\left(\frac{\alpha}{\kappa}\left(\frac{\omega\beta}{2\kappa} - 1\right) - 1\right)}{\eta f(s)^2} - \frac{k\beta}{2\eta} - k - \frac{\alpha\left(\frac{\omega\beta}{\kappa}\right)}{2f(s)}\right\}C(\xi,s) \tag{A49}$$

$$\frac{-1}{f(s)}\frac{dC(\xi,s)}{d\eta} = \left\{k\left(1 + \frac{\beta}{2\eta}\right) + \frac{\alpha\left(\frac{\omega\beta}{\kappa}\right)}{2f(s)} - \frac{\frac{\omega\alpha}{2\kappa}\left(\frac{\alpha}{\kappa}\left(\frac{\omega\beta}{2\kappa} - 1\right) - 1\right)}{\eta f(s)^2}\right\}C(\xi,s) \tag{A50}$$

$$\frac{-1}{f(s)}\frac{dC(\xi,s)}{d\eta} = \left\{k\left(1 + \frac{\beta}{2\eta}\right) + \frac{\frac{1}{2}\left(\frac{\alpha\omega\beta}{\kappa}\right)}{f(s)} - \frac{\frac{\omega\alpha}{2\eta\kappa}\left(\frac{\alpha}{\kappa}\left(\frac{\omega\beta}{2\kappa} - 1\right) - 1\right)}{f(s)^2}\right\}C(\xi,s). \tag{A51}$$

Rearranging:

$$\left(1 + \frac{\beta}{2\eta}\right)C(\xi,s) = \frac{-1}{kf(s)}\frac{dC(\xi,s)}{d\eta} - \left\{\frac{\frac{1}{2}\left(\frac{\alpha\omega\beta}{\kappa}\right)}{kf(s)} - \frac{\frac{\omega\alpha}{2\eta}\left(\frac{\alpha}{\kappa}\left(\frac{\omega\beta}{2\kappa} - 1\right) - 1\right)}{k^2 f(s)^2}\right\}C(\xi,s) \tag{A52}$$

$$kf(s) = \sqrt{\left(\frac{\alpha^2\omega^2}{4k^2} + \frac{\varphi(k_r + s)}{k}\right)}$$

$$= \sqrt{\frac{\varphi}{k}\sqrt{\left[\left(\frac{\alpha^2\omega^2}{4\varphi k} + k_r\right) + s\right]}}. \tag{A53}$$

The Inverse Laplace Transform of $f(s)^{-1}$ is:

$$L^{-1}\left[\frac{1}{\left(\sqrt{\frac{\varphi}{k}}\sqrt{\left(\left(\frac{\alpha^2\omega^2}{4\varphi k} + k_r\right) + s\right)}\right)}\right] = \frac{e^{\left(-\left(\frac{\alpha^2\omega^2}{4\varphi k} + k_r\right)t\right)}}{\sqrt{\frac{\varphi}{\kappa}\pi t}}. \tag{A54}$$

Similarly, the inverse Laplace transform of $f(s)^{-2}$ is:

$$L^{-1}\left[\frac{1}{\left(\sqrt{\frac{\varphi}{k}}\sqrt{\left(\left(\frac{\alpha^2\omega^2}{4\varphi k} + k_r\right) + s\right)}\right)^2}\right] = \frac{e^{\left(-\left(\frac{\alpha^2\omega^2}{4\varphi k} + k_r\right)t\right)}}{\frac{\varphi}{\kappa}}. \tag{A55}$$

Evaluating the inverse Laplace transform of the general solution:

$$(1 + \frac{\beta}{2\eta})C(\xi, t) = -\sqrt{\frac{\kappa}{\varphi}} \int_0^t \left(\frac{dC(\xi, \tau)}{d\eta}\right) \frac{e^{\left(-\left(\frac{a^2\omega^2}{4\varphi\kappa} + \kappa_r\right)(t-\tau)\right)}}{\sqrt{\pi(t-\tau)}} d\tau$$

$$- \sqrt{\frac{\kappa}{\varphi}} \left(\frac{a\omega\beta}{\kappa}\right) \int_0^t C(\xi, t) \frac{e^{\left(-\left(\frac{a^2\omega^2}{4\varphi\kappa} + \kappa_r\right)(t-\tau)\right)}}{\sqrt{\pi(t-\tau)}} d\tau$$

$$+ \frac{a\omega\kappa}{2\eta\varphi} \left(\frac{a}{\kappa}\left(\frac{\omega\beta}{2\kappa} - 1\right) - 1\right) \int_0^t C(\xi, \tau) e^{\left(-\left(\frac{a^2\omega^2}{4\varphi\kappa} + \kappa_r\right)(t-\tau)\right)} d\tau \quad (A56)$$

$$\kappa = D_m + \beta\omega$$
$$= D_m\left(1 + \frac{\beta\omega}{D_m}\right), \quad (A57)$$

$$f(s) = \frac{1}{\kappa}\sqrt{\left(\frac{a^2\omega^2}{4\kappa^2} + \frac{\varphi(\kappa_r + s)}{\kappa}\right)}$$

$$= \frac{1}{D_m + \beta\omega}\sqrt{\left(\frac{a^2\omega^2}{4(D_m + \beta\omega)^2} + \frac{\varphi(\kappa_r + s)}{D_m + \beta\omega}\right)}. \quad (A58)$$

Equations (39) and (40) respectively become:

$$\mu = -\frac{\frac{\alpha}{D_m}}{2(1 + \frac{D_0 v_d}{D_m})^2}, \quad (A59)$$

and:

$$K = \frac{(1 + \frac{D_0 v_d}{D_m})^2 - \left(\frac{\alpha}{D_0 v_d}\right)\left[\frac{D_0 v_d}{D_m} + 2\phi\left(1 + \frac{D_0 v_d}{D_m}\right)(\kappa_r + s)\right]}{4\left(1 + \frac{D_0 v_d}{D_m}\right)^2 \sqrt{1 + 4\phi\left(\frac{D_m}{D_0 v_d}\right)^2\left(1 + \frac{D_0 v_d}{D_m}\right)(\kappa_r + s)}}, \quad (A60)$$

here:

$$\phi' = \phi\frac{D_0^2}{D_m}. \quad (A61)$$

Appendix C. Weak-Form Solution—Separation Constant and Extended Confluent Hypergeometric Functions

Rewriting Equation (88) as:

$$(\beta + xd) X'' - \frac{v_0}{D_0}(\beta + xd) X' - \frac{\mu}{D_0}xX = 0, \quad (A62)$$

and expressing in terms of the variable x defined in relation to the drift ratio d as $x = \frac{x_1 - \beta}{d}$ and also defining the 1st-order and 2nd-order differentials in terms of x_1:

$$(\beta + xd) X'' - \frac{v_0}{D_0}(\beta + xd) X' - \frac{\mu}{D_0}xX = 0, \quad (A63)$$

$$x_1 X'' - \frac{v_0}{D_0 d}x_1 X' - \left(\frac{\mu}{d^3 D_0}x_1 - \frac{\mu\beta}{d^3 D_0}\right) X = 0. \quad (A64)$$

Similarly, defining $x_1 = \frac{D_0 d}{v_0} x_2$ in Equation (A64) and the 1st-order and 2nd-order differential in relation to x_2:

$$\frac{v_0}{D_0 d} x_2 X'' - \frac{v_0}{D_0 d} x_2 X' - \left(\frac{\mu}{v_0 d^2} x_2 - \frac{\mu \beta}{v_0 d^2} \right) X = 0, \tag{A65}$$

multiplying Equation (A65) by $\frac{D_0 d}{v_0}$ and expressing in terms of x_2 :

$$x_2 X'' - x_2 X' - \left(\frac{\mu D_0}{v_0 d^2} x_2 - \frac{\mu D_0}{v_0^2 d} \right) X = 0. \tag{A66}$$

Equation (A66) is comparable to the extended confluent hypergeometric equation of degree N [35]:

$$x X'' + (\gamma - x) X' - \left(\alpha + \sum_{n=1}^{N} \alpha_n x^n \right) X = 0, \tag{A67}$$

with one of its solutions being the extended confluent hypergeometric function:

$$X_1(x) =_1 F_1^N(\alpha; \gamma; A_a, \ldots, A_N; x), \tag{A68}$$

where the lower-case subscripts are the same as for the original confluent hypergeometric function:

$$\begin{aligned} {}_1F_1^N(\alpha; \gamma; x) &= {}_1F_1^0(\alpha; \gamma; x) \tag{A69} \\ &= {}_1F_1^N(\alpha; \gamma; 0, \ldots, 0; x), \tag{A70} \end{aligned}$$

and the upper case subscript, $N = 0$ in this case, and an arbitrary positive integer in general [35].

Equation (A66) is comparable to the extended confluent hypergeometric equation of degree 1 (Equation (A67)) thus:

$$x X'' + (\gamma - x) X' - (\alpha + \alpha_1 x) X = 0, \tag{A71}$$

where:

$$n = 1 \tag{A72}$$
$$\gamma = 0 \tag{A73}$$
$$\alpha = -\frac{\mu \beta}{v_0 d^2} \tag{A74}$$
$$\alpha_1 = \frac{\mu D_0}{v_0^2 d}. \tag{A75}$$

The extended confluent hypergeometric function of degree one can be obtained by introducing:

$$\begin{aligned} \sigma_\pm &= 1/2 \pm \sqrt{1/4 + \alpha_1} \tag{A76} \\ &= \sqrt{1 - \frac{4 \omega^2 D_0}{v_0^2 d}}, \tag{A77} \end{aligned}$$

and the first kind leads to the integral representation:

$$\mathrm{Re}(\gamma) > \mathrm{Re}(\alpha) > 0: \quad F(\alpha; \gamma; x) = D^+ \int_0^1 z^{\alpha-1} (z-1)^{\gamma-\alpha-1} e^{xz} dz, \tag{A78}$$

and the second kind leads to the integral representation:

$$Re(\gamma) > 0, Re(\alpha) > 0: \quad G(\alpha; \gamma; x) = D^- e^{i\pi\gamma} \int_0^\infty z^{\alpha-1}(z+1)^{\gamma-\alpha-1} e^{-zx} dz, \tag{A79}$$

where the constants $D^\pm$ chosen are:

$$D^- = \frac{e^{-ir\gamma i}}{\Gamma(\alpha)}, \tag{A80}$$

and D^+ takes the value unity at the origin. The solution of Equation (A71) is in the form of extended confluent hypergeometric function of degree one and 2nd kind:

$$Re(x_2) > 0, Re(\alpha) > 0:$$

$$X(x_2) = {}_1G_1^1(\alpha; \gamma; \alpha_1; x_2)$$

$$= \left(\frac{e^{-\frac{1}{2}\left(\frac{1}{\sigma}-1\right)x_2}}{\Gamma(\alpha)}\right) \int_0^\infty z^{\frac{\alpha}{\sigma}-1}(z+1)^{-\left(\frac{\alpha}{\sigma}+1\right)} e^{-\sigma z x_2} dz$$

$$x_1 = \beta + xd \tag{A81}$$

$$x_2 = \frac{v_0}{D_0 d} x_1 \tag{A82}$$

$$\mu = -\omega^2 \tag{A83}$$

$$\omega^2 = |\mu| \tag{A84}$$

Defining the Tricomi–Kummer function [36,37]:

$$U(a,b,x) = \frac{1}{\Gamma(\alpha)} \int_0^\infty z^{\alpha-1}(z+1)^{-(\alpha+1-b)} e^{-zx} dz, \tag{A85}$$

and rearranging:

$$\Gamma(\alpha)U(a,b,x) = \int_0^\infty z^{\alpha-1}(z+1)^{-(\alpha+1-b)} e^{-zx} dz. \tag{A86}$$

For the Y- component, substituting Equation (A83) in Equation (89) and multiplying by $\frac{y}{\lambda D_0}$ gives:

$$Y'' - \frac{v_0}{D_0} Y' - \frac{Rs + R\kappa + \omega^2}{D_0\lambda} y Y(y,s) = 0. \tag{A87}$$

The following variables are hereby defined:

$$Y = \psi e^{\frac{v_0 y}{2D_0}} \tag{A88}$$

$$Y' = \psi' e^{\frac{v_0 y}{2D_0}} + \frac{v_0}{2D_0} \psi e^{\frac{v_0 y}{2D_0}} \tag{A89}$$

$$Y'' = \psi'' e^{\frac{v_0 y}{2D_0}} + \frac{v_0}{D_0} \psi' e^{\frac{v_0 y}{2D_0}} + \frac{1}{4}\left(\frac{v_0}{D_0}\right)^2 \psi e^{\frac{v_0 y}{2D_0}}, \tag{A90}$$

and transform Equation (A87) thus:

$$\psi'' - \left(\frac{D_0\lambda}{Rs + R\kappa + \omega^2}\right)^2 y_1\psi = 0, \tag{A91}$$

which is a general form of the Airy equation.

References

1. Baldwin, D.E. Prediction of Tracer Performance in a Five-Spot Pattern. *J. Pet. Technol.* **1966**, *18*, 513–517. [CrossRef]
2. Rubin, J.; James, R.V. Dispersion affected transport of reacting solutes in saturated porous media: Galerkin Method applied to equilibrium-controlled exchange in unidirectional steady water flow. *Water Resour. Res.* **1973**, *9*, 1332–1356, doi:10.1029/WR009i005p01332. [CrossRef]
3. Moreno, L.; Neretnieks, I.; Klockars, C.E. *Evaluation of Some Tracer Test in the Granitic Rock at FinnsjöRn*; SKBF/KBS Tech; Technical Report 83-38; Nuclear Fuel Safety Project; SKB: Stockholm, Sweden, 1983.
4. Herbert, A.W.; Hodgkinson, D.P.; Lever, D.A.; Rae, J.; Robinson, P.C. Mathematical modelling of radionuclide migration in groundwater. *Q. J. Eng. Geol.* **1986**, *19*, 109–120. [CrossRef]
5. Mabuza, S.; Čanić, S.; Muha, B. Modeling and analysis of reactive solute transport in deformable channels with wall adsorption-desorption. *Math. Methods Appl. Sci.* **2016**, *39*, 1780–1802, doi:10.1002/mma.3601. [CrossRef]
6. Vetter, O.J.; Zinnow, K.P. *Evaluation of Well-to-Well Tracers for Geothermal Reservoirs*; Lawrence Berkeley National Laboratory: Berkeley, CA, USA, 1981; Volume 14.
7. Tomich, J.F.; Dalton, R.L.J.; Deans, H.A.; Shallenberger, L.K. Single-Well Tracer Method To Measure Residual Oil Saturation. *J. Pet. Technol.* **1973**, *25*, 211–218. [CrossRef]
8. Bear, J. *Dynamics of Fluids in Porous Media*; Dover Civil and Mechanical Engineering; Dover Publications: Mineola, NY, USA, 2013.
9. Zhou, R.; Zhan, H. Reactive solute transport in an asymmetrical fracture-rock matrix system. *Adv. Water Resour.* **2018**, *112*, 224–234, doi:10.1016/j.advwatres.2017.12.021. [CrossRef]
10. Landa-Marbán, D.; Radu, F.A.; Nordbotten, J.M. Modeling and Simulation of Microbial Enhanced Oil Recovery Including Interfacial Area. *Transp. Porous Media* **2017**, *120*, 395–413, doi:10.1007/s11242-017-0929-6. [CrossRef]
11. Noetinger, B.; Roubinet, D.; Russian, A.; Borgne, T.L.; Delay, F.; Dentz, M.; de Dreuzy, J.; Gouze, P. Random Walk Methods for Modeling Hydrodynamic Transport in Porous and Fractured Media from Pore to Reservoir Scale. *Transp. Porous Media* **2016**, *115*, 345–385, doi:10.1007/s11242-016-0693-z. [CrossRef]
12. Zlotnik, V.A.; Logan, J.D. Boundary conditions for convergent radial tracer tests and effect of well bore mixing volume. *Water Resour. Res.* **1996**, *32*, 2323–2328. [CrossRef]
13. Carslaw, H.S.; Jaeger, J.C. *Conduction of Heat in Solids*; Clarendon Press: Oxford, UK, 1959.
14. Falade, G.K.; Brigham, W.E. Analysis of Radial Transport of Reactive Tracer in Porous Media. *SPE Reserv. Eng.* **1989**, *4*, 85–90. [CrossRef]
15. Attinger, S.; Abdulle, A. Effective velocity for transport in heterogeneous compressible flows with mean drift. *Phys. Fluids* **2008**, *20*, 016102. [CrossRef]
16. Vergassola, M.; Avellaneda, M. Scalar transport in compressible flow. *Phys. D Nonlinear Phenom.* **1997**, *106*, 148–166, doi:10.1016/S0167-2789(97)00022-5. [CrossRef]
17. Moench, A.F.; Ogata, A. A numerical inversion of the Laplace transform solution to radial dispersion in a porous medium. *Water Resour. Res.* **1981**, *17*, 250–252, doi:10.1029/WR017i001p00250. [CrossRef]
18. Stehfest, H. Algorithm 368: Numerical inversion of Laplace transforms. *Assoc. Comput. Mach. J.* **1970**, *13*, 47–49.
19. Abramowitz, M.; Stegun, I.A. *Handbook of Mathematical Functions, Appl. Math. Ser. 55*; National Bureau of Standards: Gaithersburg, MD, USA, 1964; p. 1046.
20. Mashayekhizadeh, V.; Dejam, M.; Ghazanfari, M.H. The Application of Numerical Laplace Inversion Methods for Type Curve Development in Well Testing: A Comparative Study. *Pet. Sci. Technol.* **2011**, *29*, 695–707, doi:10.1080/10916460903394060. [CrossRef]

21. De-Hoog, F.R.; Knight, D.H.; Stokes, A.N. An Improved Method for Numerical Inversion of Laplace Transforms. *SIAM J. Sci. Stat. Comput.* **1982**, *3*, 357–366. [CrossRef]
22. Dubner, H.; Abate, J. Numerical Inversion of Laplace Transforms by Relating Them to the Finite Fourier Cosine Transform. *Assoc. Comput. Mach. J.* **1968**, *15*, 115–123, doi:10.1145/321439.321446. [CrossRef]
23. Zakian, V. Numerical inversion of Laplace transform. *Electron. Lett.* **1969**, *5*, 120–121. [CrossRef]
24. Schapery, R.A. Approximate methods of transform inversion for viscoelastic stress analysis. In Proceedings of the 4th U.S. National Congress on Applied Mechanics, Berkeley, CA, USA, 18–21 June 1962; pp. 1075–1085.
25. Brzeziński, D.W.; Ostalczyk, P. Numerical calculations accuracy comparison of the Inverse Laplace Transform algorithms for solutions of fractional order differential equations. *Nonlinear Dyn.* **2016**, *84*, 65–77, doi:10.1007/s11071-015-2225-8. [CrossRef]
26. Fick, A. Ueber Diffusion. *Ann. Phys.* **1855**, *94*, 59–86. (In German) doi:10.1002/andp.18551700105. [CrossRef]
27. Skellam, J. Random Dispersal in Theoretical Populations. *Biometrika* **1951**, *38*, 196–218. [CrossRef] [PubMed]
28. Whittaker, E.T. An expression of certain known functions as generalized hypergeometric functions. *Bull. Am. Math. Soc.* **1903**, *10*, 125–134. [CrossRef]
29. Gaver, D.P. Observing stochastic processes and approximate transform inversion. *Oper. Res.* **1966**, *14*, 444–459. [CrossRef]
30. Talbot, A. The accurate numerical inversion of Laplace transforms. *IMA J. Appl. Math.* **1979**, *23*, 97–120. [CrossRef]
31. Abate, J.; Whitt., W. The Fourier-series method for inverting transforms of probability distributions. *Queueing Syst.* **1992**, *10*, 5–88. [CrossRef]
32. Abate, J.; Whitt., W. Numerical inversion of Laplace transforms of probability distributions. *Oper. Res. Soc. Am. J. Comput.* **1995**, *7*, 36–43. [CrossRef]
33. McClure, T. Numerical Inverse Laplace Transform, **2018**. Available online:https://uk.mathworks.com/matlabcentral/fileexchange/39035-numerical-inverse-laplace-transform (accessed on 16 August 2017).
34. Avdis, E.; Whitt, W. Power Algorithms for Inverting Laplace Transforms. *INFORMS J. Comput.* **2007**, *19*, 341–355. [CrossRef]
35. Campos, L. On some solutions of the extended confluent hypergeometric differential equation. *J. Comput. Appl. Math.* **2001**, *137*, 177–200, doi:10.1016/S0377-0427(00)00706-8. [CrossRef]
36. Spanier, J.; Oldham, K.B. *An Atlas of Functions*; Taylor & Francis/Hemisphere: Bristol, PA, USA, 1987.
37. Slater, L.J. *The Second Form of Solutions of Kummer's Equations*; Confluent Hypergeometric Functions; Cambridge University Press: Cambridge, UK, 1960.

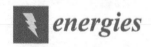

Article

Mechanism Reduction and Bunsen Burner Flame Verification of Methane

Haitao Lu [1,2], Fuqiang Liu [1,2,*], Yulan Wang [1,2], Xiongjie Fan [1,2], Jinhu Yang [1,2], Cunxi Liu [1,2] and Gang Xu [1,2]

[1]   Institute of Engineering Thermophysics, Chinese Academy of Sciences, Beijing 100190, China; luhaitao@iet.cn (H.L.); wangyulan@iet.cn (Y.W.); fanxiongjie@iet.cn (X.F.); yangjinhu@iet.cn (J.Y.); liucunxi@iet.cn (C.L.); xug@iet.cn (G.X.)

[2]   School of Engineering Science, University of Chinese Academy of Sciences, Beijing 100049, China

*   Correspondence: liufuqiang@iet.cn; Tel.: +86-010-8254-3156

Received: 5 December 2018; Accepted: 24 December 2018; Published: 29 December 2018

Abstract: Based on directed relation graph with error propagation methods, 39 species and 231 reactions skeletal mechanism were obtained from Mech_56.54 (113 species and 710 reactions) mechanism of methane. The ignition delay times, laminar flame propagation speed, and important species were calculated using the simplified mechanism at different pressures and equivalence ratios. The simulation results were in good agreement with that of detailed mechanisms and experimental data. The numerical simulation of the Bunsen burner jet flame was carried out using the simplified methane mechanism, and the simulation results well reproduced the temperature, flow fields and distribution of important species at flame zone. The compact methane reduced mechanism can not only correctly respond to its dynamic characteristics, but also can be well used for numerical simulation, which is of great significance in engineering applications.

Keywords: methane; combustion mechanism; mechanism reduction; skeletal mechanism; Bunsen burner

1. Introduction

The depletion of coal and crude oil has forced people to start looking for alternative energy sources, but in a short period of time, hydrocarbon combustion is still the most important way to generate energy. The burning of fossil fuels inevitably produces pollutants and causes a greenhouse effect, while biofuels can be produced from renewable energy sources and can reduce emissions from conventional fossil fuels. As the main component of natural gas, methane (CH_4) is a relatively clean fossil energy source. Since biomethane can be produced by biomass anaerobic reaction, it is considered as a renewable energy source and widely used in industry and life [1].

In order to reduce pollutant emissions and improve combustion efficiency, researchers have conducted a lot of research on the chemical reaction mechanism of fuels. As the simplest hydrocarbon fuel, the study of the kinetic characteristics of methane can improve the understanding of the combustion mechanism of larger hydrocarbon fuels. As early as 1958, Enikolopyan constructed the CH_4/Air oxidation model by increasing the oxidation reaction of CO and H_2 [2]. By the 1980s, precise oxidation model of CH_4/Air was close to 30 components, and then gradually contained more macromolecular components [3,4]. Since then, researchers have used the experimental equipment such as shock tubes and rapid compression machines to obtain more accurate reaction rate constants, and obtained accurate thermodynamic parameters by means of quantum chemical calculation, which makes the simulation results of the reaction mechanism more accurate. After the 1990s, methane combustion mechanism kept emerging, but due to the differences in experimental and computational conditions, the applicable scope of each mechanism was quite different [5–8]. For obtaining a wide range of CH_4/O_2 combustion mechanism model, the researchers developed and verified the experimental data in the existing public literature, and finally formed the GRI-Mech mechanism. After 20 years of development, a relatively perfect GRI-Mech 3.0 mechanism has been formed [9]. In recent years, Curran et al. developed the Mech_56.54 reaction mechanism based on experimental data and quantum chemical calculation, and the simulated results have higher accuracy than the GRI-Mech 3.0 mechanism [10,11].

Although the detailed chemical reaction mechanism of methane has been obtained, the complex chemical reaction kinetics has put forward extremely high requirements on the numerical calculation capabilities for CFD. For meeting the engineering design requirements and obtaining acceptable calculation results within a small error range, it is an effective method to use the reduced mechanism of fuel for calculation [12,13]. Qiao et al. simplified the GRI-Mech 3.0 mechanism and achieved good results by using the optimal simplification and sensitivity analysis method [14]. Based on the eigenvalue analysis method, Wen et al. simplified the GRI-Mech 2.1 mechanism and obtained the 21 species and 83 reactions skeletal mechanism. They also simplified the GRI-Mech 3.0 mechanism and obtained a 26 species and 120 reactions mechanism and a 30 species 140 reactions mechanism. The reduced mechanisms of GRI-Mech 3.0 can well predict the flame propagation speed and concentrations of main components [15,16]. Liu et al. simplified the GRI-Mech 3.0 mechanism by using a perfect stirred reactor model and a laminar flame propagation speed model. The 14 species and 18 reactions can predict methane/air premixed combustion over a wide range [17]. Gou et al. used the three-generation path flux analysis method to simplify the combustion mechanism of GRI-Mech 3.0, obtained 36 species and 208 reactions mechanism, and improved the accuracy of path flux analysis [18,19]. Wu et al. used the CSP (computational singular perturbation) algorithm to simplify the methane mechanism and obtained the 23 species and 18 reactions mechanism, which was verified by data such as ignition delay times and laminar flame propagation speed. The reduced mechanism is applied to the simulation of 6114 engine, and the heat release rate is higher than the experimental data [20]. Hu et al. constructed the skeleton mechanism which contains 24 species and 126 reactions from GRI-Mech 3.0 mechanism for $CH_4/O_2/CO_2$ mixtures based on the direct relation graph method. Meanwhile, on the basis of the skeletal mechanism, a quasi-steady state assumption method was used to obtain the 17 species and 14 reactions package mechanism. The simulation results of the skeletal mechanism are close to the results of the package mechanism, while the results of the package mechanism are largely different from the experimental data of the plug flow reactor, and this package mechanism is not suitable for quantitative calculation [21].

At present, although many scientists have simplified the detailed mechanism of methane, they have basically modeled on the GRI-Mech 3.0 mechanism instead of the latest methane combustion mechanism. In this paper, Mech_56.54 methane reaction mechanism was adopted, and the DRGEP (directed relation graph with error propagation) method was used to construct the 39 species and 231 steps reaction reduced mechanism, and compared and verified with the existing literature on ignition delay times, laminar flame propagation speed and important components. The simulation results were all in good agreement with the experiment. On this basis, the Bunsen burner premixed jet flame was simulated by the reduced mechanism, and the experimental data of flow parameters, temperature and important components were compared. The simulation results accurately reproduced the combustion characteristics of methane.

2. Mechanism Reduction

The directed relation graph with error propagation is a mechanism reduction method developed on the directed relation graph method. This method was proposed by Law C.K., in 2005, to remove the secondary components and reactions and obtain the skeletal mechanism [22–24].

The identification and removal of unimportant components in a detailed mechanism are quite complicated due to the mutual coupling between the components. DRG method gives an effective way to this problem. General idea of DRG is shown in Figure 1, and the first thing is to map a graph structure for the reaction system. Every species in Figure 1 denotes a vertex. If the removal of species

B directly results in a significant error in the rate of formation of species A, species A is associated with species B, so A and B will be connected by an edge. And, the primary issue with the DRG method is how to define parameters to evaluate the direct impact of one species on the other. For the convenience of simplification, the DRG method does not use parameters related to the reaction system, and directly removes unimportant species by reaction mechanism analysis. The direct species A, introduced by the removal of another species B, form the mechanism. Such immediate error, noted as r_{AB}, can be expressed as:

$$r_{AB} = \frac{\sum\limits_{i=1,I} |v_{A,i}\omega_i\delta_{Bi}|}{\sum\limits_{i=1,I} |v_{A,i}\omega_i|}. \tag{1}$$

$$\delta_{Bi} = \begin{cases} 1, & \text{if the } i\text{th reaction involves species B} \\ 0, & \text{otherwise} \end{cases} \tag{2}$$

$$\omega_i = \omega_{f,i} - \omega_{b,i}. \tag{3}$$

where A and B indicate the species, i indicates the ith reaction of the mechanism, and $v_{A,i}$ indicates the stoichiometric coefficient of species A in the ith reaction. $\omega_i, \omega_{f,i}, \omega_{b,i}$ represent the net reaction rate, positive reaction rate and reverse reaction rate of the ith reaction, respectively. The denominator represents the absolute contribution of all reaction mechanisms containing species A, and the numerator represents the contributions of both species A and B.

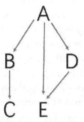

Figure 1. Scheme of general idea of the directed relation graph (DRG) methods. In the figure, species A is the target species, and species A generates species B, and species B regenerates into species C. Species E can be generated form species A or form species D.

After defining the immediate error, a search begins with the target species (e.g., fuel, oxidant) selected by the user to find the relevant path of all species relative to the target species. The program iterates until it no longer has significant species to choose from, and the reduced mechanism contains only significant species and elementary reactions. When the detailed mechanism is reduced, the existence of the vertex is decided by a threshold value (ε), which controls the size of the reduced mechanism.

However, the standard DRG method assumes that all species retained by the mechanism are of equal importance and ignores the connection between the coupled components. If A is an initial species, and species A generates species B, and species B regenerates into species C, it is assumed that species C is as important as species B in its simplification. In DRG, if r_{AB} is larger than the user-defined error value, species B needs to be retained. In addition, if r_{BC} is larger than the defined error value, the species C also needs to be retained due to the coupling of species B to the A and C [25–27]. As can be seen, the DRG method totally neglects the longer connection way to reach species A.

To overcome this shortcoming, Pepiot–Desjardins and Pitsch proposed the concept of propagation error. In DRGEP, if species A is retained, other species can be judged by R value, which is defined as [25]:

$$R_{AB} = \max_s \{r_{ij}\}. \tag{4}$$

where S is the set of all possible paths leading from species A to species B, and r_{ij} is the cumulative error of the given path. For example, if A is connected to B through a reaction and B is connected to C through a reaction, there is a path connecting from A to C via B and the R-value of this particular path is $r_{AC} = r_{AB} \cdot r_{BC}$. And in Figure 1, r_{AE} is calculated by $r_{AE} = \max(r_{AE}^{dir}, r_{AD} \cdot r_{DE})$. Based on this definition, a species B must be kept in the mechanism if there is at least one path connecting from A to B whose R-value is larger than the user-specified threshold, and a species C will be removed if r_{AC} is smaller than the user-specified threshold. The test cases of the Reaction Design have shown that, under many conditions, DRGEP method can generate a skeletal mechanism that contains approximately 10% fewer species than one generated using the DRG method [28].

Not only can the ignition delay time reflect the low-temperature combustion properties, but it can also include the high temperature combustion chemistry, so the ignition delay time is selected as the parameter in mechanism reduction, and the error of the reduced mechanism are calculated as:

$$\text{error}_{rel} = \frac{\left| \tau_{ign,red} - \tau_{ign,det} \right|}{\tau_{ign,det}} \times 100\%. \tag{5}$$

where the ignition delay times of reduced and detailed mechanisms are presented as $\tau_{ign,red}$ and $\tau_{ign,det}$, respectively. In this paper, the ignition delay time is defined as the time form the start to the maximum rate of temperature rise as shown in Figure 2.

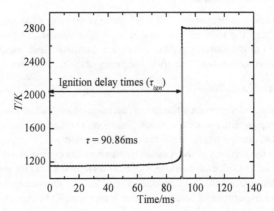

Figure 2. Typical temperature trace used to determine ignition delay time for a $\phi = 1.0$ of CH_4/Air mixture at $p = 1$ atm, T = 1150 K, τ_{ign}.

In this paper, the detailed mechanism of Mech_56.54 (including 113 species and 710 reactions) developed by Curran et al. was simplified with DRGEP method using ANSYS Chemkin Reaction Workbench software (Vision 17.0, ANSYS Reaction Design, San Diego, CA, USA) [10,28]. Ignition delay times were selected as the target parameter, and the homogeneous constant volume reactor is used to solve the solution, and the solution error error_{rel} is controlled within 5%. The simplified temperature range is 1100–1650 K, the pressure range is 1–40 atm, and the equivalent ratio is 0.3–2.0.

From Figure 3, it can be found that with the increase of the threshold value, the size of the reduced mechanism decreases, and the maximum relative error increases. In order to get a smaller skeletal mechanism with smaller relative error, the threshold value ε is set to 0.185, and the maximum relative error ε_{rel} is 5.88%, and 39 species reduced mechanism was obtained. And retained species in the corresponding reduced mechanisms with different threshold value are shown in Appendix A, Table A1.

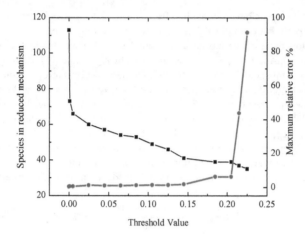

Figure 3. The change of the size of the reduced mechanism and the change of the maximum relative error with the increase of the threshold value. When the threshold value is 0.185, the reduced mechanism contains 39 species, and the maximum relative error is 0.588.

3. Verification of Reduced Mechanism

With the reduced procedure, the detailed mechanism of methane of Mech_56.54 was simplified by DRGEP method, and the reduced mechanism of 39 species and 231 reactions was obtained. The experimental data of the existing publications were simulated and compared using detailed mechanism and reduced mechanism to verify the accuracy of the mechanism.

3.1. Verification of Ignition Delay Times

Ignition delay times are important attributes of fuel combustion, and they have great influence on fuel ignition, flame propagation, and development. Therefore, accurate simulation of methane ignition delay times is crucial to capture methane combustion characteristics.

In order to verify the accuracy of the reduced mechanism, the experimental data of Bowman and Curran et al. were used [10,29]. Due to the lack of heat dissipation data for the rapid compression machine, it is considered as an adiabatic process. Therefore, the simulating results will be slightly smaller than Curran's experimental results at lower temperature. In Figure 4a,b are the ignition delay times of the $CH_4/O_2/Ar$ mixture at different equivalent ratios at pressures of 1.9 atm and 3.9 atm, respectively. In these two figures, as the temperature increases, ignition delay times decrease continuously; as the equivalence ratio increase, ignition delay times increase. At high temperature conditions, the ignition delay times of CH_4 change almost linearly with the reciprocal of temperature. The simulation results well capture the variation trend of ignition delay times with temperature and equivalent ratio, but there is still small error with the experimental data. In Figure 4, when equivalent ratios are smaller than 2.0, the relative errors between reduced model and full model are smaller than 4.36%. When equivalent ratio is 5.0, the relative errors become large, and the biggest relative errors are 20%. Figure 5 shows experimental data and simulation results of ignition delay times of CH_4/Air mixture at different equivalent ratios at high pressures (10 atm and 25 atm). At low temperature, ignition delay times become shorter as the temperature increases. While in the middle temperature, ignition delay times are less affected by the temperature. When the temperature exceeds 1100 K, the ignition delay times are exponentially negatively correlated with the temperature reciprocal. In the low temperature range, as the equivalence ratios increase, ignition delay times decrease, which is opposite to the change in the high temperature range. Both detailed mechanism and reduced mechanism well capture the variation trend of the ignition delay times with temperature in different

temperature intervals, and the errors between the reduced mechanism and the detailed mechanism
are extremely small.

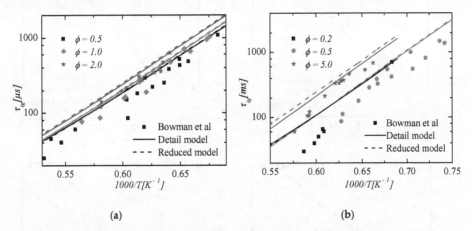

(a)   **(b)**

Figure 4. Influence of equivalent ratios on ignition delay times of $CH_4/O_2/Ar$ mixture. Solid line
shows results of the full model, dash line shows results of the reduced model and discrete point shows
experimental data from [29]. (**a**) $p = 1.9$ atm and $\phi = 0.5$, 1.0, and 2.0 in black, red, and blue, respectively;
(**b**) $p = 3.9$ atm and $\phi = 0.2$, 0.5, and 5.0 in black, red, and blue, respectively. Not all data are visible due
to overlapping profiles.

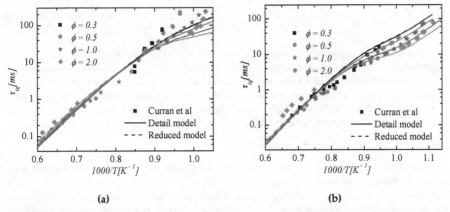

(a)   **(b)**

Figure 5. Influence of equivalent ratios on ignition delay times of CH_4/Air mixture. Solid line shows
results of the full model, dash line shows results of the reduced model and discrete point shows
experimental data from [10]. (**a**) $p = 10$ atm and $\phi = 0.3$, 0.5, 1.0, and 2.0 in black, red, blue, and magenta,
respectively; (**b**) $p = 25$ atm and $\phi = 0.3$, 0.5, 1.0, and 2.0 in black, red, blue, and magenta, respectively.
Not all data are visible due to overlapping profiles.

Figure 6 shows the effect of different pressures on the ignition delay times of the CH_4/Air mixture.
In Figure 6a–d, ignition delay times decrease with the increase of pressure at the equivalent ratio of
0.3/0.5/1.0/2.0, respectively. In the low temperature range, the calculation result of ignition delay
times is slightly lower because the heat dissipation in the experiments of Curran et al. is not considered.
In the high temperature range, all simulation results fitted the experimental results well, and the
reduced mechanism is almost coincident with the detailed mechanism. This result again demonstrates
that the reduced mechanism reflects well the ignition delay characteristics of CH_4 over a wide range.

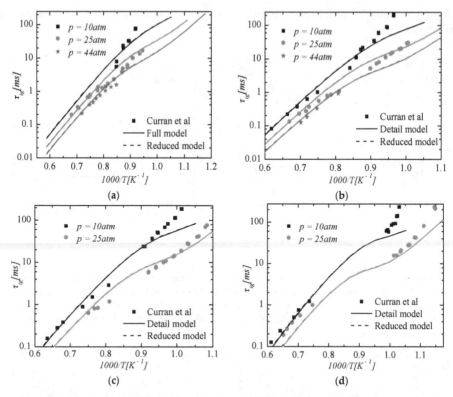

Figure 6. Influence of pressure on ignition delay times of CH$_4$/Air mixture. Solid line shows results of the full model, dash line shows results of the reduced model and discrete point shows experimental data from [10]. (**a**) $\phi = 0.3$ and $p = 10$ atm, 25 atm, and 44 atm in black, red, and blue respectively; (**b**) $\phi = 0.5$ and $p = 10$ atm, 25 atm, and 44 atm in black, red, and blue respectively; (**c**) $\phi = 1.0$ and $p = 10$ atm and 25 atm, in black and red, respectively; (**d**) $\phi = 2.0$ and $p = 10$ atm and 25 atm, in black and red, respectively. Not all data are visible due to overlapping profiles.

3.2. Verification of Laminar Flame Propagation Speed

As an important characteristic parameter of fuel, laminar flame propagation speed can be used to characterize the chemical reaction kinetics of fuel [30,31].

In this paper, the Laminar Flame Speed Calculation Solver of the Chemkin software (Vision 17.0, ANSYS Reaction Design, San Diego, CA, USA) was used to calculate the propagation speed of the CH$_4$/Air mixture. The laminar flame propagation speed of CH$_4$ at different pressures is shown in Figure 7, wherein the temperature of the unburned mixture is 298 K. At the same pressure condition, the flame propagation speed of CH$_4$ increases first and then decreases with the increase of the equivalence ratios, and reaches a peak value between the equivalent ratio of 1–1.1. When the pressure is low, the flame propagation speed between different equivalent ratios varies greatly, while when the pressure is high, the difference is small. The solid line and the dashed line in Figure 7 represent the calculated results of the detailed mechanism and the reduced mechanism, respectively. The simulated curve accurately captures the change in the laminar flame propagation speed of CH$_4$ as a function of pressures and equivalence ratios. At all equivalent ratios and pressures conditions, the results calculated by the reduced mechanism and the detailed mechanism have little error and are very close to the experimental results [32–35]. Average relative error between reduced model and full model of all conditions is 0.88%, and the biggest relative error between two models is 2.99%.

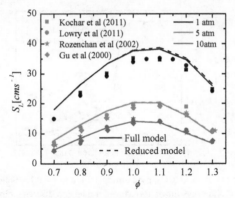

Figure 7. Influence of pressure on laminar flame speed of CH_4/Air mixture. Solid line shows results of the full model, dash line shows results of the reduced model and discrete point shows experimental data from [32–35]. Kochar et al.'s data are shown as square points; Lowry et al.'s data are shown as circle points; Rozenchan et al.'s data are shown as star points; and Gu et al.'s data are shown as diamond points.

3.3. Verification of Important Species

In order to further verify the accuracy of the mechanism, the experimental data of the jet-stirred reactors of Dagaut et al. and Lecong et al. were simulated and compared using a reduced mechanism and a detailed mechanism [36,37]. In the Dagaut's experiment, the proportion of $CH_4/O_2/N_2$ at 1 atm was 0.3%/6%/93.7%, and the residence time was 0.1 s. The species concentration distribution of the CH_4/air mixture at different temperatures is shown in Figure 8. As can be seen from the figure, the concentration of CH_4 gradually decreases with the increase of temperature, while the concentrations of CO and CO_2 keep increasing. Figure 9 shows the intermediate product of CH_4/air mixture. At approximately 1140 K, the concentrations of C_2H_4 and C_2H_6 reach a maximum, and then the concentrations decrease as the temperature increased. Detailed mechanism and reduced mechanism can simulate well the trend of formation and consumption of important components in combustion. For relatively stable components, such as $CH_4/CO/CO_2$, the error between simulation results and experimental values is small. And for the intermediate component C_2H_4, the simulation results of the reduced mechanism are slightly larger. C_2H_4 is consumed by reaction $C_2H_4 + H + M = C_2H_5 + M$ in both reduced mechanism and detailed mechanism. However, the producing rate of C_2H_4 is bigger than the consumption rate of C_2H_4 in detailed mechanism, so the mole fraction of C_2H_4 is small. While, the producing rate of C_2H_4 is even smaller than the consumption rate of C_2H_4 in reduced mechanism, so the mole fraction of C_2H_4 is relatively large.

The discrete points in Figures 10–12 are experimental data obtained by Lecong et al. using a jet stir reactor. The solid and dashed lines are the results of a detailed mechanism and a reduced mechanism, respectively [37]. In the experiment, pressure of 10 atm, equivalent ratio of 0.3, and residence time of 0.25 s were selected. In this experiment, the variations of components are similar to those in Figures 8 and 9. The reactants are continuously consumed as the temperature increases, and the products accumulate continuously, and the concentrations of the intermediate products first increase and then decrease. Compared with the experiment when the pressure was 1 atm and the equivalent ratio was 0.1, the components change relatively slowly at each temperature. It can be seen from the figure that both the simulation results and the experimental results can well capture the variation characteristics of each component at different temperatures. For the intermediate components, such as CH_2O and C_2H_4, the errors are relatively large, and all the other components are in good agreement.

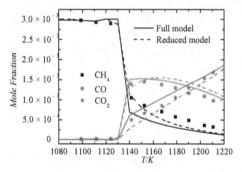

Figure 8. Mole fraction of $CH_4/CO/CO_2$. $p = 1$ atm, $\phi = 0.1$, and $\tau = 0.1$ s. Solid line shows results of the full model, dash line shows results of the reduced model and discrete point shows experimental data from [36]. CH_4 mole fraction is shown in black; CO mole fraction is shown in red; and CO_2 mole fraction is shown in blue.

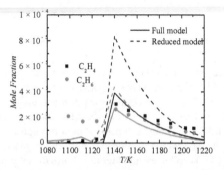

Figure 9. Mole fraction of C_2H_4/C_2H_6. $p = 1$ atm, $\phi = 0.1$, and $\tau = 0.1$ s. Solid line shows results of the full model, dash line shows results of the reduced model, and discrete point shows experimental data from [36]. C_2H_4 mole fraction is shown in black and C_2H_6 mole fraction is shown in red.

In order to make the reduced mechanism analysis more effective, the reduced mechanism is also compared with other mechanisms, as shown in Appendix B.

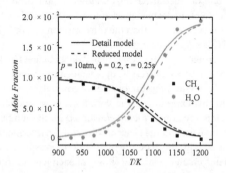

Figure 10. Mole fraction of CH_4/H_2O. $p = 10$ atm, $\phi = 0.3$, and $\tau = 0.25$ s. Solid line shows results of the full model, dash line shows results of the reduced model, and discrete point shows experimental data from [37]. CH_4 mole fraction is shown in black and H_2O mole fraction is shown in red.

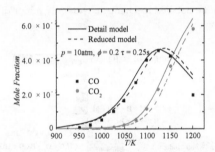

Figure 11. Mole fraction of CO/CO_2. p = 10 atm, ϕ = 0.3, and τ = 0.25 s. Solid line shows results of the full model, dash line shows results of the reduced model, and discrete point shows experimental data from [37]. CO mole fraction is shown in black and CO_2 mole fraction is shown in red.

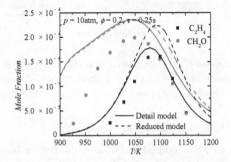

Figure 12. Mole fraction of C_2H_4/CH_2O. p = 10 atm, ϕ = 0.3, and τ = 0.25 s. Solid line shows results of the full model, dash line shows results of the reduced model, and discrete point shows experimental data from [37]. C_2H_4 mole fraction is shown in black and CH_2O mole fraction is shown in red.

4. Sensitivity Analysis

Sensitivity analysis is of great significance to understand the key reactions and mechanism simplification of chemical reaction kinetics of fuel combustion [38]. In order to find the important reaction to control the ignition in methane combustion, it is necessary to analyze sensitivity of the ignition delay times. In general, the sensitivity coefficient (S) can be expressed by the following equation:

$$S = \frac{ln\left(\frac{\tau_+}{\tau_-}\right)}{ln\left(\frac{k_+}{k_-}\right)} = \frac{ln\left(\frac{\tau_+}{\tau_-}\right)}{ln\left(\frac{2.0}{0.5}\right)}. \tag{6}$$

When the sensitivity coefficient (S) is negative, the reaction is the promoting reaction (the ignition delay time will be reduced with the increase of pre-exponential factor). When the sensitivity coefficient (S) is positive, the reaction is an inhibitory reaction (the ignition delay will increase with the increase of pre-exponential factor increases) [10].

As shown in Figure 13, it is the sensitivity analysis of the ignition delay times of the detailed mechanism and the reduced mechanism of the CH_4/Air stoichiometric mixture at 1470 K and pressure 10 atm. As can be seen from the figure, $CH_3 + O_2 = CH_2O + OH$ promotes the combustion of CH_4, which rapidly converts CH_3 produced by the pyrolysis of CH_4 into CH_2O and OH radical. For the $2CH_3(+M) = C_2H_6(+M)$ reaction, CH_3 radicals are combined to form ethane, which reduces the reaction rate and therefore inhibits the reaction. It can be seen from Figure 13 that the formation of small molecule active radicals ($HO_2/O/OH$) usually promotes the ignition, while the reaction of consumption of active radicals and the formation of relatively stable components is usually not conducive to ignition. From the figure, we can find that the analysis results of the reduced mechanism are basically identical with the detailed mechanism, with only minor differences in the sensitivity coefficients of partial reactions.

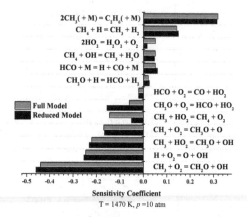

Figure 13. Sensitivity analyses of CH_4/Air Mixture. $T = 1470$ K, $p = 10$ atm, and $\phi = 1.0$. Red bars show results of the full model and black bars show results of the reduced model. When the sensitivity coefficient is negative, the reaction is the promoting reaction; when the sensitivity coefficient is positive, the reaction is an inhibitory reaction.

5. Flame Simulation and Verification of Bunsen Burner

In order to further verify the reduced mechanism, the Bunsen burner premixed flame was simulated using a two-dimensional model and compared with the experiments of Chen Y.C. et al [39]. The calculating geometry is shown in Figure 14. The diameter of inlet 1 is $D = 12$ mm, the average inlet velocity is $u_0 = 50$ m/s, the average turbulent energy is $k_0 = 10.8 \text{m}^2/\text{s}^2$, and the Reynolds number is 40,000. In the experiment, gases of the inlet 1 are CH_4/Air stoichiometric mixture and the temperature was 298 K. The gases of the inlet 2 are burned gases, the temperature is 1900 K, and the gases of inlet 3 and the inlet 4 are air. During the simulation, the boundary conditions of the inlet are given according to the experimental measurement data given the velocity and the turbulent energy, and the outlet is the pressure boundary condition. The calculation domain is 900 mm × 450 mm, and the independence of the grid is verified during simulation. The software ANSYS FLUENT (Version 17.0, ANSYS Inc, Pittsburgh, PA, USA.) was used for simulation, and the eddy-dissipation-concept model (referred to as EDC model) was selected [40]. The EDC model can simulate the turbulent-chemical kinetic interaction using a detailed chemical reaction mechanism with high simulation accuracy [41].

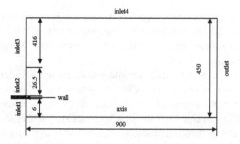

Figure 14. The Bunsen burner geometry. Radius of the inlet 1 is 6 mm, outer radius of the inlet 2 is 34 mm. The calculation domain is 900 mm × 450 mm. Gases of the inlet 1 are CH_4/Air stoichiometric mixture and the temperature was 298 K. The gas of the inlet 2 is burned gas, the temperature is 1900 K, and the gases of inlet 3 and the inlet 4 are air.

5.1. Comparison of Axis Velocity and Temperature

In order to better show the simulation results, the contour image of the simulation results is symmetrically processed. Figure 15 shows the axial velocity distribution of the Bunsen burner flame. The airflow expands slightly outward after flowing through the outlet, and the axis velocity gradually decreases with the increase of jet depth. Figure 16 shows the contour of flame temperature distribution of Bunsen burner. Outside the outlet of the premixed gas, the surrounding temperature is around 1900 K, and the temperature in the downstream gradually increases due to the heat release caused by chemical reaction. The flame temperature is lower in the downstream of the air outlet due to the larger velocity, but the flame temperature increases rapidly when it reaches the reaction zone.

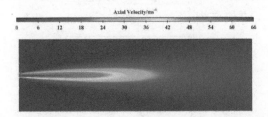

Figure 15. Axial velocity contour of Bunsen burner. The image is mirrored at the axis of symmetry. The airflow expands slightly outward after flowing through the outlet, and the axis velocity gradually decreases with the increase of jet depth.

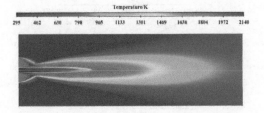

Figure 16. Temperature contour of Bunsen burner. The image is mirrored at the axis of symmetry. The flame temperature is lower in the downstream of the outlet due to the larger velocity, but the flame temperature increases rapidly when it reaches the reaction zone.

In order to quantitatively compare the calculated values with the experimental results, three axial positions ($X/D = 2.5$, $X/D = 4.5$, and $X/D = 6.5$) are selected for analysis. The axial velocity is dimensionless using the average velocity U_0, and the dimensionless temperature is defined as $C = (\tilde{T} - T_u)/(T_b - T_u)$, where K and $T_u = 298K$.

Figure 17 shows the comparison of the dimensionless axial velocity and the dimensionless temperature with the experimental results of different axial positions. The solid points are experimental results of Chen Y.C. et al., and the lines are the calculation results. As can be seen from Figure 17a, when $r/D < 0.5$, the calculated values of axial velocity almost coincide with the experimental values, while when r/D is between 0.5 and 1.1, the calculated results are slightly larger than the experimental values. In Figure 17b,c, when $r/D > $ is 0.75, the simulation results are slightly larger than the experimental values, while at $r/D < 0.75$, the simulation results are slightly smaller than the experimental results. In the three graphs, the dimensionless temperature values are basically the same as the distribution trend of the experimental results. With the increase of r/D, the values increase first and then decrease. At $X/D = 2.5$, the dimensionless temperature error was the largest, followed by $X/D = 4.5$ and $X/D = 6.5$ is the smallest. This is mainly because the adiabatic model is adopted in the simulation calculation without considering the influence of heat dissipation at the inlet, which makes the calculation temperature high. As the axial position increases, the influence of the non-adiabatic effect of the inlet becomes smaller, making the calculation result more accurate.

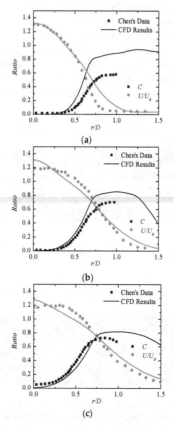

Figure 17. Non-dimensional axial velocity and temperature at different axial positions. Solid line shows CFD results and discrete point shows experimental data from [39]. Dimensionless temperature C is shown in black; dimensionless velocity U/U_0 is shown in red; (a) X/D = 2.5; (b) X/D = 4.5; and (c) X/D = 6.5.

5.2. Comparison of Important Species

Figure 18 shows the distribution of significant species of Bunsen burner flame, and Figure 18a shows the mass fraction distribution of CH_4. As the depth of the jet increases, the concentration of CH_4 decreases continuously. This distribution is similar to the axial velocity in Figure 15, but the distribution is narrower than the axial velocity and presents a tendency to converge continuously. In Figure 18b, since the CH_4/Air mixture is a stoichiometric premixed gas, the oxygen concentration in the jet is high. In the reaction zone, as the chemical reaction occurs, the oxygen concentration gradually decreases, and the concentration of oxygen in the active reaction zone is almost zero. It can be seen from Figure 18c,d that a large amount of H_2O and CO_2 are generated in the active reaction zone.

Figure 19 shows the distribution of important species in different axial positions of Bunsen burner flame, and the simulation results in all locations perfectly capture the CH_4 consumption. At the same time, when r/D < 0.75, the change of oxygen concentration with radius is also consistent with the experiment; Since the heat dissipation at the outlet wall is not considered, the oxygen concentration is closer to the experiment as the axial position increases at r/D > 0.75; at X/D = 2.5, CO_2 and H_2O are in good agreement with the experimental results, while at X/D = 4.5 and 6.5, the overall change trend is basically consistent with the experimental results, but there is a certain error in the numerical value.

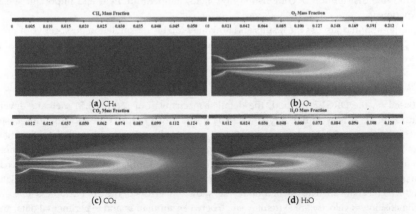

Figure 18. Important species contours of Bunsen burner flame. (**a**) Distribution of CH_4 mass fraction; (**b**) Distribution of O_2 mass fraction; (**c**) Distribution of CO_2 mass fraction; and (**d**) Distribution of H_2O mass fraction.

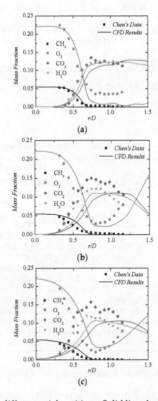

Figure 19. Important species at different axial positions. Solid line shows CFD results and discrete point shows experimental data from [39]. CH_4 mass fraction is shown in black; O_2 mass fraction is shown in red; CO_2 mass fraction is shown in blue; H_2O mass fraction is shown in magenta; (**a**) X/D = 2.5; (**b**) X/D = 4.5; and (**c**) X/D = 6.5.

By using the reduced mechanism to simulate the premixed flame of Bunsen burner, the reduced mechanism can well capture the combustion characteristics of CH_4/Air premixed flame and truly reproduce the distribution of combustion flow field, temperature field and important species in the experiment.

Authors should discuss the results and how they can be interpreted in perspective of previous studies and of the working hypotheses. The findings and their implications should be discussed in the broadest context possible. Future research directions may also be highlighted.

6. Conclusions

Based on the DRGEP method, the detailed mechanism of Mech_56.54 methane developed by Curran et al. was systematically simplified, and finally a reduced mechanism with 39 species and 231 reactions was obtained. Using the current published experimental data, the methane ignition delay times, laminar flame propagation speed, and important components obtained by the reduced mechanism were compared and verified. The simulation results of this mechanism are in good agreement with the experimental results. The Bunsen burner flame was simulated by the reduced mechanism. The calculation results show that errors of the dimensionless temperature and dimensionless velocity are extremely small between simulation and experimental data, and the distribution of important species are basically consistent with the experiment, indicating that the mechanism can be applied to CFD simulation accurately.

Author Contributions: Conceptualization, H.L., F.L., C.L., G.X., and Y.W.; methodology, H.L., X.F., Y.W., and J.H.; software, H.L., Y.W., and J.H.; Validation, H.L., X.F., F.L., and G.X.; Formal Analysis, H.L.; Investigation, H.L.; Resources, F.L. and G.X.; Data Curation, G.X.; Writing—Original Draft Preparation, H.L.; Writing—Review and Editing, H.L.; F.L., and C.L.; Visualization, H.L. and C.L.; supervision, F.L. and G.X.; Project Administration, G.X.; Funding Acquisition, J.H.

Funding: This research was funded by the National Natural Science Foundation of China, grant number 51806219.

Conflicts of Interest: The authors declare no conflict of interest.

Appendix A.

Table A1. Retained species in the corresponding reduced mechanisms with different threshold value.

Threshold value (ε)	Species	Different Species
0	113	Ar, $HOCH_2O_2H$, O_2CH_2CHO, CH_3CO_3H, CH_3CO_3, CH_3CO_2, C_2H_5OH, SC_2H_4OH, $O_2C_2H_4OH$, CH_3COCH_3, $CH_3COCH_2O_2$, C_3KET_{21}, C_2H_5CHO, C_2H_5CO, CH_3OCH_2, $CH_3OCH_2O_2$, $CH_2OCH_2O_2H$, $CH_3OCH_2O_2H$, CH_3OCH_2O, CH_3OCHO, CH_3OCO, CH_2OCHO, He, $C_3H_6OOH_1$-2, $C_3H_6OOH_1$-3, CC_3H_4, $CJ^*CC^*CC^*O$, $C^*CC^*CCJ^*O$, CJ^*CC^*O, C^*CC^*CCJ, C^*CC^*CC, C^*CC^*CCOH, HOC^*CC^*O, HOC^*CCJ^*O, HOCO
0.001	78	C_2H_2OH, HO_2CH_2CO, PC_2H_4OH, $C_2H_4O_2H$, CH_3COCH_2, C_2H_3CHO, $O_2CH_2OCH_2O_2H$, HO_2CH_2OCHO, OCH_2OCHO, $HOCH_2OCO$, C_3H_8, IC_3H_7, NC_3H_7, C_3H_6, C_3H_5-A, C_3H_5-S, C_3H_5-T, C_3H_5O
0.025	60	$HOCH_2O_2$, C, CH_3CO, $C_2H_5O_2H$, $C_2H_5O_2$, $C_2H_4O_1$-2, $C_2H_3O_1$-2, C_2H_3CO, C_3H_4-P, C_3H_4-A, C_3H_3, H_2CC, $H_2CCC(S)$, $C\#CC^*CCJ$
0.125	46	OCH_2O_2H, CH^*, C_2H, CH_3CHO, C_2H_3OH, CH_2CHO, C_2H_5O
0.185	39	H, H_2, O, O_2, OH, OH^*, H_2O, N_2, HO_2, H_2O_2, CO, CO_2, CH_2O, HCO, HO_2CHO, HCOH, O_2CHO, HOCHO, OCHO, $HOCH_2O$, CH_3OH, CH_2OH, CH_3O, CH_3O_2H, CH_3O_2, CH_4, CH_3, CH_2, $CH_2(S)$, CH, C_2H_6, C_2H_5, C_2H_4, C_2H_3, C_2H_2, CH_2CO, HCCO, HCCOH, CH_3OCH_3

Appendix B. Comparison of the Reduced Mechanism with Other Simplified Mechanisms

In order to make the reduced mechanism analysis effective, the reduced mechanism is compared with other mechanisms. Since many simplified mechanisms cannot be obtained directly from the literature, authors chose two mechanisms which are contained in literature. Dong et al. adopted genetic algorithm to simplify the GRI-MECH 3.0 mechanism with the component concentration as the goal, and finally obtained the 17 species and 24 steps reaction mechanism [42]. Kee et al. proposed a 17 species and 58 steps reaction mechanism. Ignition delay times, laminar flame speed and important species are compared with these mechanisms [43].

Comparison of Ignition Delay Times

As can be seen in Figure A1, results of Dong's model is smaller than experimental data and results of Kee's model is larger than experimental data. And results of the reduced model are in good agreement with results of the full model and experimental data. In Dong's model, with the increase of the equivalent ratios, ignition delay times decrease, this phenomenon is different from experimental data and other models.

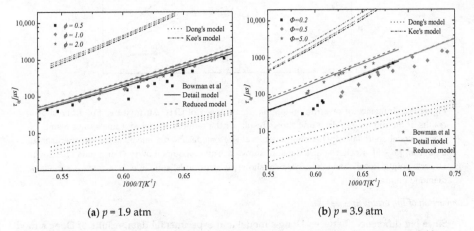

(**a**) $p = 1.9$ atm   (**b**) $p = 3.9$ atm

Figure A1. Influence of equivalent ratios on ignition delay times of $CH_4/O_2/Ar$ mixture. Solid line shows results of the full model, dash line shows results of the reduced model, dot line shows results of Dong's model, short dash dot line shows results of Kee's model, and discrete point shows experimental data from [29]. (**a**) $p = 1.9$ atm and $\phi = 0.5$, 1.0, and 2.0 in black, red, and blue, respectively; (**b**) $p = 3.9$ atm and $\phi = 0.2$, 0.5, and 5.0 in black, red, and blue, respectively. Not all data are visible due to overlapping profiles.

Comparison of the Laminar Flame Propagation Speed

As can be seen in Figure A2, when $p = 1$ atm, equivalent ratios are smaller than 1.1, all models can capture changes of laminar flame speed with the increase of equivalent ratios. There are great errors in Dong's model when equivalent ratios are larger than 1.1. When $p = 5$ atm or $p = 10$ atm, both Kee's model and Dong's model have great errors with experimental data. Overall, the reduced model by authors are in good agreement with experimental data and full model.

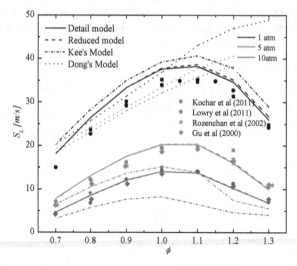

Figure A2. Influence of pressure on laminar flame speed of CH_4/Air mixture. Solid line shows results of the full model, dash line shows results of the reduced model, dot line shows results of Dong's model, short dash dot line shows results of Kee's model and discrete point shows experimental data from [32–35]. Kochar et al.'s data are shown as square points; Lowry et al.'s data are shown as circle points; Rozenchan et al.'s data are shown as star points; and Gu et al.'s data are shown as diamond points.

Comparison of Important Species

Since big differences between Dong's model and experimental data, results of Dong's model are not shown in Figures A3–A6, when comparing the important species. For relatively stable components, such as CH_4/CO/CO_2, results of reduced model and full model can well match with experimental data, however results of Kee's model doesn't even show the trend. For the intermediate component of CH_2O, results of Kee's model are totally different from experimental data.

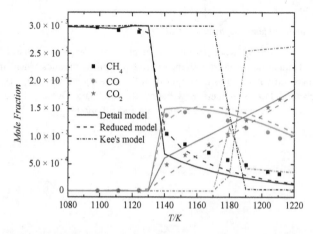

Figure A3. Mole fraction of CH_4/CO/CO_2. $p = 1$ atm, $\phi = 0.1$, and $\tau = 0.1$ s. Solid line shows results of the full model, dash line shows results of the reduced model, short dash dot line shows results of Kee's model, and discrete point shows experimental data from [36]. CH_4 mole fraction is shown in black and; CO mole fraction is shown in red; and CO_2 mole fraction is shown in blue.

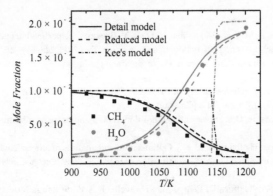

Figure A4. Mole fraction of CH_4/H_2O. $p = 10$ atm, $\phi = 0.3$, and $\tau = 0.25$ s. Solid line shows results of the full model, dash line shows results of the reduced model, short dash dot line shows results of Kee's model, and discrete point shows experimental data from [37]. CH_4 mole fraction is shown in black and H_2O mole fraction is shown in red.

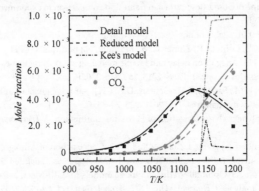

Figure A5. Mole fraction of CO/CO_2. $p = 10$ atm, $\phi = 0.3$, and $\tau = 0.25$ s. Solid line shows results of the full model, dash line shows results of the reduced model, short dash dot line shows results of Kee's model, and discrete point shows experimental data from [37]. CO mole fraction is shown in black and CO_2 mole fraction is shown in red.

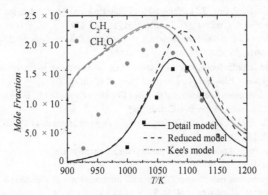

Figure A6. Mole fraction of C_2H_4/CH_2O. $p = 10$ atm, $\phi = 0.3$, and $\tau = 0.25$ s. Solid line shows results of the full model, dash line shows results of the reduced model, short dash dot line shows results of Kee's model, and discrete point shows experimental data from [37]. C_2H_4 mole fraction is shown in black and CH_2O mole fraction is shown in red.

References

1. Molino, A.; Nanna, F.; Migliori, M.; Iovane, P.; Ding, Y.; Bikson, B. Experimental and simulation results for biomethane production using peek hollow fiber membrane. *Fuel* **2013**, *112*, 489–493. [CrossRef]
2. Enikolopyan, N.S. Kinetics and mechanism of methane oxidation. *Symp. Combust.* **1958**, *7*, 157–164. [CrossRef]
3. Westbrook, C.K. An Analytical Study of the Shock Tube Ignition of Mixtures of Methane and Ethane. *Combust. Sci. Technol.* **1979**, *20*, 5–17. [CrossRef]
4. Frenklach, M.; Bornside, D.E. Shock-initiated ignition in methane-propane mixtures. *Combust. Flame* **1984**, *56*, 1–27. [CrossRef]
5. Frenklach, M.; Wang, H.; Rabinowitz, M.J. Optimization and analysis of large chemical kinetic mechanisms using the solution mapping method—combustion of methane. *Prog. Energy Combust. Sci.* **1992**, *18*, 47–73. [CrossRef]
6. Ranzi, E.; Sogaro, A.; Gaffuri, P.; Pennati, G.; Faravelli, T. A Wide Range Modeling Study of Methane Oxidation. *Combust. Sci. Technol.* **1994**, *96*, 279–325. [CrossRef]
7. Barbe, P.; Battin-Leclerc, F.; Côme, G.M. Experimental and modelling study of methane and ethane oxidation between 773 and 1573 K. *J. Chim. Phys. Physico-Chim. Biol.* **1995**, *92*, 1666–1692. [CrossRef]
8. Marinov, N.M.; Pitz, W.J.; Westbrook, C.K.; Lutz, A.E.; Vincitore, A.M.; Senkan, S.M. Chemical kinetic modeling of a methane opposed-flow diffusion flame and comparison to experiments. *Symp. Combust.* **1998**, *27*, 605–613. [CrossRef]
9. GRI-Mech, Release 3.0. Available online: http://www.me.berkeley.edu/gri_mech (accessed on 18 July 2018).
10. Burke, U.; Somers, K.P.; O'Toole, P.; Zinner, C.M.; Marquet, N.; Bourque, G.; Petersen, E.L.; Metcalfe, W.K.; Serinyel, Z.; Curran, H.J. An ignition delay and kinetic modeling study of methane, dimethyl ether, and their mixtures at high pressures. *Combust. Flame* **2015**, *162*, 315–330. [CrossRef]
11. Tingas, E.A.; Manias, D.M.; Sarathy, S.M.; Goussis, D.A. CH₄/Air homogeneous autoignition: A comparison of two chemical kinetics mechanisms. *Fuel* **2018**, *223*, 74–85. [CrossRef]
12. Wang, T.S. Thermophysics characterization of kerosene combustion. *J. Thermophys. Heat Transf.* **2001**, *15*, 76–80. [CrossRef]
13. Xu, J.Q.; Guo, J.J.A.; Liu, K.J.; Wang, L.; Tan, N.X.; Li, X.Y. Construction of Autoignition Mechanisms for the Combustion ofRP-3 Surrogate Fuel and Kinetics Simulation. *Acta Phys. Chim. Sin.* **2015**, 643–652. [CrossRef]
14. Qiao, Y.; Xu, M.H.; Yao, H. Optimally-reduced kinetic models for GRI-Mech 3. 0 combustion mechanism based on sensitivity analysis. *J. Huazhong Univ. Sci. Technol. (Nat. Sci. Ed.)* **2007**, *35*, 85–87. [CrossRef]
15. Wen, F. The Reduction Method Based on Eigenvalue Analysis for Combustion Mechanisms and Its Applications. Ph.D. Dissertation, Tsinghua University, Beijing, China, 2012.
16. Wen, F.; Zhong, B.J. Skeletal Mechanism Generation Based on Eigenvalue Analysis Method. *Acta Phys. Chim. Sin.* **2012**, *28*, 1306–1312. [CrossRef]
17. Liu, H.; Chen, F.; Liu, H.; Zheng, Z.H.; Yang, S.H. 18-Step Reduced Mechanism for Methane/Air Premixed Supersonic Combustion. *J. Combust. Sci. Technol.* **2012**, *18*, 467–472.
18. Gou, X.L.; Wang, W.; Gui, Y. Methane reaction using paths flux analysis of three generations method. *J. Eng. Thermophys.* **2014**, *35*, 1870–1873.
19. Wang, W. Studies on the Efficient Reduction Methods for the Combustion Chemical Kinetic Mechanism of Fuel. Ph.D. Dissertation, Chongqing University, Chongqing, China, 2016.
20. Wu, Z.Z. Study on Mechanism Reduction for Detailed Chemical Kinetics of IC Engine Fuel. Ph.D. Dissertation, Shanghai Jiaotong University, Shanghai, China, 2015.
21. Hu, X.Z.; Yu, Q.B.; Li, Y.M. Skeletal and Reduced Mechanisms of Methane at O₂/CO₂ Atmosphere. *Chem. J. Chin. Univ.* **2018**, *39*, 95–101. [CrossRef]
22. Lu, T.; Law, C.K. A directed relation graph method for mechanism reduction. *Proc. Combust. Inst.* **2005**, *30*, 1333–1341. [CrossRef]
23. Lu, T.; Law, C.K. Linear time reduction of large kinetic mechanisms with directed relation graph: N-Heptane and iso-octane. *Combust. Flame* **2006**, *144*, 24–36. [CrossRef]
24. Lu, T.; Law, C.K. On the applicability of directed relation graphs to the reduction of reaction mechanisms. *Combust. Flame* **2006**, *146*, 472–483. [CrossRef]

25. Pepiot-Desjardins, P.; Pitsch, H. An efficient error-propagation-based reduction method for large chemical kinetic mechanisms. *Combust. Flame* **2008**, *154*, 67–81. [CrossRef]
26. Liang, L.; Stevens, J.G.; Farrell, J.T. A dynamic adaptive chemistry scheme for reactive flow computations. *Proc. Combust. Inst.* **2009**, *32*, 527–534. [CrossRef]
27. Wang, Q.D. Skeletal Mechanism Generation for Methyl Butanoate Combustionvia Directed Relation Graph Based Methods. *Acta Phys. Chim. Sin.* **2016**, *32*, 595–604. [CrossRef]
28. ANSYS. *ANSYS Chemkin Reaction Workbench 17.0 (15151)*; ANSYS Reaction Design: San Diego, CA, USA, 2016.
29. Seery, D.J.; Bowman, C.T. An experimental and analytical study of methane oxidation behind shock waves. *Combust. Flame* **1970**, *14*, 37–47. [CrossRef]
30. Yao, T.; Zhong, B.J. Chemical kinetic model for auto-ignition and combustion of n-decane. *Acta Phys. Chim. Sin.* **2013**, *29*, 237–244. [CrossRef]
31. Zheng, D.; Yu, W.M.; Zhong, B.J. RP-3 Aviation Kerosene Surrogate Fuel and the Chemical Reaction Kinetic Model. *Acta Phys. Chim. Sin.* **2015**, *31*, 636–642. [CrossRef]
32. Kochar, Y.; Seitzman, J.; Lieuwen, T.; Metcalfe, W.; Burke, S.; Curran, H.; Krejci, M.; Lowry, W.; Petersen, E.; Bourque, G. Laminar Flame Speed Measurements and Modeling of Alkane Blends at Elevated Pressures With Various Diluents. *ASME Proc. Combust. Fuels Emiss.* **2011**, *2*, 129–140. [CrossRef]
33. Lowry, W.; Vries, J.D.; Krejci, M.; Petersen, E.; Serinyel, Z.; Metcalfe, W.; Curran, H.; Bourque, G. Laminar Flame Speed Measurements and Modeling of Pure Alkanes and Alkane Blends at Elevated Pressures. In Proceedings of the ASME Turbo Expo 2010, Power for Land, Sea, and Air, Glasgow, UK, 14–18 June 2010; pp. 855–873.
34. Rozenchan, G.; Zhu, D.L.; Law, C.K.; Tse, S.D. Outward propagation, burning velocities, and chemical effects of methane flames up to 60 ATM. *Proc. Combust. Inst.* **2002**, *29*, 1461–1470. [CrossRef]
35. Gu, X.J.; Haq, M.Z.; Lawes, M.; Woolley, R. Laminar burning velocity and Markstein lengths of methane–air mixtures. *Combust. Flame* **2000**, *121*, 41–58. [CrossRef]
36. Dagaut, P.; Boettner, J.C.; Cathonnet, M. Methane Oxidation: Experimental and Kinetic Modeling Study. *Combust. Sci. Technol.* **1991**, *77*, 127–148. [CrossRef]
37. Cong, T.L.; Dagaut, P.; Dayma, G. Oxidation of Natural Gas, Natural Gas/Syngas Mixtures, and Effect of Burnt Gas Recirculation: Experimental and Detailed Kinetic Modeling. *J. Eng. Gas Turbines Power* **2008**, *130*, 635–644. [CrossRef]
38. Yu, W.M. Research on Flame Propagation Speed and Reaction Dynamic Mechanism of Aviation Kerosene Alternative Fuel. Ph.D. Dissertation, Tsinghua University, Beijing, China, 2014.
39. Chen, Y.C.; Peters, N.; Schneemann, G.A.; Wruck, N.; Renz, U.; Mansour, M.S. The detailed flame structure of highly stretched turbulent premixed methane-air flames. *Combust. Flame* **1996**, *107*, 223–244. [CrossRef]
40. ANSYS. *ANSYS Fluent Theory Guide*; ANSYS Inc.: San Diego, CA, USA, 2017.
41. Duan, Z.Z. *Fluid Analysis and Engineering Example of Ansys Fluent*; Publishing House of Electronics Industry: Beijing, China, 2015.
42. Dong, Q.L.; Jiang, Y.; Qiu, R. Reduction and optimization of methane combustion mechanism based on PCAS and genetic algorithm. *Fire Saf. Sci.* **2014**, *23*, 43–49. [CrossRef]
43. Kee, R.J.; Grcar, J.F.; Smooke, M.D.; Miller, J.A. *A Fortran Program for Modeling Laminar One-Dimensional Premixed Flames*; Sandia Report SAND 85-8240; Sandia N Laboratories: Albuquerque, NM, USA, 1985.

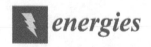

Article

Genetic Mechanism and Evolution of the Covert Fault Zone and Its Oil-Controlling Mode in Qikou Sag, Eastern China

Shuai Jiang [1,2,*], Weifeng Wang [1,2], Aizhu Zhang [1,2] and Weiwei Zhou [3]

1   School of Geosciences, China University of Petroleum (East China), Qingdao 266580, China;
    weifeng_w@yeah.net (W.W.); zhangaizhu789@163.com (A.Z.)
2   Laboratory for Marine Mineral Resources, Qingdao National Laboratory for Marine Science and Technology,
    Qingdao 266071, China
3   South China Sea Institute of Oceanology Chinese Academy of Sciences, Guangzhou 510301, China;
    weiwei_z@yeah.net
*   Correspondence: jshuai1988@163.com; Tel.: +86-15376750858

Received: 16 November 2018; Accepted: 22 December 2018; Published: 29 December 2018

Abstract: Covert fault zone is an important type of geological phenomenon that is closely related to hydrocarbon formation and distribution but has often been overlooked because it lacks obvious fault displacement and fault plane. To meet this challenge, a novel cognitive framework is proposed in this study, in which criteria for identifying the existence of covert fault zone are developed based on the regional tectonic backgrounds and geophysical data. The Riedel shear model is then utilized to analyze the genetic mechanism of the covert fault zone. The Mohr-Coulomb theory is also introduced to conduct a structural physical simulation to interpret the evolution process of the covert fault zone. Information about the genetic mechanism and evolution of the covert fault zone is finally combined to determine the oil-controlling mode. The study site is Qikou Sag in Eastern China. It is found that the covert fault zone in Qikou Sag meets four recognition criteria and is generated by the stress transferred from the strike-slip activity of the basement fault. Moreover, it can be concluded that the covert fault zone in Qikou Sag contains five evolution stages and controls the reservoir mainly via three aspects, that is, sedimentary sand, subtle traps and oil accumulation mode.

Keywords: covert fault zone; genetic mechanism; Qikou Sag; structure evolution; oil-controlling mode; Riedel shear; Mohr–Coulomb theory

1. Introduction

Fault in sedimentary strata is an important reservoir-controlling factor [1]. Especially for the petroliferous basins in China, the control effects of faults on oil and gas reservoirs is more obvious [2]. Thereby the deeply research of the fault is essential for oil and gas exploration [3–5]. Nowadays, the increase in exploration difficulty makes oil and gas exploration in deep, covert and unconventional fields desirable [6–8]. In particular, the effects of covert tectonic activity on the overlying layers requires further interpretation [9,10]. Moreover, researchers have found some small weak deformation structures in the overlying layers that without obvious fault displacement and fault plane [11,12]. The geological phenomenon formed by these structures are referred to as covert fault zone in this study.

In the past two decades, although few researches taken covert fault zone as an independent geological phenomenon, the small weak deformation structures have shown highly correlations with the distribution of oil and gas [1,10,12–18]. For example, the Ordos Basin in the central parts of China was previously believed to be no fault in the caprocks and its reservoir was controlled by the anticline, lithology and stratigraphic overlap [11]. However, many small weak deformation structures were recently discovered in the overlying layers and these structures control the distribution and migration of hydrocarbon to a great degree [1,9]. Similarly, the string distribution of the reservoirs, sags and domes in the Jinhu and Gaoyou Sag of the Subei Basin, Eastern China is found controlled by the small faults in caprocks [12]. These small faults were caused by the underlying basement faults in North East (NE), North West (NW) and South North (SN) directions [12,19].

Nevertheless, few of the previous studies performed a systematical analysis of the covert fault zone. Its genetic mechanism, evolution and oil-controlling mode are undefined. Fortunately, some referable researches have been presented in the structural geology field. For example, Morley derives a penetrating structure on the basis of the Mohr-Coulomb criterion and Byerlee's legislation [13]. This structure does not have a uniform rupture surface, which is very similar to the covert fault zone. In addition, according to the detailed analysis of the simulation of the strike-slip structure in the plate, Dooly et al. found that there are many small faults (hidden faults) scattered in the en-echelon fault zone under various regional stress in the early stages of fracture development [14]. The deformation difference of these small faults are mainly shown as variance in the maximum principle stress direction, the Riedel (R) shear angle and the construction and evolution modes. Moreover, Hardy studied the overlay deformation (defined as the discrete fault zone) features on the steep basement normal fault in the continuous increasing fault distance by using 2D discrete element modeling and concluded that the activity in the discrete fault zone affects trap formation and fluid migration [17,18].

The aforementioned studies all indicate that although the covert fault zone usually occurs as small faults in linear and discontinuous arrays, its genetic mechanism, evolution and oil-controlling mode can be studied based on the adjacent structures, such as the basement fault, en-echelon distributed small faults, regional stress and so forth. [13,14,17,18]. This was also confirmed by the simulation experiments of Bellahsen [16].

Qikou Sag is the largest hydrocarbon bearing sag of the Bohai Bay Basin in Eastern China [20]. To make clear geological structure of Qikou Sag is crucial for exploring the distribution of oil and gas in the Bohai Bay Basin. In the previous researches, the regional tectonic backgrounds of the Qikou Sag, such as stratigraphic distribution, sedimentary characteristics and basic tectonic characteristics, have been analyzed [21–24]. Many geological structure data are thus summarized. Additionally, some studies on the generation and accumulation of hydrocarbon in the Qikou Sag also have been performed [20,25–30]. These works have accumulated many geophysical data such as seismic profiles and horizontal slices of Qikou Sag. The interpretation of the seismic profiles and horizontal slices can provide the profile and planar characteristics of geological structure patterns, respectively [31–37]. However, scarcely any study has paid attention to the covert fault zone in this sag.

In this paper, to perform a deep analysis of the genetic mechanism, evolution and oil-controlling mode of covert fault zone in Qikou Sag, a novel cognitive framework for the covert fault zone is proposed. In the cognitive framework, the first aim is to establish a set of recognition criteria for identifying the existence of covert fault zone. Then for a covert fault zone, to build a genetic mechanism analysis method on the basis of the R shear model. Additionally, to design a structural physical simulation method based on the Mohr-Coulomb theory for interpreting the evolution process of the covert fault zone. Finally based upon the genetic mechanism and evolution process, to construct the oil-controlling mode of the covert fault zone.

2. Study Area

The Qikou Sag is the largest hydrocarbon bearing sag of the Bohai Bay Basin in Eastern China, located in the center of the Huanghua Depression with an area of about 5280 km^2 [26]. It is divided into land areas on the west side and sea areas on the east side by the Boxi coastline, which is about 250 km long [26]. The sag is controlled by the faults in North NE (NNE), NE and East West (EW) directions and has the characteristics of extensional fault depression. Its strike-slip activity is very active because its basement from the south to the north is passed by the Lanliao strike-slip fault, as shown in

Figure 1 [1,26]. The structural units of Qikou Sag include four negative structural units, namely the Qikou main sag, the Banqiao sub-sag, the Qibei sub-sag and the Qinan sub-sag and four positive structural units named the Beidagang buried hill, the Nandagang buried hill, the along-coastline basement involved fault belt and the Chengbei step-fault belt. The sedimentary caprocks of Qikou Sag have experienced two evolutionary stages: the Paleogene rifting stage and the Neogene depression stage. Seismic and drilling data show that the strata of Qikou Sag are composed of a preceding Paleogene basement and Cenozoic caprocks, which include Shahejie and Dongying formations in Paleogene and Guantao and Minghuazhen formations in Neogene [21].

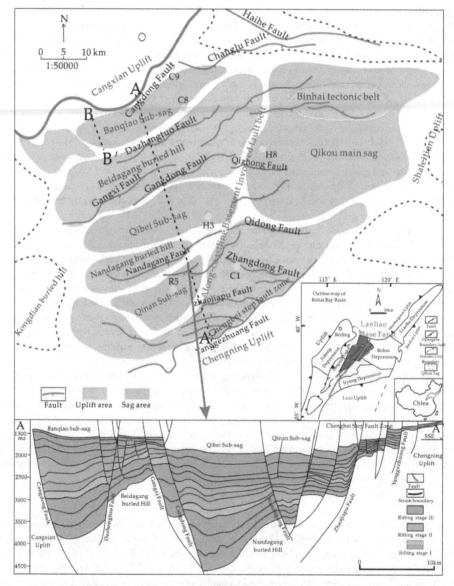

Figure 1. Regional geological map and structural framework of the Qikou Sag, Eastern China (modified from Chen et al., [21]).

3. Methods

In this paper, the genetic mechanism and evolution of the Qikou Sag were analyzed by comprehensive utilization of the regional tectonic backgrounds and geophysical data using the *R* shear model and the Mohr-Coulomb theory. The oil-controlling mode was thereby constructed for the further analysis of reservoir formation. The whole work is named as a novel cognitive framework of covert fault zone. The specific methods of the cognitive framework are introduced in the following subsections.

3.1. Establishment of Recognition Criteria

Following the previous studies, a covert fault zone can be identified by referring to the adjacent structures. The analysis of the adjacent structures in this paper was mainly based on four recognition criteria, that is, single geological element, geophysical data including seismic profiles and horizontal slices of coherent bodies and basement faults.

The single geological element adjacent to the covert fault zone can be determined based on the regional tectonic backgrounds. This is mainly because the features of the adjacent single geological element are always influenced by the covert fault zone. In this paper, the single geological element were classified into nine type, that is, small dominant structures distributed in the en-echelon discontinuous direction, fault structures discretely distributed along a fixed direction, laterally distributed buried hills and depressions, discretely zonal distributed sedimentary facies and sand bodies, discretely directional arranged traps and reservoirs, small faults and small cracks distributed in the dense zone, bead distributed volcanic rock, tectonic abrupt zones and structural separation zones. If one or more types of elements are found in a fault, it can be preliminary regarded as a covert fault zone.

The seismic profiles and horizontal slices of coherent bodies can provide the combined characteristics of the structures [32–37]. Thereby they were also used as recognition criteria in this paper. Based on the preliminary identification, if weak flower-like or semi-flower-like structures were found in the seismic profiles and if small faults or dense fracture zones were found in horizontal slices of coherent bodies, the existence of covert fault zone could be basically determined.

Additionally, basement faults are the fundamental control factors for the development of the covert fault zone in basin cap faults. Consequently, basement faults were used as one of the recognition criteria here. In practice, geophysical data such as the aforementioned seismic profiles and horizontal slices can be used to study whether there are corresponding basement faults under the basin cap. That is to say, mutual validation of these recognition criteria is possible.

On the basis of the four established recognition criteria, identification of the covert fault zone can be achieved, as illustrated in Figure 2.

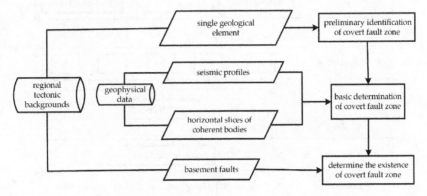

Figure 2. Recognition criteria-based identification method of covert fault zone.

3.2. R Shear Model Based Genetic Mechanism Analysis

The R shear model, which is produced by the stress from different tectonic activities, can be used to analyze the genetic mechanism of covert fault zone. This mainly because shear stress is usually expressed as small faults, such as the small dominant structures distributed in the en-echelon discontinuous direction, as discussed in Section 3.1. These small faults are important recognition criteria for identifying of the covert fault zone.

Specifically, influenced by the stress produced by the tectonic activities, such as strike-slip, the shear faults can be classified into five types as shown in Figure 3. As illustrated in Figure 3, R shear fault and antithetic R shear (R' shear) fault are firstly arisen in sequence. As the displacement increases, the amount of the R shear fault increases along the basement fault and low-angle synthetic shear (P shear) fault occurs. Then, discontinuous Y shear fault parallels to the basement fault begins to appear along with the local tensional (X shear) fault that intersects Y shear fault at a large angle.

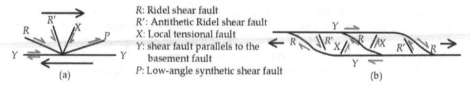

R: Ridel shear fault
R': Antithetic Ridel shear fault
X: Local tensional fault
Y: shear fault parallels to the basement fault
P: Low-angle synthetic shear fault

(a)   (b)

Figure 3. Illustration of the five types of shear fault (R, R', P, X, Y): (**a**) Mechanic principle of R shear; (**b**) Distribution of shear faults (modified from Dooley et al., [14]).

Accordingly, the genetic mechanism of covert zone are closely related to tectonic activities. Specifically, as shown in Figure 4, actives of different tectonic can generate various stress and thus produce five types of shear faults following the R shear model. Then the combination of shear faults forms the geological structures of the covert fault zone [1,32]. The genetic mechanism of a cover fault zone can thereby be analyzed by the interpretation of its geological structures. The flowchart of R shear based genetic mechanism analysis is given in Figure 4. The genetic mechanism provides the cognitive basis for the following analysis of evolution.

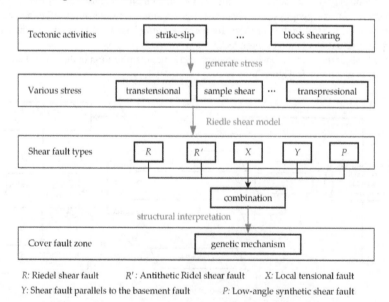

R: Riedel shear fault   R': Antithetic Ridel shear fault   X: Local tensional fault
Y: Shear fault parallels to the basement fault   P: Low-angle synthetic shear fault

Figure 4. Flowchart of *Riedel* shear based genetic mechanism analysis.

3.3. Mohr-Coulomb Theory Based Evolution Interpretation

Generally, the migration and accumulation of hydrocarbon are accompanied by the evolution of covert fault zone. The shear faults appear in the evolution of covert fault zone can provide reservoir for oil storage, be channels for oil migration or damage those pre-existing oil reservoirs. Therefore, in this paper, a physical simulation method designed based on the Mohr-Coulomb theory is utilized to interpret the evolution of covert fault zone in Qikou Sag.

The Mohr-Coulomb theory is a mathematical model that describes the responses of brittle materials, such as concrete or rubble piles, to shear stress as well as normal stress. In the Mohr-Coulomb theory, Coulomb's friction hypothesis [38] is used to determine the combination of shear and normal stress that will cause fracture of the material. Mohr's circle [39] is used to discover the principal stress that produced the aforementioned combination and find the angle of the plane in which the combination will occurs. Figure 5 displays the diagram of Mohr circle and envelope under triaxial stress. In Figure 5, envelope *a* represents fracture with uniform rupture surface belonging to the dominant stage. Envelope *b* represents the fracture line of the weak zone. Envelope *c* is the initial activity line of basement covert fault, from which the caprocks start to break and the cracks begin to form. The rectangular region between envelopes *a* and *c* represents the weak zone formed by the increasing activity strength of the basement faults. After this period, the fracture and small faults inside the fault zone begin to expand and they begin to be connected with each other to form large-scale fractures, which tend to be dominant [10].

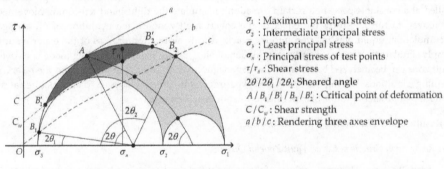

σ_1 : Maximum principal stress
σ_2 : Intermediate principal stress
σ_3 : Least principal stress
σ_n : Principal stress of test points
τ/τ_s : Shear stress
$2\theta/2\theta_1/2\theta_2$: Sheared angle
$A/B_1/B_1'/B_2/B_2'$: Critical point of deformation
C/C_w : Shear strength
$a/b/c$: Rendering three axes envelope

Figure 5. The diagram of the Mohr circle and envelope under triaxial stress (modified from Labuz et al., [40]).

As is shown in Figure 5, the reactivity condition of basement covert fault is the shear stress τ_n along with the strike fault must exceed the shear strength C. That is to say, the formation and evolution of the covert fault zone under the stress field generated by the basement fault is essentially a mechanical process. Therefore, the evolution process of the covert fault zone can be interpreted by simulating the mechanical process.

In this study, as shown in Figure 6, the principal stress, shear stress, sheared angle and shear strength were incorporated to design the physical simulation experiment. As displayed in Figure 6a, the size of the experimental device is 46 cm long and 33 cm wide, in which a basement fault with a 45°dip angle is set. Moreover, two active plates are embedded to simulate the motion of faults. As the value of the angle (α) between strip-slip and displacement, which corresponds to the θ shown in Figure 5, has a direct influence on the failure process of rocks, six different angles were designed to interpret the evolution process of the covert fault zone, as shown in Figure 6b.

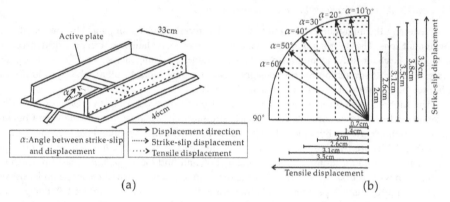

Figure 6. Schematic diagrams of the designed simulation experiment: (a) Sketch of the experimental device (modified by Dooley et al., [14]); (b) Details of the designed experimental groups.

In the designed physical simulation experiment, over-materials with a thickness of 2 cm were uniformly laid on the top of the model and were manually compacted. In addition, to facilitate the observation of deformation, 2.7 × 2.7 cm grids were printed on the surface of the cover layer in parallel and equispaced as markers during the experiment and the moving wall was pushed and pulled at a constant speed in a certain direction to simulate the right-hand tensioning movement. Moreover, to further study the effect of the covert fault zone on the evolution of oil and gas accumulation, liquid was injected into the cracks during simulation process of the experiment to simply simulate the migration and accumulation of oil. Thereby, the oil-controlling mode of a covert fault zone can be analyzed based upon the genetic mechanism, evolution process and the reservoir forming conditions. During the simulation experiment, the camera was used to take pictures at the same slip of different groups of experiments as shown in Section 4.3.

4. Results and Discussion

4.1. Recognition Criteria of Covert Fault Zone in Qikou Sag

From the comprehensive analysis of the regional tectonic backgrounds and geophysical data in Qikou Sag, four recognition criteria are found inside the sag, namely, the small dominant structures distributed in the en-echelon discontinuous direction (as shown in Figure 7), the linear zone reflected by the slice data, the flower structure reflected by the seismic profiles and the basement faults under the basin cap. Details of these four recognition criteria are given below and they confirm the existence of the covert fault zone in Qikou Sag.

Small dominant structures are found distributed in the en-echelon discontinuous direction. As introduced in Section 3.1, these structures is a type of the single geological element that can be used to preliminary identify the existence of the covert fault zone in the Qikou Sag. As shown in Figure 7, the small dominant structures (faults) on both sides of the covert fault zone have the following characteristics: the fault dip is in the opposite direction and the strike of faults causes mutation. That is, the small dominant structures in the covert fault zone are distributed in the en-echelon discontinuous direction. A series of medium and low amplitude bulges (4–8 km) along the trend of the covert fault zone are also spread out in an echelon pattern.

The linear zone reflected by the slice data is the second criterion found in Qikou Sag. As shown in Figure 8, the coherent slices in Figure 8a show that the deep part, which has a dark band with a low correlation coefficient, has an s-shaped strike-slip fault, while the shallow slices in Figure 8b reflect a linear and intermittent distribution of dark bands. These correspondences between the deep and shallow slices indicate the high possibility of the existence of a covert fault zone.

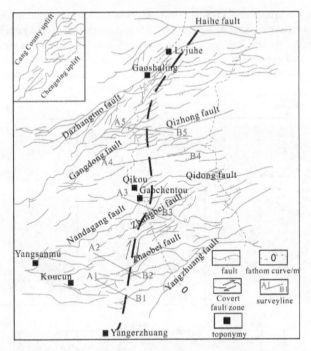

Figure 7. Distribution of small dominant Fault and tectonic location in Qikou Sag (modified from Qi et al., [28]).

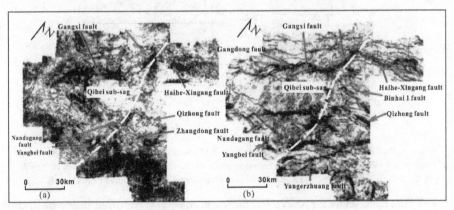

Figure 8. Coherent slices of the covert fault zone in Qikou Sag: (**a**) 3320 ms slices of the coherence cube; (**b**) 1320 ms slices of the coherence cube (modified from Zhou et al., [1]).

The flower structure reflected by the seismic profiles is the third criterion of covert fault zone in Qikou Sag. Figure 9a–d show the five profiles, that is, the profile A1–B1, the profile A2–B2, the profile A3–B3, the profile A4–B4 and the profile A5–B5 whose positions are provided in Figure 7. In the Southern section, as is shown in Figure 9a–c, the typical negative flower-like structure develops in the basement in the Paleogene era. Moreover, the flower-like structure gradually converges into a steep downward fault in the deep part. In the middle section, as shown in Figure 9d, the deep trunk strike-slip fault with an overall steep locally curved and irregular section is inserted into the base. Moreover, the deep trunk strike-slip fault is generally spread upward in a flower shape which is a normal fault on the upside and a reverse fault downwards. The covert fault zone formed in the Paleogene era and is characterized by a relatively wide and gentle flower structure. For the Northern section, the deformation of the Paleogene formation is generally confined to the regional bottom slip interface as illustrated in Figure 9e. The upper plate of the main slip zone shows folded deformation, while the tectonic features of the late Mesozoic are maintained below, with almost no deformation.

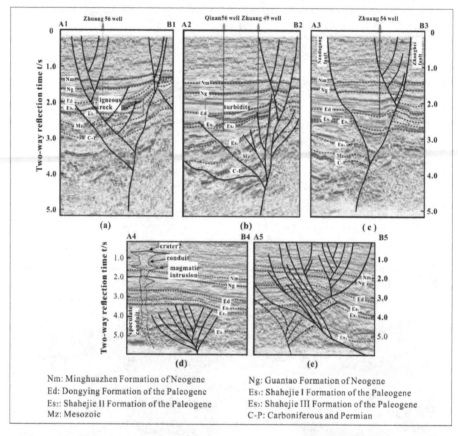

Figure 9. Profile characteristics of covert fault zone in Qikou Sag: (**a**) Profile A1–B1 of the Southern section; (**b**) Profile A2–B2 of the Southern section; (**c**) Profile A3–B3 of the Southern section; (**d**) Profile A4–B4 of the middle section; (**e**) Profile A5–B5 of the Northern section (modified from Qi et al., [28]).

Basement faults under the basin cap is the fourth criterion of the covert fault zone in Qikou Sag. The crystalline basement structure of Qikou Sag is presented in Figure 10. The covert fault zone of Qikou Sag shows completely different characteristics in terms of the aeromagnetic anomalies on both sides: the Luxi basement on the left presents a highly abnormal area with complex changes and the geological block has an overall NE strike. It can be concluded that the Taihang mountain basement on the right is characterized by an open positive and negative alternation of NNE and NE. Further, the area below the Qikou Sag is shown as the "triple point" part of the three crystalline basements of Luxi, Taihang and Yanshan. This special structural position is just like the weak zone described by Morley [10], which is prone to wiggle under the regional stress field and the covert fault zone can be easily formed by long-term weak activity or local lithologic unevenness. The buried hills and depressions on both sides of the covert fault zone are distributed in hidden places.

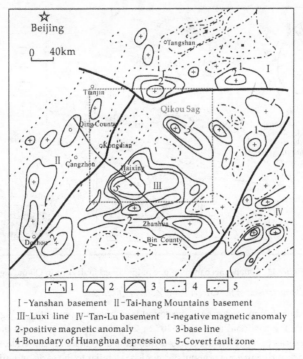

Figure 10. Crystalline basement plot of the Huanghua depression which contains the Qikou Sag.

4.2. Genetic Mechanism of the Covert Fault Zone in Qikou Sag

Through a comprehensive analysis of the tectonic activities of Qikou Sag and the four recognition criteria described in Section 4.1, it can be concluded that in the Cenozoic era, the Pacific plate subducted from NW to East West (EW), causing the Eastern part of the Chinese mainland to be under the stress field of right-handed tension. Therefore, the NNE deep basement fault near the coastline of Qikou depression has strong dextral strike-slip under this background, which is the basis for the formation of the Cenozoic structural diversity of the coastal zone. Based on the *R* shear model, *R* shear and its conjugate, *R'* shear, formed first and *P* shear formed later, according to the sequence of brittle strike-slip fracture formation. In addition, the Luxi basement and Taihang mountain basement are in contact with Qikou Sag and the splicing zone is shown as a weak zone, which is most prone to form shear fracture, as shown in Figure 11.

Overall, in the south section of the covert fault zone, the basement uplift is large and the buried depth of the basement strike-slip fault has a significant influence on the overburden deformation. The middle section is the rhombic region between the Qidong fault and the Gangdong fault. The root of the deep main strike-slip fault in this rhombic region is inserted into the basement, forming a flower-like structure and the strike of faults on both sides of the fault zone changes significantly. In the Northern section, the dense fault zone enters the slip zone, resulting in the upper displacement of the fault being adjusted so that it disappears. In general, the covert fault zone in Qikou Sag formed under the control of the strike-slip activity of the basement.

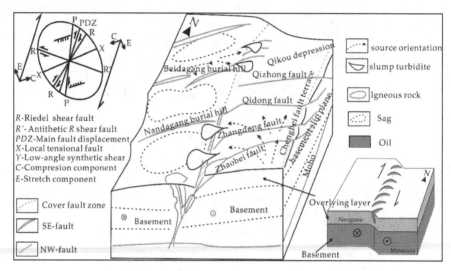

Figure 11. Genetic mechanism mode of the covert fault zone in Qikou Sag.

4.3. Evolutionary Characteristic of the Covert Fault Zone in Qikou Sag

The simulated evolution processes of the covert fault zone are presented in Figure 12. As shown in Figure 12, when the torsional angle α is 10° or 20°, the caprock tectonic deformation style is characterized by a series of relatively flat cracks arranged in an echelon. Meanwhile, the angle between the direction of crack and fault zone is about 35°, which is similar to the early and middle construction styles of simple shear. When the angle α is 30–50°, R shear faults still are arranged in an echelon but the plane morphology of most fractures will no longer be straight. Additionally, the angle between R shear faults and the direction of the fault changes from 30–40° to 5–10°, showing an arc or "S" shape. When angle α increases to 60 °, the angle between R shear faults and the direction of the fault reduces significantly to about 13° and the echelon feature of the fracture zone is no longer obvious.

To further investigate the evolution process of the covert fault zone, physical simulation experiments were performed under two stress conditions: transtensional stress and transpressional stress. The evolution processes are presented in Figure 13. As shown in Figure 13, the basement, stress and strain are the three conditions that control the formation and evolution of covert faults. Under the condition of increasing activity strength of basement faults, the caprock fault experiences a multi-stage change from "covert" to "dominant."

The first stage of the covert fault zone evolution is the induced breakage stage as shown in Figure 13. In this stage, the EW rupture is small-scale, scattered and isolated, with no break distance and poor regularity. In the covert fault zone, the induced fracture zone is roughly parallel to the strike direction and presents a dendritic shape. There is no main slip surface in the induced fractures. The second stage of the covert fault zone evolution is the localized fault stage. Under the continuous action of external stress, the activity of the basement fracture intensifies. Most R shears change into small faults with fault displacement. Due to the increases in the sizes of the above R shears, small localized rupture surfaces develop in the covert fault zone. However, they still fail to cut through covers. In this stage, P shears start to appear. This stage occurs under the condition of transtensional stress, while it does not occur under transpressional stress, as shown in Figure 13. The third stage of the covert fault zone evolution is the major fault stage. As the strike-slip effect of basement fault is further intensified, the writhing and tearing effect of cap layer is strengthened. As the en-echelon R shear increases, short P shear faults gradually appear, which are limited to the area between en-echelon faults. The P shear faults may occur at intervals or continuously develop to form local dominant faults. The fourth stage of the covert fault zone evolution is the major slip stage. As shown in Figure 13, the discontinuous partial main faults of the last stage interconnect with each other and develop into larger *Y* faults. Early-formed en-echelon *R* faults are cut into branch faults on both sides of the main faults. The last stage of the covert fault zone evolution is the slide and breakage stage. All *Y* shear faults merge with basement strike-slip faults and are shown as typical negative flower structures in the profile. In addition, many small branch faults develop along both sides of the main faults, and, combined with main faults, form a plumose structure. In this stage, the en-echelon fault are fully connected and destroyed to form the main fault plane.

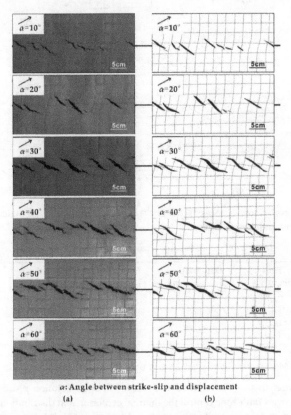

α: Angle between strike-slip and displacement
(a)   (b)

Figure 12. The evolution processes of the covert fault zone: (a) Experimental photos of different *α* angles; (b) Sketches of the photos in (a).

Based on the aforementioned analysis and oil filling simulation, the oil accumulation degree in different evolution stages is shown in Table 1. As shown, in the induced breakage stage, the structural deformation of the covert fault zone is in the form of a weak echelon, in which 20% of the trap area is filled and only small amounts of oil and gas accumulate in the fault traps close to the oil source. In the localized fault stage, small fractures expand to form larger fractures with an en-echelon distribution and small fractures began to appear at the edge of the *R* shear. In the covert fault zone, as a whole, oil and gas show an en-echelon fault block aggregation pattern and 50% of the trap area is filled. In the major fault stage, secondary *P* shear cracks begin to appear and the trap filling degree reaches 75%. In the major slip stage and slide and breakage stage, the covert fault zone is penetrated by *Y* shear and the fault properties of the covert fault zone are very obvious. The whole fault zone is rich in oil and gas and the trap is as full as 90%.

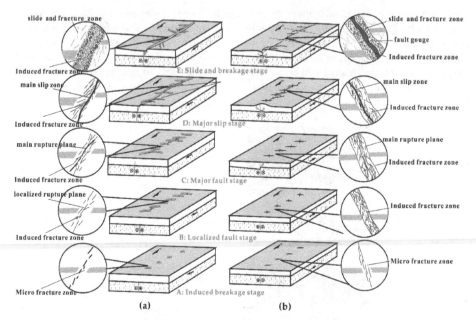

Figure 13. The evolution processes of the covert fault zone in Qikou Sag: (a) Under transtensional stress; (b) Under transpressional stress.

Table 1. Oil accumulation degree in different evolution stages.

Stage	Time Sequence	Deformation Strength (cm)	Oil Filling Level (mL)	Cumulative Time (min)	Trap Filled Area (%)
A	Early	1.56	12	3	20
B	Early-to-mid	3.76	20	7	55
C	Middle	7.32	24	9	75
D/E	End	11.6	24	10	90

4.4. Oil-Controlling Mode of the Covert Fault Zone in Qikou Sag

The genetic mechanism (as described in Section 4.2) and evolution process (as described in Section 4.3) of the covert fault zone control the thermal evolution and distribution of the source rock, reservoir physical properties, trap development and transport conditions and thus affect the law of hydrocarbon accumulation distribution [41]. Considering the geological conditions of the Qikou Sag, the distribution of the sedimentary sand, the formation and distribution of subtle traps and the oil accumulation mode are utilized to analyze the oil-controlling mode of the covert fault zone in Qikou Sag.

The controlling effect of sedimentary sand is mainly expressed as the cutting and joining action on sedimentary sand. The covert fault zone controls the formation and distribution of a large number of echelon and discontinuous small fault zones, which form the combined relationship between cutting and connection. As shown in Figure 14, the small structures shown in profile, such as fault terraces and graben, form fracture zones and the river channel can easily use its strike to wash down and cut and form sediment unloading zones along the covert fault zone. The covert fault zone of Qikou Sag has an obvious controlling effect on sedimentary facies, among which the basement fault in the Southern section is strong. The dominant fault in Zhangbei is formed in the caprock and the covert fault zone is spread in the left echelon pattern to the north. The fan delta of the Shasan Segment extends into the lake basin [42,43] and the source water system carries a large number of sand bodies deposited along the covert fault zone in the echelon pattern.

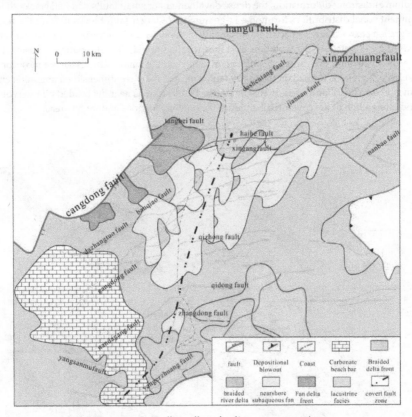

Figure 14. Controlling effect of sedimentary reservoirs.

There are primarily four kinds of subtle trap in the covert fault zone in Qikou Sag. One kind is NNE minor faults which are a kind of intersecting occlusion formed by basement fault activity in the cover. The second type is the combination of NE and NNE covert faults, which produce the arc-shaped covert fault block traps. Note that although these two traps have a tectonic background controlled by the covert fault zone, the genesis of traps is controlled by both lithology and structure. The third kind is the microstructure traps formed under the condition of different compactions of sandstone or low strength tectonic activity. The trap range of it is low, while its oil–bearing height is generally small. The fourth subtle trap includes lithologic updip pinchout traps controlled by low-amplitude slope break zones and the faults in both directions have not been developed. It is, in fact, a kind of subtle trap formed by the reconstruction of paleogeomorphology of covert fault zone.

The oil accumulation mode is the third aspect controlled by the covert fault zone in Qikou Sag. There are primarily four kinds of oil accumulation mode in the covert fault zone of Qikou Sag. As shown in Figure 15a, in the early stages of evolution under the condition of weak deformation, there are only a small number of faults, which form a few small fault block traps and they appear to be intermittent echelon or isolated and dispersed. The strike is inconsecutive and the connectivity of vertical channels is poor. In the early-to-mid stage of the evolution, as shown in Figure 15b, it can be concluded that under the conditions of medium and strong deformation, the number of faults in the covert fault zone is relatively small and the faults show an echelon arrangement. A series of small snout and fault block traps constitute the trap belt, with no connection in the strike and good connectivity in the vertical channel. In the middle stage of the evolution, as shown in Figure 15c, under the condition of medium and strong deformation, the dense development of small faults or small cracks shows an intermittent zonal distribution, forming a large number of larger fault blocks to form the trap belt. The migration channels are connected intermittently in the strike and the vertical channel connectivity is good. At the end stage of the evolution, as is shown in Figure 15d, the deformation intensity of faults is large, the small faults are densely distributed in zones and the main fault surface is almost complete. The fault block group is formed on the covert fault zone and the three-dimensional migration channel is formed in the covert fault zone. In this mode, the blank sections near the beaded oil reservoirs and the grid intersection of the covert fault zone are the potential regions and target areas.

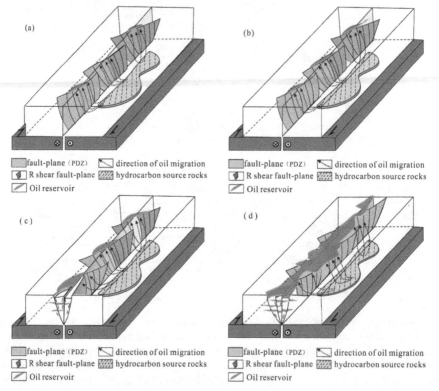

Figure 15. Oil accumulation model of the covert fault zone: (**a**) Early stage; (**b**) Early-to-mid stage; (**c**) Middle stage; (**d**) End stage.

According to the aforementioned controlling effects of the distribution of the sedimentary sand, the formation and distribution of subtle traps and the oil accumulation mode in the south, middle and north segments of the covert fault zone in Qikou Sag, three types of oil-controlling mode are established, as shown in Figure 16.

For the three-layer type shown in Figure 16a, the roots and branches of the floral structure can be used as an oil source fault with good trap, transmission and dynamic conditions. In deep series of strata, the bedrock buried hill reservoirs, fractured reservoirs and unconformity reservoirs form. The faulted noses and fault block reservoirs that formed in the Paleogene era and Neogene era have an en-echelon fault block oil trap accumulation mode. The complex oil and gas accumulation zone can be called a sandwich type vertical distribution mode. While for the mezzanine gathered type shown in Figure 16b, weak strike-slip activity of basement faults connects the oil source faults formed in the Neogene era. The Neogene caprock has relatively good sealing conditions and the oil and gas accumulate in the formations from the Paleogene era. The pattern can also be called a pie vertical distribution mode. By contrast, for the cover layer type shown in Figure 16c, basement faulting leads to the good source rock conditions and reservoir conditions. With good transmissibility in the vertical direction, oil and gas are mainly concentrated in formations from the Neogene era, which can also be called a pizza type vertical distribution mode.

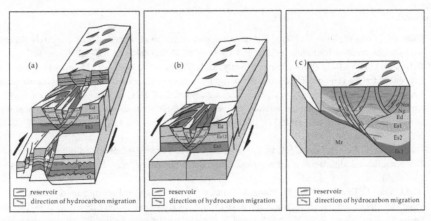

Figure 16. Oil-controlling mode of the covert fault zone in Qikou Sag: (a) Three-layer type; (b) Mezzanine gathered type; (c) Caprock type.

5. Conclusions

A novel cognitive framework of the covert fault zone is proposed in this paper. In the framework, a method for the establishment of recognition criteria is firstly presented for the identification of the covert fault zone. In this method, both the regional tectonic backgrounds and geophysical data of the adjacent structures are considered for the identification of the covert fault zone. Accordingly, the genetic mechanism of the covert fault zone is analyzed based on combination patterns of shear faults generated by the stress transferred from the activities of the basement fault following the R shear model. For the evolution interpretation of the covert fault zone, a structural physical simulation method is designed based on the Mohr-Coulomb theory. On the basis of the information about the genetic mechanism and evolution of the covert fault zone and the reservoir forming condition in the study site, the oil-controlling mode of the covert fault zone is investigated.

Studies on the Qikou Sag located in Eastern China have shown that it has four main recognition criteria, that is, en-echelon distributed small dominant structures, horizontal slices of coherent bodies, seismic profile and basement faults. Moreover, it is found that the covert fault zone in Qikou Sag is generated by the stress transferred from the strike-slip activity of the basement fault. Besides, the evolutionary characteristics show that the evolution process of the covert fault zone in Qikou Sag contains five stages: the induced breakage stage, the localized fault stage, the major fault stage, the major slip stage and the slide and breakage stage. In addition, it is concluded that the oil-controlling mode of the covert fault zone in Qikou Sag includes three types, namely the three-layer type, the mezzanine gathered type and the caprock type. Overall, the proposed cognitive framework can also be expanded to other study areas based on the corresponding regional tectonic background and geophysical data. For further improvement of the cognitive framework, variation in stress will be considered in future studies.

Author Contributions: Conceptualization, S.J. and W.Z.; Methodology, S.J., W.Z. and W.W.; Software, S.J.; Formal analysis, S.J. and W.Z.; Writing—original draft preparation, S.J. and W.Z.; Writing—review and editing, W.W. and A.Z.; Project administration, W.W.; funding acquisition, A.Z.

Funding: This research was funded by the National Natural Science Foundation of China, grant number 41801275; the Shandong Provincial Natural Science Foundation, China, grant number ZR2018BD007; the Fundamental Research Funds for the Central Universities, grant number 18CX02179A; and the Postdoctoral Application and Research Projects of Qingdao, grant number BY20170204.

Conflicts of Interest: The authors declare no conflict of interest.

References

1. Zhou, W.; Wang, W.; Shan, C.; Wang, G. Formation and evolution of concealed fault zone in sedimentary basin and its reservoir-controlling effect. *Int. J. Oil Gas Coal Technol.* **2016**, *12*, 335–358. [CrossRef]
2. Xi, K.; Cao, Y.; Haile, B.G.; Zhu, R.; Jahren, J.; Bjørlykke, K.; Zhang, X.; Hellevang, H. How does the pore-throat size control the reservoir quality and oiliness of tight sandstones? The case of the lower cretaceous quantou formation in the Southern Songliao Basin, China. *Mar. Petrol. Geol.* **2016**, *76*. [CrossRef]
3. Sun, T.; Lv, Y.; Liu, Z. Quantitative evaluation of control of faults on hydrocarbon accumulation and play fairway prediction: A case from Es3 (1) in Qijia-Yuanyanggou area, the Liaohe Depression. *Oil Gas Geol.* **2013**, *34*, 790–796. [CrossRef]
4. Zhao, X.; Jin, F.; Cui, Z.; Han, C.; Zeng, J.; Wang, Q.; Guo, K. Types of subtle buried-hill oil reservoirs and their accumulation simulation in Jizhong Depression, Bohai Bay Basin. *Petrol. Explor. Dev.* **2012**, *39*, 147–154. [CrossRef]
5. Lei, Q.; Yang, L.; Duan, Y.; Weng, D.; Wang, X.; Guan, B.; Wang, Z.; Guo, Y. The "fracture-controlled reserves" based stimulation technology for unconventional oil and gas reservoirs. *Petrol. Explor. Dev.* 2018. [CrossRef]
6. Zheng, M.; Jia, C.; Feng, Z.; Li, X.; Qu, H.; Liang, Y. Potential replacement regions of hydrocarbon reserves in exploration domain of foreland basins. *Acta Petrol. Sin.* **2010**, *5*. [CrossRef]
7. Rutqvist, J.; Rinaldi, A.P.; Cappa, F.; Moridis, G.J. Modeling of fault reactivation and induced seismicity during hydraulic fracturing of shale-gas reservoirs. *J. Petrol. Sci. Eng.* **2013**, *107*, 31–44. [CrossRef]
8. Jia, B.; Tsau, J.-S.; Barati, R. A review of the current progress of CO_2 injection eor and carbon storage in shale oil reservoirs. *Fuel* **2019**, *236*, 404–427. [CrossRef]
9. Wang, Z.; Zhao, W.; Men, X.; Zheng, H.; Li, X. Control of basement fault minor-activity on gas pool formation of upper paleozoic, Ordos Basin. *Petrol. Explor. Dev.* **2005**, *32*, 9–13. [CrossRef]
10. Morley, C.; Haranya, C.; Phoosongsee, W.; Pongwapee, S.; Kornsawan, A.; Wonganan, N. Activation of rift oblique and rift parallel pre-existing fabrics during extension and their effect on deformation style: Examples from the rifts of Thailand. *J. Struct. Geol.* **2004**, *26*, 1803–1829. [CrossRef]
11. Di, L. Controlling of petrophysical fractures on extra-low permeability oil and gas reservoirs in Ordos Basin. *Petrol. Explor. Dev.* **2006**, *33*, 667–670. [CrossRef]
12. Wang, W.; Zhou, W.; Zhou, J.; Li, S. Formation mechanism and distribution of buried fault zones in the Jinhu Sag. *J. Jilin Univ. Earth Sci. Ed.* **2014**, *44*, 1395–1405. [CrossRef]
13. Morley, C. How successful are analogue models in addressing the influence of pre-existing fabrics on rift structure? *J. Struct. Geol.* **1999**, *21*, 1267–1274. [CrossRef]
14. Dooley, T.P.; Schreurs, G. Analogue modelling of intraplate strike-slip tectonics: A review and new experimental results. *Tectonophysics* **2012**, *574*. [CrossRef]
15. Mollema, P.; Antonellini, M. Compaction bands: A structural analog for anti-mode I cracks in aeolian sandstone. *Tectonophysics* **1996**, *267*, 209–228. [CrossRef]
16. Bellahsen, N.; Daniel, J.M. Fault reactivation control on normal fault growth: An experimental study. *J. Struct. Geol.* **2005**, *27*, 769–780. [CrossRef]
17. Hardy, S. Cover deformation above steep, basement normal faults: Insights from 2D discrete element modeling. *Mar. Petrol. Geol.* **2011**, *28*, 966–972. [CrossRef]
18. Hardy, S. Propagation of blind normal faults to the surface in basaltic sequences: Insights from 2D discrete element modelling. *Mar. Petrol. Geol.* **2013**, *48*, 149–159. [CrossRef]
19. Liu, C.; Gu, L.; Wang, J.; Si, S. Reservoir characteristics and forming controls of intrusive-metamorphic reservoir complex: A case study on the diabase-metamudstone rocks in the Gaoyou Sag, Eastern China. *J. Petrol. Sci. Eng.* **2019**, *173*, 705–714. [CrossRef]

20. Pu, X.; Zhou, L.; Han, W.; Chen, C.; Yuan, X.; Lin, C.; Liu, S.; Han, G.; Zhang, W.; Jiang, W. Gravity flow sedimentation and tight oil exploration in lower first member of Shahejie Formation in slope area of Qikou Sag, Bohai Bay Basin. *Petrol. Explor. Dev.* **2014**, *41*, 153–164. [CrossRef]

21. Chen, S.; Wang, H.; Wu, Y.; Huang, C.; Wang, J.; Xiang, X.; Ren, P. Stratigraphic architecture and vertical evolution of various types of structural slope breaks in Paleogene Qikou Sag, Bohai Bay Basin, Northeastern China. *J. Petrol. Sci. Eng.* **2014**, *122*, 567–584. [CrossRef]

22. Zhao, X.; Pu, X.; Zhou, L.; Shi, Z.; Han, W.; Zhang, W. Geologic characteristics of deep water deposits and exploration discoveries in slope zones of fault lake basin: A case study of Paleogene Shahejie Formation in Banqiao-Qibei slope, Qikou Sag, Bohai Bay Basin. *Petrol. Explor. Dev.* **2017**, *44*, 171–182. [CrossRef]

23. Zhou, L.; Fu, L.; Lou, D.; Lu, Y.; Feng, J.; Zhou, S.; Santosh, M.; Li, S. Structural anatomy and dynamics of evolution of the Qikou Sag, Bohai Bay Basin: Implications for the destruction of North China Craton. *J. Asian Earth. Sci.* **2012**, *47*, 94–106. [CrossRef]

24. Pu, X.; Zhou, L.; Xiao, D.; Hua, S.; Chen, C.; Yuan, X.; Han, G.; Zhang, W. Lacustrine carbonates in the southwest margin of the Qikou Sag, Huanghua Depression, Bohai Bay Basin. *Petrol. Explor. Dev.* **2011**, *38*, 136–144. [CrossRef]

25. Pu, X.; Zhou, L.; Wang, W.; Han, W.; Xiao, D.; Liu, H.; Chen, C.; Zhang, W.; Yuan, X.; Lu, Y.; et al. Medium-deep clastic reservoirs in the slope area of Qikou Sag, Huanghua Depression, Bohai Bay Basin. *Petrol. Explor. Dev.* **2013**, *40*, 38–51. [CrossRef]

26. Yu, Z.; Liu, K.; Liu, L.; Qu, X.; Yu, M.; Zhao, S.; Ming, X. Characterization of paleogene hydrothermal events and their effects on reservoir properties in the Qikou Sag, Eastern China. *J. Petrol. Sci. Eng.* **2016**, *146*, 1226–1241. [CrossRef]

27. Huang, C.; Wang, H.; Wu, Y.; Wang, J.; Chen, S.; Ren, P.; Liao, Y.; Zhao, S.E.; Xia, C. Genetic types and sequence stratigraphy models of palaeogene slope break belts in Qikou Sag, Huanghua Depression, Bohai Bay Basin, Eastern China. *Sediment. Geol.* **2012**, *261*, 65–75. [CrossRef]

28. Qi, P.; Ren, J.; Shi, S. Features of the cenozoic structure of the coastal zone in Qikou Sag and its formation mechanism. *Acta Petrol. Sin.* **2010**, *31*, 900–905. [CrossRef]

29. Zhou, L.; Lu, Y.; Xiao, D.; Zhang, Z.; Chen, X.; Wang, H.; Hu, S. Basinal texture structure of Qikou Sag in Bohai Bay Basin and its evolution. *J. Nat. Gas Geosci.* **2011**, *22*, 373–382. [CrossRef]

30. Wu, Y.; Fu, J.; Zhou, J.; Xu, Y. Evaluation of hydrocarbon system in Qikou Sag. *Acta Petrol. Sin.* **2000**, *6*. [CrossRef]

31. Fossen, H. *Structural Geology*, 2nd ed.; Cambridge University Press: Cambridge, UK, 2016; pp. 20–45, ISBN 978-1-107-05764-7.

32. Zhang, H. New concept of petroleum system and its history genetic classification. *J. Chengdu Univ. Technol.* **1999**, *1*, 14–16. [CrossRef]

33. Jackson, C.A.-L.; Rotevatn, A. 3D seismic analysis of the structure and evolution of a salt-influenced normal fault zone: A test of competing fault growth models. *J. Struct. Geol.* **2013**, *54*, 215–234. [CrossRef]

34. Du, W.; Wu, Y.; Guan, Y.; Hao, M. Edge detection in potential filed using the correlation coefficients between the average and standard deviation of vertical derivatives. *J. Appl. Geophys.* **2017**, *143*, 231–238. [CrossRef]

35. Duffy, O.B.; Bell, R.E.; Jackson, C.A.-L.; Gawthorpe, R.L.; Whipp, P.S. Fault growth and interactions in a multiphase rift fault network: Horda platform, Norwegian North Sea. *J. Struct. Geol.* **2015**, *80*, 99–119. [CrossRef]

36. Mortimer, E.; Paton, D.; Scholz, C.; Strecker, M. Implications of structural inheritance in oblique rift zones for basin compartmentalization: Nkhata Basin, Malawi rift (EARS). *Mar. Petrol. Geol.* **2016**, *72*, 110–121. [CrossRef]

37. Li, C.; Wang, S.; Wang, L. Tectonostratigraphic history of the Southern Tian Shan, Western China, from seismic reflection profiling. *J. Asian Earth. Sci.* **2018**. [CrossRef]

38. Renard, Y. A uniqueness criterion for the signorini problem with coulomb friction. *SIAM J. Math. Anal.* **2006**, *38*, 452–467. [CrossRef]

39. Dieter, G.E.; Bacon, D.J. *Mechanical Metallurgy*, 3rd ed.; McGraw-Hill Education: New York, NY, USA, 1986; pp. 213–256, ISBN 978-0070168930.

40. Labuz, J.F.; Zang, A. Mohr-coulomb failure criterion. *Rock Mech. Rock Eng.* **2012**, *45*, 975–979. [CrossRef]

41. Parnell, J.; Watt, G.R.; Middleton, D.; Kelly, J.; Baron, M. Deformation band control on hydrocarbon migration. *J. Sediment. Res.* **2004**, *74*, 552–560. [CrossRef]

42. Bonini, M. Passive roof thrusting and forelandward fold propagation in scaled brittle-ductile physical models of thrust wedges. *J. Geophys. Res.-Solid Earth* **2001**, *106*, 2291–2311. [CrossRef]
43. Zhou, W.; Wang, W.; An, B.; Hu, Y.; Dong, M. Genetic types of potential fault zone and its significance on hydrocarbon accumulation. *J. Nat. Gas Geosci.* **2014**, *25*, 1727–1734. [CrossRef]

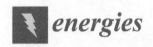

Article

Igniting Soaring Droplets of Promising Fuel Slurries

Alexander Bogomolov [1,2,*], Timur Valiullin [3], Ksenia Vershinina [3], Sergey Shevyrev [1] and Nikita Shlegel [3]

[1]   T.F. Gorbachev Kuzbass State Technical University, Kemerovo 650000, Russia; ssa.pmahp@kuzstu.ru
[2]   Institute of Thermophysics Siberian Branch, Russian Academy of Sciences, Novosibirsk 630090, Russia
[3]   National Research Tomsk Polytechnic University, Tomsk 634050, Russia; radisovich@yandex.ru (T.V.);
     vershininaks@gmail.com (K.V.); nik.shlegel.ask@gmail.com (N.S.)
*   Correspondence: barom@kuzstu.ru; Tel.: +7-(3842)-396-354 (ext. 3044)

Received: 24 November 2018; Accepted: 31 December 2018; Published: 10 January 2019

Abstract: High rates of environmental pollution by boilers and thermal power plants burning coal of different grades are the main reason for active research in the world aimed at the development of alternative fuels. The solution to the formulated problem acceptable in terms of environmental, technical and economic criteria is the creation of composite slurry fuels with the use of fine coal or coal processing and enrichment waste, water of different quality, and oil sludge additive. This study considers modern technologies of burning slurry fuels as well as perspective research methods of the corresponding processes. A model combustion chamber is developed for the adequate study of ignition processes. The calculation of the basic geometric dimensions is presented. The necessity of manufacturing the combustion chamber in the form of an object of complex geometry is substantiated. With its use, several typical modes of slurry fuel ignition are determined. Principal differences of ignition conditions of a single droplet and group of fuel droplets are shown. Typical vortex structures at the fuel spray injection are shown. A comparison with the trajectories of fuel aerosol droplets in real combustion chambers used for the combustion of slurry fuels is undertaken.

Keywords: slurry fuel; ignition; combustion; combustion chamber; soaring of fuel droplets; trajectories of fuel droplets

---

1. Introduction

1.1. Reasons for the Increased Interest in Water-Containing Slurry Fuels

Water-containing slurry fuels based on fine coal or coal processing waste are called coal-water slurry (CWS), or composite liquid fuel (CLF). Their prospects were justified more than 40 years ago. In recent years, in the world scientific community, there have been more and more arguments in favor of the active use of water-containing slurry fuel, since it could serve to solve a number of global problems.

In particular, the main attractiveness of CLF in comparison with traditional energy resources (gas, coal, and oil) derives from the following essential advantages (based on the analysis of the research results [1–11]):

(i)   cost-effectiveness of CLF; compared to fuel oil and coal, the cost of 1 ton of CLF is 1.5–2.5 times lower; operating costs for storage, transportation and combustion are 20–30% lower;

(ii)   due to almost complete burnout of coal particles in CLF, gaseous anthropogenic emissions into the atmosphere are minimal and comparable to emissions from gas combustion;

(iii) in terms of technological effectiveness, CLF is fire- and explosion-proof; it can be stored at a wide temperature range; slagging is reduced; there is no dependence on the properties and origin of the CLF components;

(iv) CLF can be used as primary and additional fuel in all operating conditions regardless of the region; virtually any coal can be used to prepare CLF;

(v) among the main combustible components of CLF, in addition to fine coal, there may be numerous wastes of coal enrichment and oil refinement; in this case, it is possible to dispose of large amounts of these wastes, to release the occupied territory, and to reduce the consumption of fossil fuels for producing thermal and electric energy.

At the same time, it is worth noting the main limitations that constrain the active use of CLF technologies [1–11]:

(i) moisture, which can make up to 40% of CLF, is a ballast, and part of the energy from fuel combustion is spent on the phase transition of water from liquid to gaseous state;

(ii) typical CLF obtained on most units (at various technologies) retains stability (does not stratify) only for 30–70 h. Modern equipment allows increasing this parameter up to 10–15 days, and even up to a year without the use of additives-stabilizers. Limited stability forces the use of additives-plasticizer or special processing methods when it is necessary to increase the CLF shelf life, which obviously increases the fuel cost. The way out of this situation is to prepare CLF immediately before combustion. This approach provides a CLF daily supply reserve in the immediate vicinity of the consumer. The main fuel supply in this case is provided by the initial coal reserves;

(iii) high abrasive wear of the injectors took place at the first stages of CLF application. For example, in Russia, China, India, Japan and Poland the first nozzles served no more than 30–40 h. In modern conditions, these issues have been resolved up to the manufacturing of serial burners for CLF.

Each of these restrictions is leveled in the process of CLF preparation by adding various additives, modifying combustion chambers compared to coal and fuel oil, and adapting technologies. It is believed traditionally that, unfortunately, the technologies of CLF burning have low efficiency, despite a large number of studies (in particular, [12,13]) on the CLF ignition mechanism, possible modes and effects.

1.2. Modern Technology of Composite Liquid Fuel (CLF) Combustion

The main ways of CLF combustion are [1–3] burning in flame (chamber) and in a fluidized bed. Flame burning is the main method, especially in boilers of medium and high power. The geometry of the boilers, as a rule, allows the torch to be organized inside the combustion chamber so that the coal particles that make up the CLF could completely burn out. From the practice of CLF burning it may be inferred that fuel ignition begins in a small vicinity of the nozzle section, which feeds it into the combustion chamber. The disadvantages of the method include rather high requirements to the burner device of the boiler (nozzle).

At CLF burning in the fluidized bed, the fuel jet is fed to the heated layer of inert material. Droplets of CLF are almost instantly ignited when they fall in the fluidized bed. The undoubted advantages of this method of combustion are relative ease of implementation in boilers of small capacity, a fairly large range of boiler capacity control (without loss of efficiency), and low requirements to the quality of the fuel supplied. The disadvantages include the capital intensity of fluidized bed organization in boilers of medium and high power, especially in the case of reconstruction of the latter.

When burning CLF in a fluidized bed, there are technological solutions for implementing the mode of slurry fuel gasification (pyrolysis), for which CLF is an ideal raw material. In this case, the combustion is carried out in two stages: gasification and direct combustion of the obtained gases. Depending on the technological features it is possible to combine the combustion methods. Synthesis gas obtained at the gasification stage increases the CLF combustion stability. The disadvantage of the method today is the lack of commercially available boilers. This is one of the reasons for the development of a large group of different methods for studying the processes of CLF ignition, as well as different reproducible conditions for the study of the relevant processes.

1.3. Limitations of Modern Methods of Investigation of CLF Ignition and Combustion

It is possible to distinguish the following main methods of experimental studies of ignition and combustion of droplets of slurry fuels [12]: on a heated substrate or any surface (to reproduce the layer combustion, for example, on stoker grates and walls of combustion chambers); in the flow of heated air using different holders (rods, wires, thermocouple junctions, etc.); and when placing CLF droplets into the heated airflow. Each of these research methods is somewhat far from the technologies described in Section 1.2. Therefore, the task of developing the experimental technique taking into account the real features of advanced combustion technologies for promising CLF is relevant.

The aim of this work is to design a model combustion chamber for recording the limiting conditions, the main characteristics of the ignition and combustion processes, as well as the trajectories of the slurry fuel droplets in conditions close to real combustion technologies.

2. Experimental Setup and Methods

2.1. Typical Industrial Layouts of Units for CLF Combustion

To date, a fairly large number of experimental setups and experimental-industrial units used for burning fuel slurries have been developed; for example, in China, there are more than 90 steam and power plants using CLF. Among the most common Russian plants burning slurry fuels are prismatic (straight-through) elongated furnace chambers of hot water and steam boilers without significant modifications, used for direct combustion of pulverized coal or fuel oil. As a rule, they are based on the principle of flame burning by fine-dispersed fuel spray through burners or nozzles [1–10]. For example, the boiler (TPE-214) of Novosibirsk thermal power plant (Russia) was operated at burning of large volumes of CWS (over 350×10^3 m^3), implemented in conjunction with the technical project of the coal pipeline (262 km) "Belovo-Novosibirsk" (Russia, from 1989 to 1997) [8]. This project was carried out on the basis of industrial research into the preparation and combustion of CWS at Belovskaya thermal power plant (Russia, from 1986 to 1987) on the basis of boilers PC-40-1 with steam capacity of 640 t/h, as well as TP-35 (Figure 1) of Min-Kush thermal power plant based on Kavak brown coal (Kyrgyzstan) [1,8].

The team of the Institute of Thermophysics (Siberian Branch of the Russian Academy of Sciences) conducted thermal calculations and changed the design of the low power boiler (KE-10-14C) for CWS combustion. For this boiler, the vortex combustion mode was used (Figure 2).

It is known that the combustion of fuel slurries based on solid (coal, coal processing waste) and liquid (water, waste oils, etc.) components is accompanied by an increase in the size of the ignition zone and a decrease in the temperature level due to the presence of liquid inert ballast. In this regard, there was a gradual transition to the vortex method of CLF burning (due to the angular circular swirl) as the most efficient combustion with the longest period of cyclicity of the soaring of fuel droplets in the reaction zone and high combustion efficiency.

For furnaces of direct-flow boiler (P-56GM) and drum boiler (BKZ-75-39FB), the tangential arrangement of burners and various forms of air blowing were applied, providing a vortex flow of combustion products and stable combustion of CWS [3].

In turn, the gas and fuel oil boiler (DKVr 6.5/13GFO) was additionally equipped with muffle (cyclone) vortex furnace extensions. They were used for pre-ignition and flame combustion of fine-sprayed fuel slurry with its subsequent burnout in the main furnace of the boiler [4]. For the reconstructed boiler DKVr-20-13 (at a transition from the layer burning of coal to the slurry one) the vortex combustion of CLF was numerically studied using the ANSYS Fluent 12 software [5].

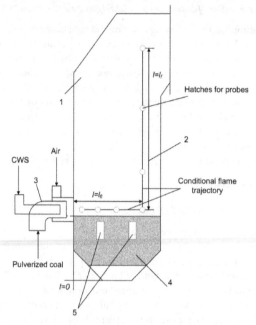

Figure 1. Variation of dimensionless parameters when firing Kavak brown coal in the TP-35 boiler in the forms of pulverized coal and coal-water slurry (CWS). (1) Furnace, (2) zone of active combustion for pulverized coal and CWS, (3) burners for firing pulverized coal and CWS, (4) and (5) design active combustion zone and burners for finely dispersed CWS (0–350 μm) [1].

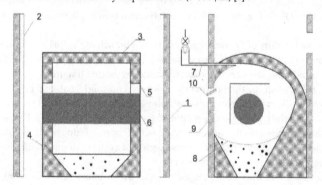

Figure 2. The scheme of the reconstructed furnace KE-10-14C: 1—cooling chamber; 2—furnace tubes; 3—muffle (combustion chamber); 4—end walls of the muffle; 5—gas-transfer windows; 6—the central base; 7—nozzle of secondary blast; 8—ash collector; 9—the front wall of the boiler; 10—CWS atomizer [2].

The study [6] considered in detail the issues of modernization of steam and hot water boilers with installing cyclone furnace extensions operating on traditional fuels: coal, gas and fuel oil (Figure 3).

The study [7,8] presents the results of designing an experimental setup (Figure 4) for CLF burning. The stable CLF ignition for this furnace was carried out in the temperature range of 600–700 °C, and the temperature of the gases at the outlet of the cyclone furnace extension was 1090–1160 °C.

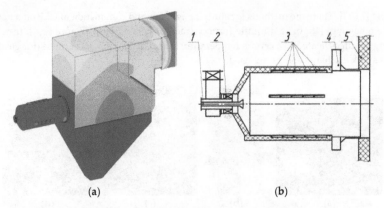

Figure 3. (**a**) Reconstructed boiler DKVR-20-13S with muffle furnace extension for composite liquid fuel (CLF) combustion [5]; (**b**) scheme of furnace extension: 1—CLF nozzle; 2—axial air supply channel; 3—channels of tangential air supply; 4—channel of tertiary air; 5—front wall of the furnace.

Figure 4. (**a**) Cyclone furnace extension for the boiler with a capacity of 6.5 MW; (**b**) overall view and section of cyclone furnace extension [7,8].

Studies [9,10] present the results of numerical calculations of coal-water fuel combustion in the adiabatic vortex combustion chamber obtained with ANSYS FLUENT software. These mathematical calculations to determine the optimal modes of combustion of CWS can improve the design accuracy of various boilers.

2.2. Typical Diagrams of Experimental Setups for Studying CLF Combustion

Efficient CLF combustion at thermal power plants or small boilers, as well as achieving the maximum efficiency of the power plant, is associated with the optimal organization of fuel ignition in the combustion chamber, stabilization of the flame combustion and achievement of the set temperature level. As a rule, it depends on the correctly chosen physical parameters (T_g^{min} is minimal ignition temperature, τ_d is the ignition delay time, τ_c is the time of complete combustion, T_d^{max} is the maximum temperature at the drop center during heating) for the relevant design calculations of furnace chambers. This requires a series of experimental studies on the combustion initiation of various CLF and the necessary conditions for their development.

From the analysis of world practice on the study of combustion initiation of slurry fuel droplets, it should be noted that the most well-known methods are [11,12]. It is believed that the most widespread is the experimental approach, implemented by suspending a single fuel droplet on the thermocouple junction, its further placing in a heated medium (heated air, combustion products, their mixture) and recording the temperature change of the droplet (Figure 5a,b). In turn, holders made of other materials (thin metal wire, quartz thread, ceramic rod) are also often used to register fuel ignition characteristics' parameters [13,14]. There are methods for studying ignition and combustion of CLF on a hot surface (conductive heating) [15] or in a muffle furnace (radiant heating [16]) (Figure 5c,d). In rare cases, local energy sources heated to a certain temperature (metal disks, etc.), laser pulse (for gasification), as well as spark discharge energy are used.

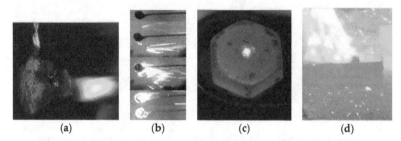

(a)　　　　　　(b)　　　　　　(c)　　　　　　(d)

Figure 5. Approaches to experimental investigation of slurry droplets combustion: (**a**,**b**) using different holders; (**c**) on a heated surface; (**d**) in a muffle furnace.

All the known aforementioned experimental approaches are far from the real combustion processes of thermal power plants because all of them use contact with the heated surface. Often holders lead to changes in the heat transfer conditions in the suspended fuel droplet [13,14]. One may observe processes of the heat sink from the droplet to the holder and an additional energy flow through the holder to it. These heat transfer processes are different from real conditions. As for experiments with fuel droplets freely falling through a cylindrical channel of heated air, there are restrictions associated with a small residence time of the fuel droplet in the combustion chamber (a cylindrical channel of limited length is used, see Figure 6a [17,18]). As noted above, the CLF burnout requires a long residence time of fuel droplets in the furnace space, i.e., directly in the active combustion zone.

(a)　　　　　　　　　(b)　　　　　　　　　(c)

Figure 6. (**a**) Ignition and combustion of a single free-falling CLF droplet; (**b**) burner device with flame burning of CLF flow in the boiler furnace; (**c**) CLF combustion fragment observed through the boiler observation window.

The study of CLF ignition and combustion in these thermal power plants (large and small boilers) is difficult and limited. This is due to rather high rates of physical and chemical processes, occurring in the boiler furnace, and the impossibility of visual recording of combustion and ignition of fuel droplets in the entire volume of the core zone (Figure 6b). Individual viewing windows allow monitoring of CLF combustion processes in a narrow region (Figure 6c).

In this regard, it is advisable to develop a model combustion chamber, which would allow, on the one hand, bringing the conditions of fuel ignition and combustion to the furnaces of real boilers, and, on the other hand, visualizing these processes for soaring CLF droplets in a swirling flow of heated air with their direct video recording in real time. It is expedient to develop and manufacture a model combustion chamber from optically transparent quartz glass based on the results of calculations of geometric dimensions.

2.3. Designing the Model Combustion Chamber

In the study of the characteristics of ignition and combustion of soaring CLF droplets, in contrast to the stationary suspended droplets [11–14], the necessary parameters for the soaring of a droplet of the fuel composition were estimated. The aim of the evaluation was to determine the geometric dimensions of the expanding part of the quartz tube (with conical inlet and outlet channels), where the droplet would be ignited by the heated air flow. A model cone-shaped combustion chamber has been developed as a promising design (in terms of manufacturing complexity, placement in the laboratory and compliance with the conditions of fuel combustion at TPP). This design allows keeping the drop in the given range of heights by changing air flow pressure (due to the pressure difference over the chamber height) in the vertical direction, and, thus, changing the residence time of the fuel droplet in the combustion core. The initial data for the calculation are presented in Table 1.

Table 1. Initial data for calculating the parameters of cone-shaped model combustion chamber.

No.	Parameter	Value	Comment
1	Initial diameter of the cone (d_{tr}), mm	80	The pipe diameter (d_{tr}) corresponds to the initial diameter of the cone
2	High pressure fan flow rate at 293 K (V_a), L/min	1200	Swirl fan Leister Robust (50 Hz)
3	Droplet diameter (d), mm	1.5	Without the sphericity coefficient
4	Initial air temperature (for igniting a fuel droplet), K	753	-
5	Nominal (maximum) air temperature, K	923	-
6	Humidity of an initial fuel droplet (W^a), %wt.	43.5	-
7	Ash content of an initial fuel droplet (A^d), %wt.	25	-

The geometric dimensions of the chamber are calculated by the method of determining the soaring of a single droplet, which assumes the equality of forces of aerodynamic drag of the droplets and the gravitational forces in the ascending air flow.

When calculating the conditions of CLF droplet soaring, the following were assumed:

1. The coefficient of droplet sphericity (spherical shape factor) $\varphi = 0.73$ [19].
2. Droplet motion in the vertical direction in the range $h = 0–120$ mm (height of the calculated cone-shaped chamber).
3. Properties of the component composition of CLF (density, ash, etc.) are subject to the additivity rule, and they can be determined using the relevant properties of the components.

The calculation method for droplet soaring is identical [20]. The air flow rate along the channel section, at which a single drop passes into a soaring state, is the rate of soaring [21]. It corresponds to the beginning of the destruction of the monodisperse soaring layer. At that,

$$\varepsilon = 1,$$

where ε is the porosity (relative fraction of volume not filled with solid phase).

$$\varepsilon = 1 - \rho_n/\rho_d, \tag{1}$$

ρ_n is the bulk density, and ρ_d is the body density (drop).

The soaring rate may be determined by Equations [20]:

$$Re_{vit} = \frac{Ar}{18 + 0.61\sqrt{Ar}}, \tag{2}$$

$$\omega_{vit} = \frac{Re_{vit}\mu_a}{d_d\rho_a}, \tag{3}$$

where Re_{vit} is the Reynolds criterion, ω_{vit} is the rate of soaring, m/s; d_d is the drop diameter, m; ρ_a, μ_a are the density (kg/m^3) and dynamic viscosity coefficient (Pa·s) of air; and Ar is the Archimedes criterion.

Density and dynamic coefficient of air viscosity (Table 2) are taken at a temperature of 823 K [22].

Table 2. Air parameters.

Density, kg/m^3	0.43
Dynamic coefficient of viscosity, Pa·s	376×10^{-7}

Archimedes' criterion is calculated by the expression [20]:

$$Ar = \frac{d_d^3 g}{v_a^2} \frac{\rho_d - \rho_a}{\rho_a} = \frac{d_d^3 g \rho_a (\rho_d - \rho_a)}{\mu_a^2}, \tag{4}$$

where v_a is the kinematic coefficient of the medium viscosity, m^2/s; and μ_a is the dynamic coefficient of the medium viscosity, Pa·s.

Droplet size (equivalent diameter):

$$d_d = d\varphi, \tag{5}$$

Considering the deviation from the spherical shape the drop size will be [19]:

$$d_d = d\varphi = 1.5 \times 0.73 = 1.095 \text{ mm.} \tag{6}$$

To calculate the Archimedes' criterion, it is necessary to determine the density of the CLF droplet. According to the reference data [23,24], the density of coal dust in the composition of CLF is 1700 kg/m^3 (water content of the initial drop is ≈43.5%).

The density of CLF droplet in the initial state:

$$\rho_d = 0.435 \times 998 + 0.565 \times 1700 = 1394 \text{ kg/m}^3. \tag{7}$$

In the future, with the known elemental composition of the used solid fuel, the droplet density is specified according to [25]:

$$\rho_d = \frac{100 \rho_{org}}{100 - A^c \left(1 - \frac{\rho_{org}}{2900}\right)}, \tag{8}$$

$$\rho_{org} = \frac{10^5}{0.344 C^p + 4.25 H^p + 23}, \tag{9}$$

where ρ_{org} is the density of the organic mass of the fuel; C^p, H^p is the percentage of carbon and hydrogen in the fuel; and A^c is the ash content per dry mass of fuel.

The criterion of Archimedes for air temperature of 823 K:

$$Ar = \frac{d_d^3 g \rho_a \cdot (\rho_d - \rho_a)}{\mu_a^2} = \frac{(1.095 \times 10^{-3})^3 \times 9.8 \times 0.43 \times (1394 - 0.43)}{(376 \times 10^{-7})^2} = 5453.68. \tag{10}$$

Reynolds criterion:

$$Re_{vit} = \frac{Ar}{18 + 0.61\sqrt{Ar}} = \frac{5453.68}{18 + 0.6 \times \sqrt{5453.68}} = 87.52. \tag{11}$$

Soaring rate:

$$\omega_{vit} = \frac{Re_{vit} \mu_a}{d_d \rho_a} = \frac{87.52 \times 376 \times 10^{-7}}{1.095 \times 10^{-3} \times 0.43} = 6.98 \text{ m/s.} \tag{12}$$

High pressure vortex fan Leister Robust provides the maximum air flow rate V_v = 1200 L/min at 293 K, which corresponds to the air velocity of 4–5 m/s in the channel with a diameter of 80 mm.

Mass air flow rate:

$$G_a = \rho_a^{20} V_a, \tag{13}$$

Air velocity in the channel:

$$w = \frac{V_a}{F} = \frac{V_a}{0.785 \cdot d_{tr}^2}, \tag{14}$$

When air temperature in the channel is 823 K, considering its density of 0.43 g/L, the mass air flow rate:

$$G_a = \rho_a^{20} V_a = 1.2 \times 1200 = 1440 \text{ g/min.} \tag{15}$$

$$V_a = \frac{G_a}{\rho_a^{550}} = \frac{1440}{0.43} = 3348 \text{ L/min or } V_a = \frac{3348 \times 10^{-3}}{60} = 0.0558 \text{ m}^3/\text{s.} \tag{16}$$

The air velocity in the channel with a diameter of 80 mm is:

$$w = \frac{V_a}{F} = \frac{V_a}{0.785 \cdot d_{tr}^2} = \frac{0.0558}{0.785 \times (80 \times 10^{-3})^2} = 11.1 \text{ m/s.} \tag{17}$$

Thus, the required slurry velocity is provided, and the CLF drop can move vertically in the chamber along the expanding part of the cone (Figure 7). In the calculation it was assumed that the smaller diameter of the cone corresponds to the diameter of the quartz tube. The maximum height of the cone will take 120 mm (due to the limitations of the size of the experimental stand). It was believed that the droplet soaring rate of 6.98 m/s corresponds to the region with a smaller cone diameter (80 mm).

Let us consider the final state of the droplet (to determine the angle of the cone opening)—complete burnout of organic mass with forming the ash envelope. For this intermediate state of the droplet, its diameter (d_d) corresponds to 1.095 mm (model of the retained ash envelope [19]). Ash content (A^d) of the initial drop of CLF is 25%.

Ash envelope density:

$$\rho_{zk} = \rho_z A^d, \tag{18}$$

The ash envelope density in a droplet with a diameter of 1.095 mm will be:

$$\rho_{zk} = \rho_z A^d = 2400 \times 0.25 = 600 \text{ kg/m}^3, \tag{19}$$

where ρ_z is the true ash density (in the range of 2100–2400 for Kuznetsk coals) [26].

The Archimedes' criterion:

$$Ar = \frac{d_d^3 g \rho_a (\rho_{zk} - \rho_a)}{\mu_a^2} = \frac{(1.095 \times 10^{-3})^3 \times 9.8 \times 0.43 \times (600 - 0.43)}{(376 \times 10^{-7})^2} = 2346.4 \tag{20}$$

Reynolds criterion:

$$Re_{vit} = \frac{Ar}{18 + 0.6\sqrt{Ar}} = \frac{2346.4}{18 + 0.6 \times \sqrt{2346.4}} = 49.85. \tag{21}$$

Soaring rate:

$$w_{vit} = \frac{Re_{vit} \mu_a}{d_d \rho_a} = \frac{49.85 \times 376 \times 10^{-7}}{1.095 \times 10^{-3} \times 0.43} = 3.98 \text{ m/s.} \tag{22}$$

Cone diameter:

$$d_c = \sqrt{\frac{V_a}{0.785 w_{vit}}}. \tag{23}$$

At this rate and air flow rate of 0.0588 m^3/s the diameter of the cone is:

$$d_c = \sqrt{\frac{V_a}{0.785 w_{vit}}} = \sqrt{\frac{0.0588}{0.785 \times 3.98}} = 0.137 \text{ m.} \tag{24}$$

Consequently, the larger diameter of the cone is 137 mm. The height of the cone was taken earlier as 120 mm. Thus, the opening angle of the cone is about 24 degrees (Figure 7).

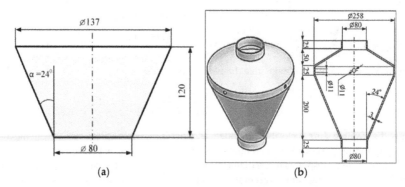

(a)   (b)

Figure 7. (a) Coned channel for creating soaring conditions of CLF droplets; (b) model of a conical combustion chamber, casted by the manufacturer.

Due to technological limitations for casting the cone-shaped chamber from transparent optical quartz glass, the dimensions of the real combustion chamber have been changed (Figure 7b). As a result, the large diameter of the cone has almost doubled (258 mm). The dimensions of the input and output channels and the cone opening angle remain unchanged. To control the temperature in the combustion chamber by chromel-aluminum thermocouple, as well as the input and discharge of CLF drops, there are two technological holes with a diameter of 11 mm in its side part. The chamber is made with a total height of 325 mm. The volume of the combustion chamber is 6 L. This allowed expanding the limits of permissible rates of soaring of CLF droplets (for a combination of a single, small group and a flow of droplets).

3. Materials

The fuels were prepared on the basis of several different components, both solid and liquid. In this study, low-grade solid fuel (lignite or lignite) and wet coal flotation waste (filter cake) were considered as the main components. The main properties of all components used are given in Tables 3 and 4. Proximate and ultimate analysis was carried out in accordance with standard procedures [27]. Below, the compositions of specific fuels are given either directly in the figures or in the explanatory tables.

Table 3. Properties of coal components [27].

Sample	W^a, %	A^d, %	V^{daf}, %	Q^a_s, MJ/kg
Brown coal	14.11	4.12	47.63	22.91
Filter-cake "T"	–	21.20	16.09	26.92
Filter-cake "K"	–	26.46	23.08	24.83

Table 4. Properties of liquid combustible components [27].

Sample	Density at 293 K, kg/m^3	W^a, %	A^d, %	T_f, K	T_{ign}, K	Q^a_s, MJ/kg
Used turbine oil	868	–	0.03	448	466	44.99
Fuel oil	1000	6.12	4.06	438	513	39.4
Used compressor oil	887	–	0.023	458	502	45.2

The choice of components is due to the fact that in this study it was necessary to test several CLF with different flammability. Figure 8 illustrates the characteristic differences in the ignition delay time for several fuel compositions. The data in Figure 8 obtained by burning fuel droplets using the designed combustion chamber (Figure 7).

The initiation of burning of CLF compositions based on brown coal, water and oil components occurs at lesser ambient temperatures than that of CLF based on wastes of enrichment of filter-cake "T" and "K" (from the coal-washing plant of the Kemerovo region, Russia) and used turbine oil. The properties are presented in Tables 1 and 2. The filter-cakes "T" and "K" had a moisture content of 43.5% and 39.1%, respectively.

An important role is played by the content of the oil component in the fuel and the minimum temperature of their ignition. The ignition temperature of fuel oil is about 513 K, which is higher relative to other liquid combustible impurities of CLF, for example, used turbine (466 K) and used compressor (502 K) oils.

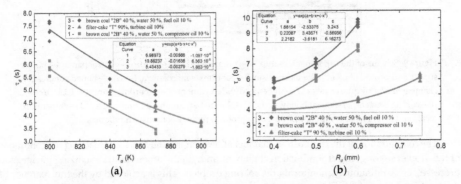

Figure 8. Ignition delay times of soaring CLF droplets depending on T_g (**a**) and R_d (**b**) at $V_g \approx 4$ m/s.

Figure 8a shows that for the composition of CLF based on cake "T" and used turbine oil, the ignition delay times are less than 1 s than for the fuel mixture based on brown coal, water and fuel oil. In addition, the determining factor is the change in the initial droplet size, which affects the ignition delay times Figure 1b. If the CLF droplet size is $R_d \approx 0.4$ mm, for the two compositions of CLF with used oil there is an identical ignition delay time of about 4.2 s within the limits of random errors. Furthermore, for the composition of cake "T" of 90% and used turbine oil of 10% there is a smooth increase in ignition delay time than for compositions with brown coal. This is most likely due to the lower moisture content in the cake "T" (about 35%). Thus, preliminary tests (Figure 8) made it possible to determine the composition of waste-derived fuel slurries of interest for further studies using different heating approaches.

4. Results and Discussion

4.1. Advantages of the Model Combustion Chamber

To study the ignition characteristics of the soaring CLF droplets, an experimental setup has been developed; its scheme is shown in Figure 9. The flow of heated air pumped by the compressor was formed in the model combustion chamber. Then a CLF droplet from a supply and discharge device was placed in the combustion chamber. To spray the flow of the fuel slurry into the combustion chamber, a T-shaped mixer was used; the compressed air was supplied in one of its channels, and slurry was fed in the second channel. The investigated processes of ignition and combustion of soaring droplets of CLF were registered with the use of high-speed video camera.

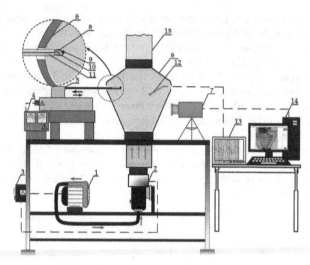

Figure 9. Scheme of the experimental setup: 1—compressor, 2—air heater, 3—control panel, 4—power supply and control of coordinate mechanism, 5—coordinate mechanism, 6—cone-shaped combustion chamber made of quartz glass, 7—high-speed video camera, 8—nichrome wire, 9—CLF droplet, 10—cutting element, 11—metal hollow rods, 12—chromel-aluminum thermocouple, 13—temperature recorder, 14—personal computer, 15—exhaust probe to remove combustion products.

The configuration of the combustion chamber allows, on the one hand, approaching the conditions of the droplet soaring in real combustion chambers, and, on the other hand, visualizing the ignition processes, i.e., continuously monitoring the soaring droplets. This is achieved by the transparency of the combustion chamber, made of heat-resistant quartz glass.

4.2. Modes of Ignition and Combustion of Slurry Fuels

In experiments with the flow of soaring CLF droplets, as well as with a single droplet, several ignition modes have been recorded. They can be divided into four modes (Figure 10a). These are characterized by characteristic trajectories of motion (mainly ellipsoidal or arbitrary), directed along the heated air motion in the combustion volume. The soaring CLF flow represents the droplets of different dispersiveness with irregular shape and different weight. Consequently, each CLF droplet has different evaporation time of moisture and flammable liquid, when interacting with hot air and in chemical reactions.

In the first mode, the fuel droplet was ignited in the upper part of the combustion chamber. If the droplet fell and was ignited directly on the metal grid at the bottom of the chamber, these conditions corresponded to the second mode. It should be noted that in the second mode, the droplet after ignition and some period of combustion on the grid changed to the soaring mode. The third and fourth modes were characterized by direct soaring of CLF droplets in the heated air flow. They differ by different areas of ignition. The third ignition mode is realized closer to the wall (along the cone-generators), and the fourth runs in the central zone of the combustion chamber. The main difference between the two modes of soaring is, most likely, due to the angle of entry into a certain area of the combustion chamber, which increases the likelihood of the modes. The most favorable is the ignition of CLF droplets in the central part (the fourth mode), which increases the completeness of fuel burn-out.

Figure 10b,c shows typical video shots of combustion of soaring CLF droplet flow based on brown coal, water and waste turbine oil, as well as CLF on the basis of filter cake and waste turbine oil. Among additional processes, it is worth noting the formation of the vortex flow of the fuel-air mixture for the period of pulse injection of CLF, which affects the rate of chemical reaction, Figure 10d.

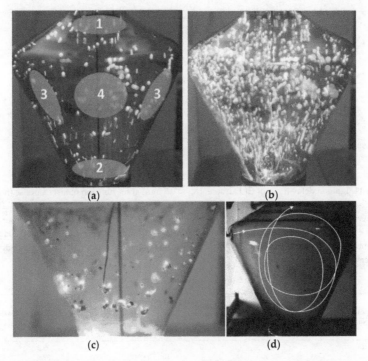

Figure 10. Video frames with four ignition modes: 1—in the upper part; 2—in the lower part; 3—near the wall; 4—in the central part (**a**); burning of soaring CLF droplets (on the basis of brown coal "2B", water and used turbine oil) (**b**); burning of soaring CLF droplets (on the basis of cake "G" and waste turbine oil) (**c**); vortex structures of the fuel-air flow at its direct injection (**d**).

In some cases, it was necessary to use the air swirler, installed in the lower part of the cone-shaped channel along the axis of symmetry. This served as an additional swirl of heated air to create the soaring conditions for the flow of fuel droplets and to reduce the consequences of their coagulation and adherence to the walls of the chamber.

4.3. Differences in the Characteristics of Ignition of Slurry Fuel Droplets

At the same component compositions of CLF, initial radius of the droplet and air temperature, the difference in the ignition delay times for a soaring droplet in contrast to the droplet suspended on a thermocouple is about 20–40%. Extrapolation of curves beyond the set values to higher temperatures allows predicting ignition delay times corresponding to real combustion processes. Figure 11 shows these dependences. The determining parameter is also the size of the soaring CLF droplet. Thus, for a soaring droplet and for a CLF droplet suspended on a thermocouple, the experimental values of τ_d can differ over 1.5 times.

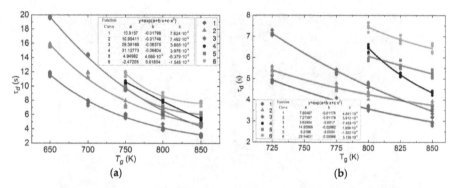

(a)   (b)

Figure 11. Ignition delay times of CLF droplet ($R_d \approx 0.5$ mm) at the junction of fast response (nominal static characteristic—platinum-rhodium-platinum, range of measured temperatures of 273–1873 K, systematic error of ± 1 K, inertia of no more than 1 s, diameter of the junction of about 0.1 s) thermocouple depending on T_g (a); ignition delay times of CLF droplets ($R_d \approx 0.5$ mm) at soaring in the combustion chamber depending on T_g (b). Numbering of fuel compositions 1–4 corresponds to the numbers of compositions from Table 5.

Table 5. Fuel compositions under study.

Composition No.	Solid Components (%)			Liquid Fuel Components (%)				Water (%)	Plasticizer (%)
	Coal "2B"	Filter-Cake "K"	Filter-Cake "G"	Waste Engine Oil	Waste Turbine Oil	Waste Compressor Oil	Fuel Oil		
1	50	-	-	10	-	-	-	39.5	0.5
2	50	-	-	-	10	-	-	39.5	0.5
3	40	-	-	-	-	10	-	50	-
4	-	-	50	-	-	-	10	39	1
5	-	50	-	10	-	-	-	39.5	0.5
6	-	50	-	-	10	-	-	39.5	0.5

Comparative analysis (Figure 12) of ignition delay times of CLF droplets on various holders shows that they are higher in relation to the soaring fuel droplet. This proves that even when using a holder with low temperature diffusivity (in this case $(2.3–2.7) \times 10^{-5}$ m^2/s), the thermocouple junction affects the heat and mass transfer conditions. Temperature diffusivity for nichrome wire is $(1.1–2.4) \times 10^{-5}$ m^2/s and for ceramic rod it is $(1–7) \times 10^{-7}$ m^2/s.

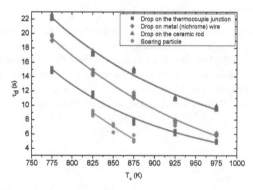

Figure 12. Ignition delay times of CLF droplets depending on air temperature ($R_d \approx 1$ mm, $V_g \approx 3$ m/s) when using different holders in comparison with the soaring CLF droplet.

When a stationary drop is streamlined by heated air, its intensive heating and subsequent ignition are realized from the side of the incoming flow at constant temperature T_g and velocity V_g. There is also an additional flow of energy through the holder. In case of a soaring CLF droplet, its ignition is realized on the entire surface (droplet rotates and warms more uniformly) with varying parameters of temperature and air velocity. Therefore, a soaring droplet requires slightly higher air temperatures for its stable ignition and subsequent combustion, as shown in Figure 13.

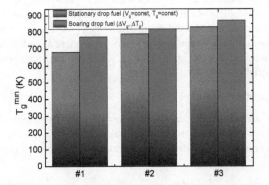

Figure 13. Limit (minimum) temperatures of stable ignition of a CLF droplet at $R_d \approx 0.5$. Compositions: No. 1—brown coal "2B" 50%, water 40%, waste turbine oil; No. 2—cake "K" 90%, waste turbine oil; No. 3—cake "K" 90%, waste turbine oil.

The most important parameter, in addition to air temperature and velocity, as well as droplet size, is the configuration of the droplet surface. Therefore, to determine differences between the integral characteristics, the experiments were held with different shapes of droplets (sphere, ellipsoid and polyhedron). It has been found that the polyhedron droplets are characterized by minimal ignition delay time, which is associated with an increased contact area due to numerous ledges [28].

4.4. Comparison of Ignition Characteristics of a Single Droplet and a Polydisperse Flow of Slurry Fuel Droplets

Spraying of fine slurry fuel into the boiler furnace is carried out by creating a non-isothermal jet with certain parameters (change in droplet mass and temperature in the jet). In the experiments, these factors are difficult to assess, because in real practice, fuel is supplied continuously. In this work, there was a short-term pulsed spray of CLF flow into the combustion chamber, associated with the transparency of the chamber walls and adequate registration of ignition of fuel droplets (Figure 14).

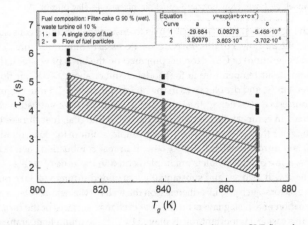

Figure 14. Ignition delay times of a single droplet and a polydisperse CLF flow depending on air temperature ($R_d \approx 0.5$ mm and $V_g \approx 4$ m/s).

Ignition delay times for the polydisperse fuel slurry flow have minimum values compared with single droplets. The main result of the work is that they differ by 20–40%. That is, all over the world, the ignition delay times predicted in experiments with one droplet are overestimated by 20–40% relative to the real polydisperse CLF flow. The shaded area is for the experimental values of ignition delay times of all droplets of the polydisperse fuel flow at identical initial parameters. In this graph, the ignition delay times for the CLF droplet flow are determined by varying the air temperature. Accordingly, the experimental points for the set of droplets are determined and random errors are indicated.

Moreover, in experiments with the flow of slurry fuels, there are additional effects of puffing, coagulation and breakup of droplets, as a result of which the fuel flow heating and subsequent combustion are intensified.

4.5. Recommendations for Applying the Research Results

The experiments have shown that the efficiency of CLF combustion in thermal power plants can be increased significantly. A schematic solution by the example of a low-temperature vortex furnace (with appropriate parameters of CLF combustion initiation) is presented in Figure 15.

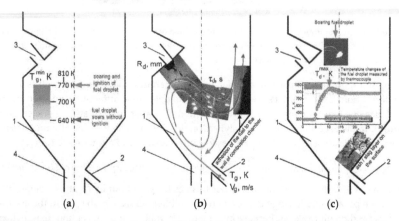

Figure 15. Combustion chamber diagram with indication of necessary parameters and conditions of ignition of a soaring CLF droplet and a number of adverse effects: typical ignition temperatures (a); typical droplet trajectories in the combustion chamber (b); typical fuel droplet combustion temperature (c). 1—cold funnel; 2—lower blow nozzle; 3—fuel-air mixture; 4—slag output.

On energy characteristics, the main problem is to ensure the conditions of soaring and ignition of a CLF droplet in the air flow. For this purpose, the necessary (minimum, threshold) conditions of soaring (followed by ignition) of CLF droplets prepared on the basis of waste-derived components have been determined: air temperature in the model chamber above 640 K, air flow velocity at the chamber inlet over 4 m/s, and droplet sizes in the range of $R_d \approx 0.4$–1.5 mm (Figure 15a). This will require fewer resources, i.e., less energy, fuel and time to warm up the combustion chamber. As far as ecology is concerned, in comparison with the combustion of fine coal, fuel slurries have much lower anthropogenic emissions (NO_x, SO_x, CO_2, CO). For example, sulfur oxide concentrations will be lower by about 30–40% and nitrogen oxides—by 50–60%. If in real combustion chambers of TPP boilers the conditions of soaring of CLF droplets are implemented in the zone of active combustion, so the depth of fuel burnout will increase and substantially less underburning will take place. In particular, the risks of adverse factors, such as CLF adherence on the walls of the heat-resistant casing of the boiler, as well as the formation of ash-slag layers on the heat exchange surfaces of the furnace, will decrease (Figure 15b,c). In this case, an important role is played by the maximum temperature and duration of fuel combustion to account for heat generation in the furnace during CLF combustion, which directly affects the steam capacity and efficiency of the TPP boiler.

The experimental data obtained on ignition characteristics of the soaring CLF droplets serve as the information base for experimental development works at designing various types of vortex furnace devices and combustion chambers (geometry, wall material, etc.). This also applies to all potentially available TPP boilers with small improvements in existing designs (for example, layer furnaces), allowing for burning of pulverized fuel.

The main practical recommendations are as follows:

(i)   a small addition (5–10% by relative mass concentration) of brown coal leads to a significant decrease in the ignition delay time and the minimum ignition temperatures of the soaring droplets of CLF; similar conclusions can be made from experiments with addition of enriched coal to the waste coal;

(ii) an increase in the mass fraction of water (from 40 to 50%) in the CLF composition leads to a significant increase in the ignition delay time (on average by 30–40%) of the soaring fuel droplets in the combustion chamber [29];

(iii) adding, for example, up to 5% of aluminum powder [30] to CLF composition has a positive effect on the combustion stabilization and the combustion temperature increase;

(iv) the use of CLF compositions with plant additives, for example, rapeseed oil, allows strengthening the main integral parameters of combustion and reducing the concentration of harmful emissions [31].

5. Conclusions

(i)   An experimental setup with a model combustion chamber has been developed to study the ignition and combustion of soaring CLF droplets in the heated air flow in the conditions close to combustion processes in thermal power plants.

(ii) Four ignition modes of CLF droplets characterized by involuntary motion trajectory and locality of ignition have been established; two of them correspond to the mode of soaring in the heated air flow.

(iii) It is shown that the ignition delay times are lower (up to 2–4 s) for the soaring CLF droplets than when initiating combustion on a fast thermocouple or even when using a material with low temperature diffusivity. In this case, the minimum (threshold) ignition temperature for the soaring CLF droplets is slightly higher than when they are heated on the holders.

(iv) It has been proved that ignition of polydisperse flow of soaring CLF droplets occurs faster than that of a single droplet with ignition delay times lower by 20–40%.

(v)   The expediency of using composite slurry fuels based on coal and oil wastes of various power plants and mechanisms as an alternative fuel for boiler combustion has been illustrated.

Author Contributions: A.B. and K.V. wrote the paper; T.V., S.S. and N.S. performed the experiments.

Funding: This research received no external funding.

Acknowledgments: This research was funded by the Russian Foundation for Basic Research and the Government of the Tomsk Region of the Russian Federation, grant number 18-43-700001.

Conflicts of Interest: The authors declare no conflict of interest.

Abbreviations

CLF	composite liquid fuels
CWS	coal-water slurries

Nomenclature

T_g	temperature in combustion chamber (K)
V_g	air flow rate (m/s)

R_d	initial droplet radius (mm)
W^a	humidity of original sample (%)
A^d	ash level of dry sample (%)
V_{daf}	yield of volatiles of filter cake converted to a dry ash-free state (s)
Q^a_s	heat of combustion (MJ/kg)
T_f	flash point (K)
T_{ign}	temperature of ignition (K)
$T_g{}^{min}$	minimum oxidizer temperature sufficient for sustainable ignition (K)
τ_d	ignition time delay (s)
d_{tr}	tube diameter corresponding to inlet diameter of the cone (mm)
V_a	flow rate of high-pressure fan (L/min)
d	droplet diameter (mm)
d_d	droplet diameter considering spherical shape factor (mm)
ε	porosity (relative share of volume not filled by solid phase)
ρ_n	bulk density (kg/m^3)
ρ_d	droplet density (kg/m^3)
Re_{vit}	Reynolds criterion
ω_{vit}	velocity of soaring (m/s)
ρ_a	air density (kg/m^3)
μ_a	dynamic viscosity of air (Pa·s)
Ar	Archimedes criterion
ν_a	kinematic viscosity coefficient of the medium (m^2/s)
φ	spherical shape factor of the droplet
ρ_{org}	density of the organic mass of fuel (kg/m^3)
C^P	carbon content in the fuel (%)
H^P	hydrogen content in the fuel (%)
A^c	ash content per dry mass of fuel (%)
G_a	mass air flow rate (g/min)
ω	air velocity in the channel (m/s)
F	cross-sectional area of the input channel (m^2)
ρ_{zk}	ash frame density (kg/m^3)
ρ_z	true ash density (kg/m^3)
d_c	large cone diameter (mm)

References

1. Osintsev, K.V. Studying flame combustion of coal-water slurries in the furnaces of power-generating boilers. *Therm. Eng.* **2012**, *59*, 439–445. [CrossRef]
2. Salomatov, V.V.; Dorokhova, U.V.; Syrodoy, S.V. Transfer of low power boilers to coal-water technology. *Polzunovskii Vestnik* **2013**, *4*, 38–46.
3. Puzyrev, E.M.; Golubev, V.A. Technology of coal-water fuel combustion in power boilers. *Vestnik Altaiskoy Nauki* **2014**, *4*, 325–331.
4. Puzyrev, E.M.; Murko, V.I.; Chernetskii, M.Y. Results of pilot tests of the fuel oil boiler DKVR 6.5/13 on coal-water fuel. *Therm. Eng.* **2001**, *2*, 69–71.
5. Tsepenok, A.I.; Ovchinnikov, Y.V.; Lutsenko, S.V.; Kvrivishvili, A.R.; Lavrinenko, A.A.; Mezhov, E.A. *Numerical Studies of the Combustion of Composite CWF in the Boiler DKVR-20-13*; Reports of VIII All-Russian Conference "Combustion of Solid Fuel"; Kutateladze Institute of Thermophysics SB RAS: Novosibirsk, Russia, 2012.
6. Shtym, A.N.; Shtym, K.A.; Vorotnikov, E.G.; Rasputin, O.V. Study and development of vortical technology of fuel burning. *Bull. Far East. Fed. Univ.* **2010**, *2*, 43–59.
7. Ovchinnikov, Y.V.; Boiko, E.E.; Serant, F.A. Problems of combustion of hydrocarbon fuels and proposals on the development of combustion technologies. *Proc. Acad. High. Sch. Russ. Fed.* **2015**, *1*, 26.
8. Ovchinnikov, Y.V.; Boiko, E.E. *The Technology of Obtaining and Studying Fine-Dispersed Coal-Water Slurry: Monograph*; NSTU Publisher: Novosibirsk, Russia, 2017; p. 308.

9. Murko, V.I.; Riestor, A.; Tsetsorina, S.A.; Fedyaev, V.I.; Karpenok, V.I. Results of numerical simulation of combustion of coal-water fuel. *Polzunovskii Vestnik* **2013**, *4*, 38–46.

10. Murko, V.I.; Senchurova, Y.A.; Fedyaev, V.I.; Karpenok, V.I. Study of combustion technology of coal slurry fuel in a vortex chamber. *Bull. Kuzbass State Tech. Univ.* **2013**, *2*, 103–105.

11. Kijo-Kleczkowska, A. Combustion of coal-water suspensions. *Fuel* **2011**, *90*, 865–877. [CrossRef]

12. Glushkov, D.O.; Strizhak, P.A.; Chernetskii, M.Y. Organic coal-water: Problems and advances (Review). *Ther. Eng.* **2016**, *63*, 707–717. [CrossRef]

13. Glushkov, D.O.; Strizhak, P.A.; Vershinina, K.Y. Minimum temperatures for sustainable ignition of coal water slurry containing petrochemicals. *Appl. Ther. Eng.* **2016**, *96*, 534–546. [CrossRef]

14. Vershinina, K.Y.; Iegorov, R.I.; Strizhak, P.A. The ignition parameters of the coal-water slurry droplets at the different methods of injection into the hot oxidant flow. *Appl. Ther. Eng.* **2016**, *107*, 10–20. [CrossRef]

15. Glushkov, D.O.; Strizhak, P.A.; Vershinina, K.Y. Hot surface ignition of a composite fuel droplet. *MATEC Web Conf.* **2015**, *23*, 1–4. [CrossRef]

16. Fedorova, N.I.; Patrakov, Y.F.; Surkov, V.G.; Golovko, A.K. Analysis of the nature of combustion of composite fuels, obtained by cavitation method. *Bull. Kuzbass State Tech. Univ.* **2007**, *4*, 38–41.

17. Atal, A.; Levendis, Y.A. Observations on the combustion behavior of coal water fuels and coal water fuels impregnated with calcium magnesium acetate. *Combust. Flame* **1993**, *93*, 61–89. [CrossRef]

18. Atal, A.; Levendis, Y.A. Combustion of CWF agglomerates from pulverized or micronized bituminous coal, carbon black, and diesel soot. *Combust. Flame* **1994**, *93*, 326–342. [CrossRef]

19. Baskakov, A.P.; Mukhlenov, I.P.; Sazhin, B.S.; Frolov, V.F. *Calculations of Fluidized-Bed Apparatuses: Reference Book*; Khimiya: Leningrad, Russia, 1986; p. 352.

20. Aerov, M.E.; Todes, O.M. *Hydraulic and Thermal Bases of Operation of Devices with Stationary and Boiling Granular Layer*; Khimiya: Leningrad, Russia, 1968; p. 247.

21. Pavlov, K.F.; Romankov, P.G.; Noskov, A.A. *Examples and Problems of the Course on Processes and Devices of Chemical Technology*; Khimiya: Leningrad, Russia, 1976; p. 552.

22. Vargaftik, N.B. *Handbook of Thermophysical Properties of Gases and Liquids*; Nauka: Moscow, Russia, 1972; p. 720.

23. Sukhorukov, V.I. *Scientific Bases of the Improvement of Equipment and Technologies for the Production of Coke*; ALLO: Yekaterinburg, Russia, 1999; p. 393.

24. Guva, A.Y. *Brief Thermophysical Handbook*; Sibvuzizdat: Novosibirsk, Russia, 2002; p. 300.

25. Antonov, P.P.; Sidorov, A.M.; Tyurkin, A.S.; Shcherbakov, F.V. Revised thermal calculation of furnaces of low-temperature fluidized bed. *Polzunovskii Vestnik* **2008**, *1–2*, 115–122.

26. Pavlenko, S.I. *Fine-Grained Concrete from Industrial Waste*; ASV: Moscow, Russia, 1997; p. 176.

27. Glushkov, D.O.; Lyrshchikov, S.Y.; Shevyrev, S.A.; Strizhak, P.A. Burning Properties of Slurry Based on Coal and Oil Processing Waste. *Energy Fuels* **2016**, *30*, 3441–3450. [CrossRef]

28. Valiullin, T.R.; Strizhak, P.A. Influence of the shape of soaring particle based on coal-water slurry containing petrochemicals on ignition characteristics. *Ther. Sci.* **2017**, *21*, 1399–1408. [CrossRef]

29. Glushkov, D.O.; Syrodoy, S.V.; Zakharevich, A.V.; Strizhak, P.A. Ignition of promising coal-water slurry containing petrochemicals: Analysis of key aspects. *Fuel Process. Technol.* **2016**, *148*, 224–235. [CrossRef]

30. Egorov, R.I.; Valiullin, T.R.; Strizhak, P.A. Energetic and ecological effect of small amount of metalline powders at the doping of the waste-derived fuels. *Combust. Flame* **2018**, *193*, 335–343. [CrossRef]

31. Nyashina, G.S.; Vershinina, K.Y.; Dmitrienko, M.A.; Strizhak, P.A. Environmental benefits and drawbacks of composite fuels based on industrial wastes and different ranks of coal. *J. Hazard Mater.* **2018**, *347*, 359–370. [CrossRef] [PubMed]

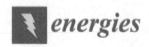

Article

Experimental Study on the Physical Performance and Flow Behavior of Decorated Polyacrylamide for Enhanced Oil Recovery

Shuang Liang [1,*], Yikun Liu [1,*], Shaoquan Hu [2], Anqi Shen [1,*], Qiannan Yu [1,*], Hua Yan [2] and Mingxing Bai [1,*]

1   EOR key lab in the Ministry of Education in Northeast Petroleum University, Daqing 163318, China
2   Daqing Oilfield Co., Ltd. No. 6 Oil Production Company, Daqing 163318, China;
     hushaoquan1@petrochina.com.cn (S.H.); yanhuajieyouxiang@163.com (H.Y.)
*   Correspondence: liangshuang21@163.com (S.L.); liuyikun111@126.com (Y.L.); anqi1986@126.com (A.S.);
     canaan184@163.com (Q.Y.); baimingxing@hotmail.com (M.B.)

Received: 21 January 2019; Accepted: 11 February 2019; Published: 12 February 2019

Abstract: With the rapid growth of energy consumption, enhanced oil recovery (EOR) methods are continually emerging, the most effective and widely used was polymer flooding. However, the shortcomings were gradually exposed. A novel decorated polyacrylamide might be a better alternative than polymer. In this work, the molecular structure and the properties reflecting the viscosity of decorated polyacrylamide, interfacial tension, and emulsification were examined. In order to better understand the interactions between decorated polyacrylamide and oil as well as the displacement mechanism, the displacement experiment were conducted in the etched-glass microscale model. Moreover, the coreflooding comparison experiments between decorated polyacrylamide and polymer were performed to investigate the displacement effect. The statistical analysis showed that the decorated polyacrylamide has excellent characteristics of salt tolerance, viscosity stability, and viscosification like polymer. Besides, the ability to reduce the interfacial tension in order 10^{-1} and emulsification, which were more similar to surfactant. Therefore, the decorated polyacrylamide was a multifunctional polymer. The displacement process captured by camera illustrated that the decorated polyacrylamide flooded oil mainly by means of 'pull and drag', 'entrainment', and 'bridging', based on the mechanism of viscosifying, emulsifying, and viscoelasticity. The results of the coreflooding experiment indicated that the recovery of decorated polyacrylamide can be improved by approximately 11–16% after water flooding when the concentration was more than 800 mg/L, which was higher than that of conventional polymer flooding. It should be mentioned that a new injection mode of 'concentration reduction multi-slug' was first proposed, and it obtained an exciting result of increasing oil production and decreasing water-cut, the effect of conformance control was more significant.

Keywords: decorated polyacrylamide; physical properties; displacement mechanism; flow behavior; enhanced recovery; injection mode

---

1. Introduction

Polymer flooding as an EOR technology boosted in the 1950s in order to ensure the sufficient oil production. Pye and Sandiford [1,2] first noted that the mobility of brine solution could be substantially reduced when the polymer was added. Until now, the researches concerning polymer flooding have matured enough [3–6], the mechanism of conformance control and mobility control has also been clearly understood by laboratory experiment [7]. The permeability of the reservoir decreased by increasing the viscosity of the water phase and the retention of polymer in the reservoir. Consequently, the mobility ratio and injection profile were improved, the swept region was enlarged. Both commercial and inhouse simulators were applied to optimize the injection process of polymer flooding. Oliveira et al. [8] studied the influence of polymer properties on economic indicators, which was helpful in the sensitivity analysis of polymer parameters. Janiga et al. [9] screened the efficient polymer injection strategy based on the nature-inspired algorithms and reservoir simulation. When the concentration of polymer was higher, the elasticity appeared, which was helpful to pull and drag the residual oil droplets [10]. Wang and Xia [11–13] stated that the high visco-elastic polymers can reduce the amount of remaining oil compared with water flooding by the etched microscale model. Zhong [14] established a mathematical model to simulate the transient flows of viscoelastic polymer, and obtained the pressure and velocity distribution in the course of flooding. Mohammad Sadegh et al. [15] took some factors into account such as effective concentration, thermal effects, dispersion, diffusion, etc., in order to improve the accuracy of viscoelastic polymer flooding model at present stage. Large-scale applications have gradually been realized in China after the 1990s, mainly in mature oil fields and offshore heavy oil fields [16,17]—for instance, Daqing Oilfield, Shengli Oilfield, and Bohai Oilfield.

Laboratory and field tests showed that recovery based polymer flooding was higher than that of water flooding by 10% [18]. However, lots of oil reserves remained underground by approximately 50% after polymer flooding. This was probably due to the shortcomings of polymer, a lower viscosity retention when located in the high temperature and salinity reservoir [19], as well as in the process of high-speed injection [20]. Additionally, water preferential channels appeared after long term displacement in the actual non-homogeneous reservoir. All those led to highly scattered distribution of remaining oil and coexisted with the water advantage channels, causing the waste and pollution of injected polymer [21,22]. Consequently, numerous studies were carried out to develop novel polymers. Zhang et al. [23] investigated the molecular structure of hydrophobically associating polyacrylamides (HAPAM) characterized by its tackifying performance. Lai et al. [24] synthesized shear-resistance hyperbranched polymers with special network, which can reduce the effect of shear. Li et al. [25] created a hydrophobically associating fluorinated polyacrylamide with amazing surface activity and thickening property. You et al. [26] studied the properties of a self-thickening polymer (STP), which was similar to gel, and can be prepared by produced water, moreover, a better effect of water control and oil increase was obtained. It was concluded that nanoparticles such as silica and titanium dioxide were applied broadly in polymer flooding and surfactant flooding to enhanced oil recovery, especially for heavy oil reservoir [27–31]. Cheraghian et al. [32] synthesized a nanopolymer used nanoclay and PAM, which has a good stability in the polymer solution. Besides, the amphiphilic polymer, biopolymers, low-tension polymer, thermally-stable, and salt-tolerant polymers have also been studied [33–35]. Overall, the mainly EOR mechanism of above polymers was still increasing the viscosity or decreasing the viscosity loss.

Nevertheless, it was unsatisfactory to further enhance oil recovery only depended the viscous property of polymers, lower interfacial tension and emulsification also played important roles in enhancing oil recovery. Polymer/surfactant binary system, alkali-surfactant-polymer (ASP) combinational flooding system, heterogeneous compound flooding system and foam flooding system have become the main direction of enhanced oil recovery technology after polymer flooding [36–39]. It was proved that the ASP flooding improved oil recovery by 17.2%, the recovery increment of SP flooding reached 14.3%, and that of the foam flooding was 13.1% [40]. However, some problems were still in existence, chromatographic separation due to the differences lied in migration velocity of components was the main problem, especially for the low permeability reservoir [41]. Which resulted in poor synergy of multi-component and waste of costs and chemicals.

Furthermore, the injection mode was important for enhanced oil recovery. Zhu et al. [42] stated that adjustable-mobility polymer flooding can enlarge the swept region of medium and low permeability layer. A variable viscosity injection was also applied in polymer flooding [43], Li et al. [44] compared the different injection pattern of polymer flooding, the result showed that the three-stage plugged injection with relatively high molecular polymer was the optimal approach to enhanced recovery.

Considering above points, single displacement agent with multi-functional groups might be a better alternative and new injection pattern needed to be explored urgently.

The primary objective of this study was to evaluate the physical properties of a novel decorated polyacrylamide, which possessed the excellent characteristic of polymer and surfactant. In addition, the oil displacement mechanism and flow behavior were uncovered by means of etching glass model. Finally, the comparison experiments were performed to verify the displacement effect of decorated polyacrylamide. Meanwhile, a new injection method has been proposed to improve the recovery.

2. Materials and Methods

2.1. Materials

The decorated polyacrylamide (DP) used in this research was provided by Chinese Academy of Sciences, the purity was more than 91%, the particle size from 0.2 mm to 1.0 mm accounted for 94%. The polymers were polyacrylamide from Daqing Refining & Chemical Company, which molecular weight were 1.5×10^7 and 3.5×10^7. The oil sample used in the experiment was from the First Oil Production Plant in Daqing Oilfield, the viscosity of simulated oil was 10 mPa·s at 45 °C. The brine water with salinity of 6778 mg/L was prepared according to the water of the First Oil Production Plant, the composition of injection water and wastewater were all presented in Table 1. Sodium chloride (NaCl) was from Shanghai Yansheng Biochemical Co. Ltd., China.

Table 1. Composition of water from First Oil Production Plant of Daqing Oilfield.

Water Type	Content (mg/L)						
	$Na^+ + K^+$	Ca^{2+}	Mg^{2+}	HCO_3^-	Cl^-	SO_4^{2-}	Total Salinity
Formation water	2186.3	14.9	52.4	2054.4	2267.7	54.1	6778
Injection water	231.2	34.1	24.3	225.1	88.7	36	729
Injection wastewater	1265	32.1	7.3	1708.56	780.12	9.61	4013

2.2. Equipment

The main equipment included an environmental scanning electron microscope (ESEM), a vacuum freeze drier, a LVDV-II + Pro viscosimeter, a HJ-6 magnetic stirrer, a thermostat, a Texas-500 interfacial tensiometer, an electronic balance, beakers, test tubes, and measuring cylinders. The coreflooding experiment equipment was from Wuxi City Petroleum Instrument Equipment Co., Ltd. China, the experimental setup was depicted in Figure 1a. The displacement experiments were carried out by artificial sinter cores, which were square with length 30 cm and the thickness 4.5 cm, and made of quartz sand and epoxy resin. The key parameters of cores were listed in Table 2. The oil displacement mechanism experiment device was presented in Figure 1b, the etching glass model was 40 cm in length and 40 cm in width, the average pore size of unevenly distributed pores and throats was 0.1 mm and the cross section was elliptical. One injection well and one production well were distributed at both ends of the diagonal line of the model.

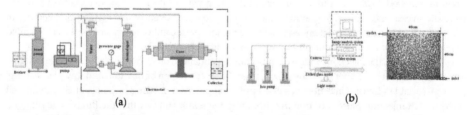

(a)　　　　　　　　　　　　　　　　(b)

Figure 1. (a) Coreflooding experimental setup; (b) oil displacement mechanism experiment device.

Table 2. Key parameters of cores.

Core Number	Permeability (μm^2)	Porosity (%)	Initial Oil Saturation (%)	Injection Chemicals and Concentrations (mg/L)
1	1.126	23.8	72.3	P(1000)
2	1.078	22.8	72.5	DP(300)
3	1.082	23.2	73.6	DP(300)
4	1.12	24.6	74.2	DP(500)
5	1.114	24.4	73.9	DP(500)
6	1.105	21.8	72.9	DP(800)
7	1.127	20.9	74.2	DP(800)
8	1.094	21.7	71.5	DP(1000)
9	1.102	23.4	72.8	DP(1000)
10	1.15	22.4	71.6	P(1000) + DP(1500) + DP(1000)
	0.153	20.1	52.9	

Note: P refers to polymer, DP refers to decorated polyacrylamide, and the concentrations were in the brackets.

2.3. Method

2.3.1. Microstructure Measurement

The polymer (molecular weight was 1.5×10^7) and DP solution with a concentration of 1000 mg/L were prepared, then the liquid sample of two chemical agents were placed on the stainless steel concave groove by a dropper respectively. Immediately, the samples solidified rapidly by the injected liquid ammonia. Meanwhile, a vacuum was created by means of vacuum freeze drier, which leaded to the water molecular located in the solution was sublimated, the dried samples were obtained. Finally, the samples were moved into the ESEM to observe and analyze the visualization images.

2.3.2. Salt Tolerance Property Measurement

Considering the Na^+ was the main cation in the formation water, the sodium chloride was selected to perform the salt tolerance property measurement. Firstly, the two chemicals' mother liquors of 5000 mg/L were simulated respectively, the NaCl solution was added into the mother liquors until the concentration up to 1000, 3000, 5000, 6000, 7000, and 9000 mg/L respectively. At last, the viscosities of different salinities were determined at 45 °C.

2.3.3. Viscosity-Improvement Measurement

In this study, the mother liquors with a concentration of 5000 mg/L were prepared by injection water, and diluted to different concentrations by injection wastewater (i.e., 500, 800, 1000, 1500, 1800, and 2000 mg/L). The viscosities of polymer (molecular weight was 3.5×10^7) and DP solution of different concentrations were tested by a LVDV-II + Pro viscosimeter at the rate of 6 rpm with magnetic stirrer.

2.3.4. Viscosity Stability Measurement

The DP solution of different concentrations (1500 mg/L and 2000 mg/L) were made up by injection wastewater, and the solution were placed in different days (0, 3, 7, 15, 30, 60, and 90 days), then the concentrations were measured by a viscosimeter to evaluate the viscosity stability.

2.3.5. Emulsification Property Measurement

Samples were prepared by mixing different concentrations of DP solution (20, 50, 100, 150, 200, 300, 500, 800, and 1000 mg/L) with oil at the volume ratio of 1:1. After that, the measuring cylinders with mixed samples were put into the thermostat for 15 min, then shaken by approximately 200 times to form an emulsion. Meanwhile, the emulsification phenomenon was observed regularly.

2.3.6. Interfacial Tension Measurement

The spinning drop method was applied to interfacial tension measurement, the experimental temperature was 45 °C. The DP of different concentrations (800, 900, 1000, 1200, 1400, 1600, and 1800 mg/L) and a 2 μL oil drop were injected into the test tubes respectively. An elliptical or cylindrical droplet will be formed at the rotation speed of 5000 r/min, under the action of gravity, centrifugal force and interfacial tension. Subsequently, the length and width of droplets were determined once every 20 min until the morphology was stable. According to Equation (1), the interfacial tension was calculated.

$$IFT = \frac{0.521 \times \Delta\rho \times D^3}{T^2} \tag{1}$$

where *IFT* refers to the interfacial tension (mN/m), *T* the rotation period (s), $\Delta\rho$ the density difference of oil and DP (g/cm^3), *D* the diameter of oil droplet (cm).

2.3.7. Cores Displacement Experiment

Experimental flowchart presented in Figure 1a was used in this study, nine single-tubes were applied to compare the oil recovery of polymer and DP (numbered 1–9). The cores parameters and injection fluids were shown in Table 2. According to the measured relative permeability curves before, the mobility ratios of water–oil and polymer–oil were approximately 55 and 0.3, respectively. The cores flooding experiment steps included that:

(1)   The casted cores were vacuumed for several hours until the weight did not decrease, then the dry weight were measured. Next, saturated the cores with brine (6778 mg/L), the porosities were attained.

(2)   The above cores were taken into the thermostat at 45 °C for more than 12 h. The simulated oil was injected until the water cut was 0%. Step (1) and (2) were called core treatment in the following text. The initial oil volume (saturation) and irreducible water saturation were obtained.

(3)   The brine was injected at a rate of 0.1 mL/min as far as the water-cut equaled 98%, the amount of displaced oil was obtained from the collector. The ratio of displaced oil volume to initial oil volume was the recovery of water flooding.

(4)   The chemicals were injected as designed in Table 2, then the chase water was injected again until the water content raised to 98%. As above, the oil volumes from chemicals flooding were attained from collector, based on the initial oil volume, the recovery were calculated. Consequently, the recovery increments were received.

In addition, the dual-tubes model (no. 10) was used to investigate the concentration reduction multi-slug injection of DP. The injection order was that: water injection, injection of polymer (1000 mg/L, 0.6 PV), succeeding water injection, injection of DP (1500 mg/L, 0.6 PV), injection of DP (1000 mg/L, 0.3 PV) and subsequent water injection. The injection rate was 1.2 mL/min.

2.3.8. Displacement Experiment in Etched-Glass Micromodel

The experiment was completed using the installations depicted in Figure 1b, the experimental procedures were similar to Section 2.3.7. The process included that water flooding, polymer flooding, chase water flooding, DP flooding, and succeeding water flooding in turn. Simultaneously, the experiment course was photographed by camera for analysis and research.

3. Results and Discussion

3.1. Microstructure of the DP

Figure 2a depicted the molecular structure of the DP. It can be seen that the chemical agent took flexible acrylamide and sodium acrylate hydrocarbon chain as the backbone, there occurred copolymerization with multiple functional monomers as the side chains, such as anionic or nonionic surfactants. A novel polymer we called decorated polyacrylamide (DP) was formed, which has the both characteristics of polymer and surfactant, coexistence with hydrophilic and hydrophobic groups. Figure 2b,c compared the microstructures of polymer and DP solution with a concentration of 1000 mg/L. It was indicated that the network structure formed in the polymer solution, which resulted in a good viscoelasticity to enlarge the sweep efficiency. However, the network structure was three-dimensinal in the DP solution and became denser. At a high concentration, the DP solution had the plugging property, and it was easier to enter inaccessible pore at low concentration. Consequently, the sweep volume was expanded.

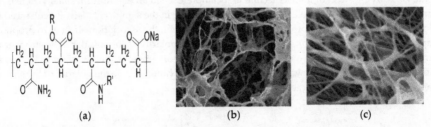

(a) (b) (c)

Figure 2. (a) Chemical structure of the DP; (b) Microstructure of polymer (1.5×10^7); (c) Microstructure of DP.

3.2. Physical Properties of the DP

3.2.1. Salt Tolerance Property Analysis

As the viscosity data showed in Table 3, with the concentration of NaCl increased, the viscosity of polymer was declined. In contrast, the viscosity of DP was increased first, and then decreased at 6000 mg/L. It indicated that the polymer has no salt tolerance property, due to there was a repulsion effect among the carboxyl ions in polymer solution. Besides the formation water was rich in metal ions, the existence of sodium and potassium ions weaken the repulsion of charges, and it made the molecules coil up and the viscosity declined. It showed a significant salt tolerance property as the concentration of NaCl increased as it concerned the DP. Because the polarity of the solution was enhanced by adding NaCl, it brought about the decrease of contact area between the DP molecule and water molecule. Meanwhile, the physical contact points among the molecule groups were increased correspondingly, it was possible to form a network structure. In addition, due to the association of hydrophobic groups, the molecules were close to each other, the hydrodynamic radius was increased and the viscosity was increased. The increment was more obvious from 3000 mg/L to 5000 mg/L, and maintained over 300 mPa·s. This increase rate after a certain NaCl concentration of 6000 mg/L off, it could be explained by the fact that the probability of intramolecular association was greater than that of intermolecular association, therefore, the viscosity decreased.

Table 3. Viscosities under different salinities of chemical agents.

Concentration of NaCl (mg/L)	Polymer Viscosity (mPa·s)	DP Viscosity (mPa·s)
1000	90.75	40.60
3000	52.54	80.00
5000	37.01	329.55
6000	32.24	349.85
7000	29.85	322.39
9000	22.69	220.90

3.2.2. Viscosification Property Analysis

The correlations of apparent viscosity and concentration of different chemical agents were shown in Figure 3a. It was indicated that the apparent viscosity increased with the concentration increased, and the increase increment of DP was larger than that of polymer. The reasons might be that the molecular weight of polyacrylamide was 3.5×10^7, which belonged to high-molecular polymer, the hydrodynamic size was larger. The molecular attraction increased when the concentration increased, meanwhile, the molecular motion became intense, the internal friction force and flow resistance increased, then the viscosity increased accordingly. In addition, as the molecular chain increased, the entanglement effect became more obvious, and the viscosity was larger. Although the molecular weight of DP was lower than the polyacrylamide, there were hydrophobic groups in the DP. When the concentration was below 1000 mg/L, the viscosity reduction effect of intramolecular association and the viscosifying effect of intermolecular association reached an equilibrium state. Whereas, the intermolecular association played a dominant role, resulting in the significant viscosity enhancement when the concentration was more than 1000 mg/L. Moreover, the association effect of hydrophobic groups was a dehydration action, which depended on the polarity of the solution. The stronger the polarity, the more compact the molecular chain, and the more obvious the association. Duo to the solution was prepared by wastewater, the existence of more cations made the polarity of the solution enhanced, and the association was intensified.

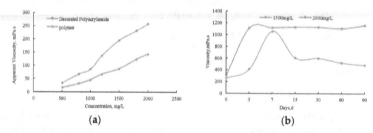

Figure 3. (a) Apparent viscosity curves of different chemical agents; (b) correlation between viscosity and days at different concentrations.

3.2.3. Viscosity Stability Analysis

Figure 3b clearly showed that the viscosity was increased substantially at the beginning and then decreased as the time. The difference was that when the concentration was 1500 mg/L, the viscosity reached a maximum value after seven days, then reduced sharply, and finally the viscosity kept stable, the viscosity retention rate was about 52%. However, the viscosity achieved maximal after three days, then gradually flatten with slight declines at 2000 mg/L, the viscosity retention rate was almost 99%. On account of the solution was prepared by injection wastewater, the existence of dissolved oxygen, bacteria, and microorganisms made the apparent viscosity of the solution instable, and the oil displacement was worse. Fu et al. [45] revealed that the polymer degraded severely in both clean water and waste water, and viscosity decreased with time. Nevertheless, the –SH and surfactants located in DP, which has an antioxidation and resistance to biodegradation. The peroxide in the solution was decomposed effectively via the self-oxidation of oxygen transfer groups, and the fracture of molecular chains under oxidation was inhibited, consequently, the viscosity did not decrease continuously. When the concentration was below a certain value, it was difficult to degrade substantially. Yet, the number of oxygen transfer radicals achieved a certain degree as the concentration increased, the viscosity loss rate was reduced under the comprehensive action.

3.2.4. Emulsification Property Analysis

It can be seen from Figure 4a that emulsification occurred in the system, which was not separated after resting for 16 days, and the emulsion type was oil-in-water when the concentration was over 800 mg/L. The emulsion was thermodynamically unstable system, it was necessary to inhibit the stratification and coalescence to maintain stability. The stratification rate can be expressed by Stockes formula (Equation (2)). The viscosity was increased with the concentration as mentioned above, which leaded to less the stratification rate. In addition, it was mentioned that wettability affected the multiphase flow, and wettability of sandstone was generally water-wet [46]. Due to the functional monomers of DP might be the alkyl sulfonate surfactant unit and quaternary ammonium Gemini surfactant unit, which led to the DP being adsorbed at the water–wet interface, the adsorption film formed as shown in Figure 4b. The stability of emulsion was improved due to the collision of emulsion droplets was prevented. Moreover, the existence of film made it difficult to wet the pore wall during the process of oil droplets migration, the displacement efficiency was improved. When the functional monomer was anionic surfactant, the charge density on the pore wall will be increased by the adsorption film. Thus, the electrostatic repulsion force between oil droplets and pore wall was increased, the oil droplets were easily to be taken away. Simultaneously, the emulsion droplets will be stuck in the three-dimensional network structure at a high concentration of DP, the droplet coalescence and floatation can be avoided.

$$v = \frac{(\rho_o - \rho_{DP})g}{18\mu}D^2 \qquad (2)$$

where v is the stratification rate (cm/s), ρ_O the density of oil (g/cm^3), ρ_{DP} the density of DP (g/cm^3), g the gravitational acceleration (981 cm/s^2), D the diameter of emulsion droplet (cm), μ the viscosity of DP (Pa·s).

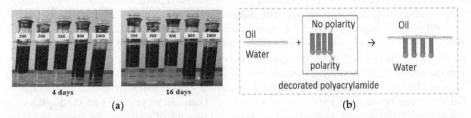

4 days   16 days

(a)   (b)

Figure 4. (**a**) Emulsification results of DP with different concentrations (200, 300, 500, 800, and 1000 mg/L); (**b**) arrangement of DP at oil–water interface.

A large number of experiments showed that emulsification was contribute to enhance oil recovery. The mechanism included entrainment and profile control. On the one hand, the emulsion flowed at a low speed, and the oil droplets of dispersed phase were extruded to kick-off the oil film at the pore wall, which has been named entrainment. On the other hand, it was easier for high viscosity emulsion to enter into high permeability layer preferentially, the large channels were plugged, and the swept region was enlarged. Besides, the emulsion can also reduce the resistance of lipophilic pore and improve the mobility of fluid.

3.2.5. Interfacial Tension Analysis

The interfacial tensions (IFT) at different concentrations were given in Figure 5. It can be determined that the IFT decreased first, and then nearly flat as the concentration increased. The IFT was in order 10 mN/m when the concentration was lower than 1000 mg/L. However, the oil-DP IFT maintained in order 10^{-1} mN/m when the concentration over 1000 mg/L. Therefore, the DP can reduce interfacial tension to some extent, exhibit the characteristic of surfactants. It was due to the hydrophilic and hydrophobic groups of DP having different directions on the interface as shown in Figure 4b. On the basis of capillary number (Ca) theory [47], the larger the Ca, the lower the residual oil saturation. In general, the Ca of water flooding was about 10^{-6}–10^{-5}, if the higher production was obtained, the Ca needed to be more than 10^{-3}. As a result, it was possible for DP to displace the remaining oil by a higher Ca.

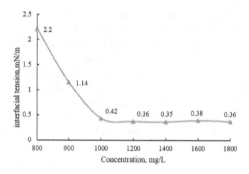

Figure 5. Correlation between interfacial tension and concentration.

3.3. Oil Displacement Mechanism and Flow Behavior in Etching Glass model

3.3.1. Displacement Mechanism of DP

Figure 6 clearly showed that the residual oil distributions in the micromodel at different stages. The dark brown sections corresponded to oil. It can be seen that there were amounts of residual oil after water flooding. Although the decrease in remaining oil after polymer treatment, some of them was still concentrated at the outlet end as shown in Figure 6c. After the DP injected as showed in Figure 6e, there was obvious effect on recovery, the swept region was enlarged significantly. The reason might be that the viscosity of DP was greater than that of polymer, the ability of expanding sweep volume was stronger. According to the Newton's law of friction (Equation (3)), the high viscosity increased the internal friction between liquid layers, which was the main force of fluid flow. Therefore, the shear stress between DP and oil was greater than that between polymer and oil, the displacement efficiency of the former was inevitably enhanced. In addition, the DP possessed the properties of both polymer and surfactant. On the one hand, it can push and pull residual oil in pores and throats with viscoelasticity like polymer flooding. On the other hand, the interfacial tension can be reduced to make the residual oil deform easily, meanwhile, accompanied by emulsification, which can enhance the carrying capacity [48].

$$\tau = \mu \frac{dv}{dy} \tag{3}$$

where τ is the shear stress (N/m^2), dv/dy the velocity gradient (s^{-1}), μ the viscosity of DP (Pa·s).

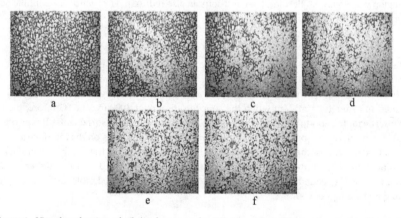

Figure 6. Visual evaluation of oil displacement for DP: (**a**) saturated by crude oil; (**b**) after water flooding; (**c**) polymer treatment; (**d**) after polymer treatment (water flooding); (**e**) DP(1000 mg/L) treatment; (**f**) water flooding again.

3.3.2. Flow Behavior of DP in Non-Homogeneous Porous Media

(1) Pull and Drag

At the beginning, the DP was adsorbed on the pore wall according to the creeping principle, the flow resistance of oil was reduced, the flow speed was accelerated, and they passed through the macropores and small pores smoothly. Concurrently, the DP carried residual oil through the throat as shown in Figure 7a-(i), then the residual oil was stretched and deformed which was similar to the amoeba effect in the small pore [49]. However, the deformation was not confined to small pores. Under the action of shear force, the band-shaped oil was extended, and transformed into slender oil filaments with large ends and slim middle shown in Figure 7a-(ii),a-(iii). Finally, the front of the filament was cut off into oil droplets. Those above illustrated that the mode was mainly based on shear stress, the remaining oil in the channel was elongated-sheared into small droplets by 'pull and drag'.

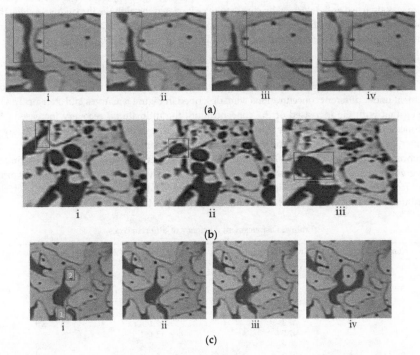

Figure 7. (a) A schematic of pull and drag mode: (i) Residual oil in the throat; (ii) Residual oil was stretched and deformed; (iii) Residual oil was drawn into filaments; (iv) Oil filaments broken into droplets. (b) A schematic of entrainment mode: (i) Oil film on pore wall; (ii) Oil film was entrained to form oil droplets; (iii) oil droplets aggregation. (c) A schematic of bridging mode: (i) Oil film enclosed pore wall; (ii) Oil film flowed along pore wall; (iii) Oil film bridging; (iv) Oil film merging together.

(2) Entrainment

Figure 7b-(i),b-(ii) exhibited that with the injection of DP, the oil film adhered to the pore wall was kicked-off and entrained by the DP solution to form small oil droplets, which was mainly depended on the emulsification. From Figure 6 we can see that the distribution of remaining oil after polymer flooding was more scattered. When the residual oil encountered the DP, the oil-in-water emulsion was formed. With the range of emulsification became larger, the flow resistance of emulsified small oil droplets was much smaller. Consequently, the original non-flowing residual oil was transformed from the bulk volume into small volume after emulsification, and then flowed again. This process was called emulsification start-up. At the intersection of channels, those small emulsified and dispersed oil droplets were entrained and accumulated into large oil droplets as shown in Figure 7b-(iii), this course was denoted as the emulsification entrainment.

(3) Bridging

Figure 7c represented the bridging mode in non-homogeneous porous media. The oil film might be absorbed on the pore wall when the rock surface was oil-wet. As shown in Figure 7c-(i), the Pore 1 and Pore 2 was enclosed by oil film, and Pore 1 was in the upstream direction of Pore 2. With the continuous injection of DP, the oil film around Pore 1 moved downstream along the pore wall as depicted in Figure 7c-(ii). Immediately, the extruded oil contacted the oil film around downstream Pore 2, this process was named bridging, after that the oil films was connected, the remaining oil in the upstream was transported downstream by the action of bridging as shown in Figure 7c-(iii). Then the oil film was stretched and deformed by shear stress as it passed through the throat. Meanwhile, the oil droplets formed by emulsification and deformation cutting were aggregated further. Consequently, the flow channel expanded, and the remaining oil flowed to the outlet end gradually.

3.4. Comparing Efficiency of Oil Displacement and Concentration Reduction Multi-Slug Injection of DP

The displacement experimental results were presented in Table 4, the variation rate of recovery increment under different concentrations were described in Figure 8. Curves and statistical analysis showed that with the increased of the concentration, the ultimate oil recovery increased. When the concentration of DP was lower than 800 mg/L, the displacement effect was not as good as polymer flooding, and the variation rate of recovery increment was below 20%. However, when the concentration was more than 800 mg/L, the recovery increment was over 11%. The ultimate oil recovery reached 63.7% at 1000mg/L, and the recovery improvement achieved 16.2%, the variation rate of recovery increment was as high as 37.87%. It proved that DP was a promising agent, which can contribute to the enhanced oil recovery further compared to polymer.

Table 4. Displacement efficiency of different cores.

Core Number	Water Flooding Oil Recovery (%)	Recovery Increment (%)	Ultimate Oil Recovery (%)
1	47.5	11.2	58.7
2	47.3	8.6	55.9
3	47.6	9.1	56.7
4	47.5	10.2	57.7
5	47.2	9.7	56.9
6	47.4	11.4	58.8
7	47.7	12.1	59.8
8	47.8	15.8	63.6
9	47.2	16.6	63.8

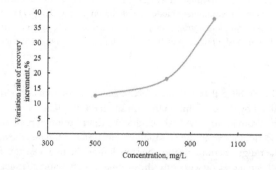

Figure 8. Relationship between variation rate of recovery increment and concentration.

Figure 9a,b showed the experimental result of concentration reduction multi-slug injection of DP. The curves can be divided into six parts, a represented the water flooding stage, b referred to polymer flooding, c stood for succeeding water flooding, d was DP flooding (1500 mg/L), e represented DP flooding (1000 mg/L), and f was subsequent water injection. It implied that with the increase of PV injected, the recovery percent showed an increasing trend, especially after polymer flooding, the recovery was improved by approximately 11% at part b. From the part d, there was a rising stage in the curve of injection pressure, which was the injection of DP (1500 mg/L), then the injected pressure was immediately declined and became nearly flat after 3.5 PV, which was the injection of DP (1000 mg/L). A reasonable explanation was that the high-permeability tube might be plugged by the chemical agent with 1500 mg/L. Afterward, the mobility was changed with the injection of a slightly low concentration DP, meanwhile, the chemical agent with 1000 mg/L entered into the low-permeability tube so that the injection pressure declined. Concurrently, the curve of water content has two descending stages, one was the injection of DP (1500 mg/L), the other was the injection of DP (1000 mg/L). This meant that the two kinds of concentrations of DP had the effect of increasing oil production and decreasing water-cut. As shown in Figure 9b, the shunt rate of high permeability layer decreased while that of the low permeability layer increased at the two injection stages of DP. Those above illustrated that the DP achieved a better effect of profile control. It should be mentioned that the recovery efficiency can be improved better by concentration reduction multi-slug injection, especially for low permeability zones. Further research concerned the optimization of slug number, slug size, injection rate under different concentrations, and the mechanism of profile control should be taken into account.

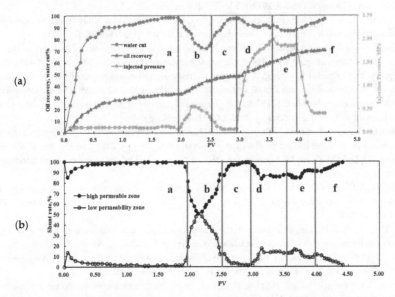

Figure 9. (**a**) Curves of oil recovery, water cut, and injection pressure versus PV injected. (**b**) Curves of shunt rate versus PV injected.

4. Conclusions

(1)   The DP simultaneously possessed characteristics of polymer and surfactant, exhibiting good salt tolerance, viscosification, viscosity stability, lower interfacial tension, and emulsification properties.

(2)   The experiment conducted in an etching glass model indicated that the displacement mechanism of DP included viscoelasticity, viscosification, lower interfacial tension, and emulsification. The flow behaviors were concluded to be 'pull and drag', 'entrainment', and 'bridging'.

(3)   Oil displacement experimental results clearly illustrated that the recovery increment of DP was about 11–16% higher than that of polymer flooding when the concentration was more than 800 mg/L. Additionally, the mode of concentration reduction multi-slug injection was favorable to enhance recovery by means of profile control, especially for a low permeability layer.

Author Contributions: Conceptualization, H.Y.; Methodology, Q.Y.; Validation, A.S.; Investigation, Y.L.; Resources, S.H. and S.L.; Data curation, S.L. and Q.Y.; Writing—original draft preparation, S.L. and Y.L.; Writing—review and editing, S.L. and M.B.; Supervision, Y.L.

Funding: This paper was funded by the National Natural Science Foundation of China, grant number 51804078 and the University Nursing Program for Young Scholars with Creative Talents in Heilongjiang Province, grant number UNPYSCT-2017031.

Conflicts of Interest: The authors declare no conflict of interest.

References

1. Pye, D.J. Improved secondary recovery by control of water mobility. *SPE J.* **1964**, *16*, 911–916. [CrossRef]
2. Sandiford, B.B. Laboratory and field studies of water floods using polymer solutions to increase oil recoveries. *SPE J.* **1964**, *16*, 917–922. [CrossRef]
3. Riahinezhad, M.; Romero-Zerón, L.; McManus, N.; Penlidis, A. Evaluating the performance of tailor-made water-soluble copolymers for enhanced oil recovery polymer flooding applications. *Fuel* **2017**, *203*, 269–278. [CrossRef]
4. Shi, L.T.; Chen, L.; Ye, Z.B.; Zhou, W.; Zhang, J.; Yang, J.; Jin, J.B. Effect of polymer solution structure on displacement efficiency. *Pet. Sci.* **2012**, *9*, 230–235. [CrossRef]
5. Salmo, I.C.; Pettersen, Ø.; Skauge, A. Polymer flooding at an adverse mobility ratio: Acceleration of oil production by crossflow into water channels. *Energy Fuels* **2017**, *31*, 5948–5958. [CrossRef]
6. Li, M.; Romero-Zerón, L.; Marica, F.; Balcom, B.J. Polymer flooding enhanced oil recovery evaluated with magnetic resonance imaging and relaxation time measurements. *Energy Fuels* **2017**, *31*, 4904–4914. [CrossRef]
7. Meybodi, H.E.; Kharrat, R.; Wang, X. Study of microscopic and macroscopic displacement behaviors of polymer solution in water-wet and oil-wet media. *Transp. Porous Media* **2011**, *89*, 97–120. [CrossRef]
8. De Oliveira, L.F.L.; Schiozer, D.J.; Delshad, M. Impacts of polymer properties on field indicators of reservoir development projects. *J. Pet. Sci. Eng.* **2016**, *147*, 346–355. [CrossRef]
9. Janiga, D.; Czarnota, R.; Stopa, J.; Wojnarowski, P.; Kosowski, P. Performance of nature inspired optimization algorithms for polymer enhanced oil recovery process. *J. Pet. Sci. Eng.* **2017**, *154*, 354–366. [CrossRef]
10. Zhang, Z.; Li, J.C.; Zhou, J.F. Microscopic roles of "viscoelasticity" in hpma polymer flooding for eor. *Transp. Porous Media* **2011**, *86*, 199–214. [CrossRef]
11. Wang, D.M.; Cheng, J.C.; Yang, Q.Y. Viscous-elastic polymer can increase micro-scale displacement efficiency in cores. In Proceedings of the SPE Annual Technical Conference and Exhibition, Dallas, TX, USA, 1–4 October 2000.
12. Xia, H.F.; Wang, D.M.; Liu, Z.C. Study on the mechanism of polymer solution with visco-elastic behavior increasing microscopic displacement efficiency. *Acta Petrolei Sin.* **2001**, *22*, 60–65.
13. Wang, D.M.; Cheng, J.C.; Xia, H.F. Improvement of displacement efficiency of cores by driving forces parallel to the oil–water interface of viscous-elastic fluids. *Acta Petrolei Sin.* **2002**, *23*, 47–52.
14. Zhong, H.Y.; Zhang, W.D.; Yin, H.J.; Liu, H.Y. Study on mechanism of viscoelastic polymer transient flow in porous media. *Geofluids* **2017**, *2017*. [CrossRef]
15. Sharafi, M.S.; Jamialahmadi, M.; Hoseinpour, S.A. Modeling of viscoelastic polymer flooding in core-scale for prediction of oil recovery using numerical approach. *J. Mol. Liq.* **2018**, *250*, 295–306. [CrossRef]
16. Maghzi, A.; Kharrat, R.; Mohebbi, A.; Ghazanfari, M.H. The impact of silica nanoparticles on the performance of polymer solution in presence of salts in polymer flooding for heavy oil recovery. *Fuel* **2014**, *123*, 123–132. [CrossRef]
17. Silva, I.P.G.; Aguiar, A.A.; Rezende, V.P.; Monsores, A.L.M.; Lucas, E.F. A polymer flooding mechanism for mature oil fields: Laboratory measurements and field results interpretation. *J. Pet. Sci. Eng.* **2018**, *161*, 468–475. [CrossRef]
18. Wang, Q.M.; Liao, G.Z.; Niu, J.G. Practice and Understanding of Polymer Flooding Technology. *Pet. Geol. Oilfield Dev. Daqing* **1999**, *18*, 3–7.

19. Manrique, E.J.; Thomas, C.P.; Ravikiran, R. EOR: Current Status and Opportunities. In Proceedings of the SPE Improved Oil Recovery Symposium, Tulsa, Oklahoma, 24–28 April 2010.

20. Howe, A.M.; Clarke, A.; Giernalczyk, D. Flow of concentrated viscoelastic polymer solutions in porous media: Effect of mw and concentration on elastic turbulence onset in various geometries. *Soft Matter* **2015**, *11*, 6419–6431. [CrossRef] [PubMed]

21. Feng, Q.H.; Qi, J.L.; Yin, X.M.; Yang, Y.; Bing, S.X.; Zhang, B.H. Simulation of fluid-solid coupling during formation and evolution of high-permeability channels. *Pet. Explor. Dev.* **2009**, *36*, 498–502.

22. Cui, C.Z.; Li, K.K.; Yang, Y.; Huang, Y.S.; Cao, Q. Identification and quantitative description of large pore path in unconsolidated sandstone reservoir during the ultra-high water-cut stage. *J. Pet. Sci. Eng.* **2014**, *122*, 10–17.

23. Zhang, P.; Huang, W.Z.; Jia, Z.F.; Zhou, C.Y.; Guo, M.L.; Wang, Y.F. Conformation and adsorption behavior of associative polymer for enhanced oil recovery using single molecule force spectroscopy. *J. Polym. Res.* **2014**, *21*, 523. [CrossRef]

24. Lai, N.J.; Li, S.T.; Liu, L.; Li, Y.X.; Li, J.; Zhao, M.Y. Synthesis and rheological property of various modified nano-sio2/am/aa hyperbranched polymers for oil displacement. *Russ. J. Appl. Chem.* **2017**, *90*, 480–491. [CrossRef]

25. Li, F.; Luo, Y.; Hu, P.; Yan, X.M. Intrinsic viscosity, rheological property, and oil displacement of hydrophobically associating fluorinated polyacrylamide. *J. Appl. Polym. Sci.* **2016**, *134*, 44672. [CrossRef]

26. You, Q.; Wang, K.; Tang, Y.C.; Zhao, G.; Liu, Y.F.; Zhao, M.W.; Li, Y.Y.; Dai, C.L. Study of a novel self-thickening polymer for improved oil recovery. *Ind. Eng. Chem. Res.* **2015**, *54*, 9667–9674. [CrossRef]

27. Cheraghian, G. Effect of nano titanium dioxide on heavy oil recovery during polymer flooding. *Pet. Sci. Technol.* **2016**, *7*, 633–641. [CrossRef]

28. Cheraghian, G. Thermal resistance and application of nanoclay on polymer flooding in heavy oil recovery. *Pet. Sci. Technol.* **2015**, *17–18*, 1580–1586. [CrossRef]

29. Cheraghian, G. Effects of titanium dioxide nanoparticles on the efficiency of surfactant flooding of heavy oil in a glass micromodel. *Pet. Sci. Technol.* **2016**, *3*, 260–267. [CrossRef]

30. Cheraghian, G. Improved Heavy Oil Recovery by Nanofluid Surfactant Flooding-An Experimental Study. In Proceedings of the 78th EAGE Conference and Exhibition, Vienna, Austria, 30 May–2 June 2016.

31. Cheraghian, G. Evaluation of clay and fumed silica nanoparticles on adsorption of surfactant polymer during enhanced oil recovery. *J. Jpn. Pet. Inst.* **2017**, *2*, 85–94. [CrossRef]

32. Cheraghian, G. Synthesis and properties of polyacrylamide by nanoparticles, effect nanoclay on stability polyacrylamide solution. *Micro Nano Lett.* **2017**, *1*, 40–44. [CrossRef]

33. Kang, W.L.; Jiang, J.T.; Lu, Y.; Xu, D.R.; Wu, H.R.; Yang, M.; Zhou, Q.; Tang, X.C. The optimum synergistic effect of amphiphilic polymers and the stabilization mechanism of a crude oil emulsion. *Pet. Sci. Technol.* **2017**, *35*, 1180–1187. [CrossRef]

34. Shu, Q.L.; Yu, T.T.; Gu, L.; Fan, Y.L.; Geng, J.; Fan, H.M. Viscosifying ability and oil displacement efficiency of new biopolymer in high salinity reservoir. *J. China Univ. Pet. (Ed. Nat. Sci.)* **2018**, *42*, 171–178.

35. Yadali Jamaloei, B.; Kharrat, R.; Torabi, F. Analysis and correlations of viscous fingering in low-tension polymer flooding in heavy oil reservoirs. *Energy Fuels* **2010**, *24*, 6384–6392. [CrossRef]

36. Zhao, G.; Dai, C.L.; You, Q. Characteristics and displacement mechanisms of the dispersed particle gel soft heterogeneous compound flooding system. *Pet. Explor. Dev.* **2018**, *45*, 481–490. [CrossRef]

37. Wei, P.; Pu, W.F.; Sun, L.; Pu, Y.; Wang, S.; Fang, Z.K. Oil recovery enhancement in low permeable and severe heterogeneous oil reservoirs via gas and foam flooding. *J. Pet. Sci. Eng.* **2018**, *163*, 340–348. [CrossRef]

38. Liu, S.H.; Zhang, D.H.; Yan, W.; Puerto, M.; Hirasaki, G.J.; Miller, C.A. Favorable attributes of alkaline-surfactant-polymer flooding. *SPE J.* **2008**, *13*, 5–16. [CrossRef]

39. Lai, N.J.; Zhang, Y.; Wu, T.; Zhou, N.; Liu, Y.Q.; Ye, Z.B. Effect of sodium dodecyl benzene sulfonate to the displacement performance of hyperbranched polymer. *Russ. J. Appl. Chem.* **2016**, *89*, 70–79. [CrossRef]

40. Han, P.H.; Su, W.M.; Lin, H.C.; Gao, S.L.; Cao, R.B.; Li, Y.Q. Evaluation and comparison of different EOR techniques after polymer flooding. *J. Xi'an Shiyou Univ. (Nat. Sci. Ed.)* **2011**, *26*, 44–48.

41. Li, K.X.; Jing, X.Q.; He, S.; Ren, H.; Wei, B. Laboratory study displacement efficiency of viscoelastic surfactant solution in eor. *Energy Fuels* **2016**, *30*, 4467–4474. [CrossRef]

42. Zhu, Y.; Gao, W.B.; Li, R.S.; Li, Y.Q.; Yuan, J.S.; Kong, D.B.; Liu, J.Y.; Yue, Z.C. Action laws and application effect of enhanced oil recovery by adjustable-mobility polymer flooding. *Acta Petrolei Sin.* **2018**, *2*, 189–200.

43. Cao, R.B.; Han, P.H.; Sun, G. Oil displacement efficiency evaluation of variable viscosity polymer slug alternative injection. *Oil Drill. Prod. Technol.* **2011**, *6*, 88–91.

44. Li, Y.Q.; Liang, S.Q.; Lin, L.H. Comparison and evaluation on injection patterns of polymer driving. *Pet. Geol. Recov. Effic.* **2010**, *6*, 58–60.

45. Fu, M.L.; Zhou, K.H.; Zhao, L.; Zhang, D.H. Research on viscosity stability of polymer solution made up by sewage. *OGRT* **2000**, *7*, 6–8.

46. Czarnota, R.; Janiga, D.; Stopa, J.; Wojnarowski, P. Wettability Investigation as a Prerequisite During Selecting Enhanced Oil Recovery Methods For Sandstone and Dolomite Formations. In Proceedings of the 17th International Multidisciplinary Scientific Geo Conference, Albena, Bulgaria, 29 June–5 July 2017; Volume 14, pp. 1013–1020.

47. Kutter, B.L. Effects of capillary number, bond number, and gas solubility on water saturation of sand specimens. *Can. Geotech. J.* **2013**, *50*, 133–144. [CrossRef]

48. Zhou, Y.Z.; Yin, D.Y.; Cao, R.; Zhang, C.L. The mechanism for pore-throat scale emulsion displacing residual oil after water flooding. *J. Pet. Sci. Eng.* **2018**, *163*, 519–525. [CrossRef]

49. Wang, Y.L.; Jin, J.F.; Bai, B.J.; Wei, M.Z. Study of displacement efficiency and flow behavior of foamed gel in non-homogeneous porous media. *PLoS ONE* **2015**, *10*, e0128414. [CrossRef] [PubMed]

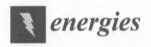

Article

Forecasting of Coal Demand in China Based on Support Vector Machine Optimized by the Improved Gravitational Search Algorithm

Yanbin Li and Zhen Li *

School of Economics and Management, North China Electric Power University, Beijing 102206, China;
liyb@ncepu.edu.cn
* Correspondence: lizhen2020@ncepu.edu.cn; Tel.: +86-188-1066-8262

Received: 1 April 2019; Accepted: 10 June 2019; Published: 12 June 2019

Abstract: The main target of the energy revolution in the new period is coal, but the proportion of coal in primary energy consumption will gradually decrease. As coal is a major producer and consumer of energy, analyzing the trend of coal demand in the future is of great significance for formulating the policy of coal development planning and driving the revolution of energy sources in China. In order to predict coal demand scientifically and accurately, firstly, the index system of influencing factors of coal demand was constructed, and the grey relational analysis method was used to select key indicators as input variables of the model. Then, the kernel function of SVM (support vector machine) was optimized by taking advantage of the fast convergence speed of GSA (gravitational search algorithm), and the memory function and boundary mutation strategy of PSO (particle swarm optimization) were introduced to improve the gravitational search algorithm, and the improved GSA (IGSA)–SVM prediction model was obtained. After that, the effectiveness of IGSA–SVM in predicting coal demand was further proven through empirical and comparative analysis. Finally, IGSA–SVM was used to forecast China's coal demand in 2018–2025. According to the forecasting results, relevant suggestions about coal supply, consumption, and transformation are put forward, providing scientific basis for formulating an energy development strategy.

Keywords: coal consumption forecasting; support vector machine; improved gravitational search algorithm; grey relational analysis

1. Introduction

The coal industry is an important basic industry related to national economic and energy security. At the stage of high-quality development, the environment of coal industry development is more complex, and the problems of unbalanced, uncoordinated, and unsustainable development are still outstanding. With the acceleration of the energy revolution, the proportion of coal in primary energy consumption will gradually decrease, but the main energy status will not change for a long time in China [1]. Coal has made irreplaceable contributions to the growth of the national economy. The shortage and surplus of coal supply will seriously interfere with the normal operation of the national economy.

As coal is a major producer and consumer of energy, due to the lack of scientific planning, the imbalance between supply and demand has been affecting the healthy development of the coal industry and national economy for a long time in China. At the same time, given the dominance of coal in China's energy structure, it is the leading source of carbon dioxide emissions. Thus, the change in coal demand is closely related to China's low-carbon and clean energy transition. Therefore, the timely adjustment of energy strategic planning with the help of scientific and accurate prediction of coal demand will not only help the coal industry to achieve the balance between supply and demand but also play an important role in promoting the establishment of modern energy systems and realizing the energy revolution.

In order to study the operation rules and characteristics of coal and guide the healthy and sustainable development of the coal industry, it is necessary to focus on the influencing factors of coal demand and coal demand forecasting. Coal demand forecasting is a very complex system. Rational estimation of each component of coal demand will be affected and restricted by many factors. The accuracy of coal demand forecasting depends on the detailed consumption data of major coal-consuming industries and the scientificness of forecasting methods and models. This paper chose the improved gravity search algorithm to optimize the parameters of the support vector machine algorithm and then used the optimized algorithm to predict the values of key factors. This improved intelligent algorithm model can greatly improve the prediction accuracy. The innovations of this article are as follows:

(1) In the analysis of the influencing factors of coal demand, combined with the actual situation of coal production and consumption, economic, social, and environmental constraints, this paper systematically selected 15 impact indicators from the four dimensions of economy, energy, industry, and environment. The grey correlation method was used to select the key indicators as the input variables of the forecasting model.

(2) Forecasting of coal demand based on the improved gravitational search algorithm–support vector machine (IGSA–SVM). Compared to the traditional optimization algorithm, GSA has the characteristics of fast convergence and strong pioneering performance when optimizing SVM. By introducing the memory function and boundary mutation strategy of particle swarm optimization, it avoids falling into the local optimum when GSA optimizes the parameters of SVM.

The remainder of this study is organized as follows. Section 2 mainly presents some literature related to coal demand forecasting and methods for energy demand forecasting. Section 3 screens the main influencing factors based on grey relational analysis and presents the IGSA–SVM adopted in this paper. The effectiveness of IGSA–SVM in predicting coal demand is further proven by comparing the errors of back propagation (BP), SVM, GSA–SVM, and IGSA–SVM in predicting coal demand, and IGSA–SVM is used to forecast China's coal demand in 2018–2025 in Section 4. Finally, we summarize the conclusions of this paper and put forward several recommendations in Section 5.

2. Relevant Literature Review

This section summarizes the relevant research from two aspects: coal demand forecasting and energy methods for energy demand forecasting.

2.1. Relevant Research on Coal Demand Forecasting

Coal demand forecasting is a complex system engineering involving economy, energy, and the environment. Scholars' research mainly focuses on two aspects: influencing factors of demand and coal demand forecasting.

In terms of influencing factors of coal demand, Michieka and Fletcher [2] used the vector autoregressive model and modified Granger causality test to study the relationship between urban population, gross domestic product (GDP), electricity output, and coal consumption. Chong et al. [3] used the logarithmic average Dirichlet decomposition method and found that rapid economic development is the biggest driving factor for the growth of coal consumption, and the improvement of coal utilization efficiency is the main factor to restrain the growth of coal consumption. Kulshreshtha [4] believes that economic development has a great impact on coal consumption. When Lin Boqiang [5] studied the deviation between China's coal demand growth and economic growth, he believed that technological progress, coal quality change, energy structure adjustment, industrial structure change, environmental governance, and carbon dioxide emissions were the important reasons for the change of coal demand.

In terms of coal demand forecasting, including direct forecasting and indirect forecasting, direct prediction is to get the change of coal consumption through the development and change of various departments which mainly consume coal resources. Reference [6] presented the formulation of forecasting models to get the total coal consumption by forecasting the coal consumption of different sectors in India, including domestic, transportation, and electricity. Reference [7] chose exponential smoothing (ETS) and the Holt–Winters model and used R language software to predict coal consumption in seven major coal-consuming industries. The indirect forecasting method is to construct the model by estimating the internal relationship between coal demand and its driving factors, so as to provide reference for coal demand forecasting. Reference [8] constructed an inverted U-shaped environmental Kuznets curve (EKC), which described the relationship between coal consumption per capita and GDP per capita in China, to predict China's coal consumption. Reference [9] predicted coal demand of electric power through the Granger causality test and predicted the trend of Chinese coal demand.

Due to the increasingly stringent constraints of climate, resources, and the environment, the complex and changeable market environment, people's awareness of energy conservation and emission reduction, and other factors, the influencing factors of coal demand should also be considered in a more comprehensive way to adapt to the new environmental requirements. Meanwhile, the forecasting method should keep pace with the times and be combined with the intelligent algorithm to realize coal demand forecasting more scientifically and efficiently.

2.2. Relevant Methods for Energy Demand Forecasting

Research on energy demand forecasting can be roughly divided into two categories: One is to use a single model for forecasting. The other is the combination forecasting method.

Among the single forecasting methods, there is the elastic coefficient method [10,11], regression method [12,13], system dynamics method [14,15], grey forecasting method [16,17], neural network method [18,19], support vector machine [20,21], etc. Damrongkulkamjorn et al. [22] introduced a new method combining ARIMA (autoregressive integrated moving average) with classical decomposition techniques. Reference [23] proposed a novel approach that incorporates ensemble empirical mode decomposition, which is widely used in time series analysis, sparse Bayesian learning for forecasting crude oil prices. Reference [24] used the system dynamics method to simulate and analyze China's energy consumption and carbon dioxide emissions under the target constraints of 2020. Reference [25] proposed a forecasting model combining the imperialist competitive algorithm with the back-propagation (BP) neural network. Cao and Wu [26] used the support vector regression machine to forecast the power demand series of China and the United States. The drosophila algorithm was used to optimize the parameters of the Support Vector Regression (SVR) model, and then the forecast results were revised according to seasonal index. Reference [27] introduced an accurate deep neural algorithm for short-term load forecasting, which displayed very high forecasting accuracy by comparing with other five artificial intelligence algorithms that are commonly used in load forecasting.

In combination forecasting methods, the prediction accuracy of the combination forecasting model is usually higher than that of single model. At present, there are three kinds of commonly used combination forecasting models. One is the combination model of classical forecasting methods. Reference [28] forecasted energy demand of transport in 2010, 2015, and 2020 based on the partial least square regression (PLSR) method. Reference [29] presented regression models of 34 customer energy sales and total energy sales based on the least squares technique, which considers economic and air temperature influencing factors. Secondly, based on the grey model, it combines with other statistical models. Reference [30] optimized the traditional Grey Model (GM (1,1)) model via the genetic algorithm, established the genetic algorithm-based remnant GM (1,1) (GARGM (1,1)) model, and used this model to forecast China's future energy demand. Reference [13] predicted China's total energy consumption through the Shapley value method, combined exponential smoothing model, system dynamics model, and GM (1,1) model.

Thirdly, the combination model was formed by various intelligent algorithms. Compared with traditional prediction methods, the intelligent algorithm has obvious advantages in speed and accuracy. Reference [31] made the adaptive bat algorithm based on exponential annealing (ABA–ESA) to establish the energy demand forecasting model. A new hybrid algorithm was formed by combining the ant colony algorithm and artificial bee colony algorithm, and it is used for probabilistic optimal allocation and classification of distributed energy resources [32]. Rahman et al. [33] combined the large data analysis technology based on Hadoop and other software with some machine learning methods such as artificial neural network (ANN) and back-propagation neural networks (BPNN) to forecast American power production. Reference [34] introduced a forecasting model combining data preprocessing with the extreme learning machine optimized by the cuckoo algorithm. Dai [35] proposed a novel model EEMD–ISFLA–LSSVM (Ensemble Empirical Mode Decomposition and Least Squares Support Vector Machine Optimized by Improved Shuffled Frog Leaping Algorithm) for forecasting the energy consumption in China. Reference [36] proposed a novel TS–PSO (particle swarm optimization)–LSSVM forecasting model. When predicting the energy consumption of China from 2017 to 2030, it was proven to be effective.

In conclusion, it can be seen that different prediction methods have their own advantages and disadvantages. In this paper, the intelligent algorithm was selected to solve the problem of coal demand forecasting. Through a literature review, it was found that BP and SVM are most widely used in prediction, while SVM operation results are more stable and reliable when the number of test samples is relatively small [37]. However, it is easy for SVM to fall into the local optimum when forecasting. Therefore, scholars adopted the genetic algorithm [38,39], particle swarm optimization [40,41], ant colony algorithm [42], gray Wolf algorithm [43], artificial swarm algorithm [44], and so on to optimize SVM. In this paper, the gravitational search algorithm (GSA) was used to optimize SVM. GSA is a new swarm intelligence optimization algorithm proposed by Professor Esmat Rashedi [45] in 2009. This algorithm searches the global optimal solution by simulating gravitation in physics. It makes use of the idea that the whole is larger than the part and has the characteristics of the whole search. These are the abilities and characteristics that some algorithms cannot achieve using a single individual to solve the optimal problem. Reference [46] introduced a mathematical model for predicting the enthalpy of steam turbine exhaust using the GSA to optimize the penalty factor of the LSSVM and two parameters of the radial range of the kernel. Reference [47] proposed a GSA–SVM network security situation prediction model, improving the accuracy and speed of network security situation prediction. In order to avoid premature convergence of GSA, Reference [48] introduced inertia decreasing weight and a local search operator for acceleration and velocity and predicted menstrual flow in dry season through SVM improved by IGSA. In a word, the improved combined prediction model can significantly improve the global search ability and avoid falling into the local optimal solution. In summary, it is not difficult to find that coal demand forecasting has been the focus of scholars' attention. However, the accuracy of the prediction results is restricted by many factors, such as the contradiction between the comprehensiveness of the variable selection process and the independence and the availability of data, and the scientific value of each variable in the prediction period, which all affect the prediction effect of the model to some extent. The intelligent algorithm has made great progress in energy demand forecasting, but the research on coal demand forecasting needs to be further studied.

3. Methods and Models

This section presents the methods and model adopted in this paper. The main influencing factors were screened according to grey relational analysis, and an IGSA–SVM forecasting model was constructed, combining the advantages of SVM and IGSA.

3.1. Screening of the Main Influencing Factors Based on Grey Relational Analysis

In order to predict coal demand scientifically, a set of scientific and reasonable index systems should firstly be established. Therefore, drawing on the experience of experts and scholars, combining with the actual development of China's coal industry, the following indicators were selected from the economy, energy, industry, and environment dimensions as alternative indicators: GDP (gross domestic product), total population, CPI (consumer price index), value added of secondary industry, urbanization rate, energy consumption intensity, the proportion of coal consumption to energy consumption, coal reserves, inventory of coal and products, quantity of imported coal, average annual price of Shanxi premium blended coal in Qinhuangdao Port, new productivity increased in raw coal mining, national railway coal freight volume, completion of investment in industrial pollution control, and sulfur dioxide emissions, and the index system is constructed, as shown in Figure 1.

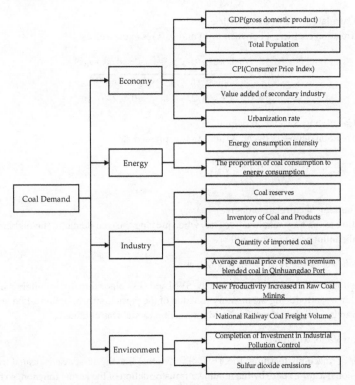

Figure 1. Influence factors system for coal demand forecasting.

In order to avoid the poor effect of the prediction model caused by too many input factors, the grey relational analysis [49] was used to screen the key factors affecting coal demand. The steps of calculating grey correlation degree are as follows:

Step 1 Determining the sequence of analysis

The time series of coal consumption is expressed as:

$$X_0 = (x_0(1), x_0(2), \cdots , x_0(n)) \tag{1}$$

The time series of factors affecting coal demand are expressed as follows:

$$\begin{cases} X_1 = (x_1(1), x_1(2), \cdots, x_1(n)) \\ X_2 = (x_2(1), x_2(2), \cdots, x_2(n)) \\ \cdots \\ X_m = (x_m(1), x_m(2), \cdots, x_m(n)) \end{cases} \tag{2}$$

Step 2 Standardized processing. Due to the different units of each influencing factor, the absolute value difference is also great, which affects the accuracy of prediction. Therefore, all influencing factors need to be standardized before calculating the correlation degree, with dimension reduction for each sequence.

$$y_i(k) = \frac{x_i(k)}{x_0(k)} \ k = 1, 2, \cdots, n, \ i = 0, 1, 2, \cdots, m \tag{3}$$

Step 3 Calculating correlation coefficient

The correlation coefficient between $y_j(k)$ and $y_0(k)$ is expressed as:

$$\zeta_j(k) = \frac{\min_j \min_k |y_0(k) - y_j(k)| + \rho \max_j \max_k |y_0(k) - y_j(k)|}{|y_0(k) - y_j(k)| + \rho \max_j \max_k |y_0(k) - y_j(k)|} \tag{4}$$

$$\rho \in (0, 1), k = 1, 2, \cdots, n, \ j = 0, 1, 2, \cdots, m$$

Step 4 Calculating correlation degree

The correlation degree between X_j and X_0 is expressed as:

$$r_j = \frac{1}{n} \sum_{k=1}^{n} \zeta_j(k) \ k = 1, 2, \cdots, n, \ j = 0, 1, 2, \cdots, m \tag{5}$$

Step 5 Ranking of correlation degree

According to the grey relational degree, when judging the coal demand, the higher the value is, the greater the influence of factors will be.

3.2. Construction of IGSA–SVM Forecasting Model

This section introduced the principle of the SVM and GSA algorithm firstly and then improved the GSA algorithm by introducing the memory function of the particle swarm optimization and boundary mutation strategy, and then the IGSA–SVM forecasting model was obtained.

3.2.1. Support Vector Machine

SVM is a machine learning theory [50]. Firstly, the input space is transformed into a high-dimensional feature space by the nonlinear transformation of the kernel function, so that it can be linearly separable, and then the optimal classification hyperplane is obtained in this feature space.

Supposing that the sample set is $T = \{(x_i, y_i)|i = 1, 2, \ldots, N\}$, the optimal classification hyperplane $(\omega \cdot x) + b = 0$ can be obtained by nonlinear mapping, then the optimal classification hyperplane can be translated into the following optimization problems:

$$\min \Phi(\omega) = \frac{1}{2}\|\omega\|^2 + C\sum_{i=1}^{N} \xi_i \tag{6}$$

$$s.t. \ (\omega \cdot x) + b \geq 1 - \xi_i, i = 1, 2, \ldots, N$$

where ω is normal vector for hyperplane, b is a deviation; C is a penalty parameter, and ξ_i is a relaxation variable. For solving this quadratic programming problem, the Lagrange function is introduced, and the dual principle is used to transform the original optimization problem into:

$$\max W(\alpha) = \sum_{i=1}^{N} \alpha_i - \frac{1}{2} \sum_{i,j=1}^{N} y_i y_j \alpha_i \alpha_j K\left(x_i, x_j\right)$$

$$s.t. \sum_{i=1}^{N} \alpha_i y_i = 0, 0 \leq C, i = 1, 2, \dots, N \tag{7}$$

where α_i is Lagrange multiplier, $K\left(x_i, x_j\right)$ is a kernel function. We chose the Gauss radial Bbsis Kernel function, which is expressed as follows:

$$K\left(x_i, x_j\right) = e^{-g\|x_i - x_j\|^2} \tag{8}$$

where g is a kernel parameter, controlling the range of action of the Gauss kernel, and is an important parameter affecting the classification performance of SVM.

The decision function obtained using the Gauss radial basis function is as follows:

$$f(x) = \text{sgn}\left(\sum_{i=1}^{N} \alpha_i y_i K(x_i, x) + b\right) \tag{9}$$

3.2.2. Gravitational Search Algorithm

GSA (gravitational search algorithm) is a new swarm intelligence optimization algorithm proposed, and the Schematic diagram of the GSA algorithm is shown as Figure 2.

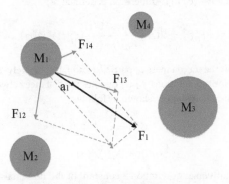

Figure 2. The schematic diagram of the gravitational search algorithm (GSA).

In Figure 2, M_1, M_2, M_3, and M_4 represent four objects with different masses. The larger the area, the greater the mass. Object M_1 is subject to gravity F_{12}, F_{13}, and F_{14} of object M_2, M_3, and M_4, respectively, producing a reasonable F_1 and corresponding acceleration a_1. Under the action of the resultant F, Object M_1 moves towards an F_1 direction. It can be seen from the figure that F_1 direction and F_{13} direction are close to each other. In the figure, object M_3 has the maximum mass. Under the action of the surrounding gravity, object M_1 will approach the object with the maximum mass nearby. This approach is actually the optimization process of the gravity search algorithm. Therefore, the optimization process of the GSA algorithm is actually the process of finding the individual with the largest mass, and the quality of the optimization result is judged by the size of the individual mass.

Assuming a population of N particles X_i, in D-dimensional search space, defining the position and velocity of the first particle are $X_i = \left(X_i^1, X_i^2, \cdots, X_i^k, \cdots X_i^D\right)$ and $V_i = \left(v_i^1, v_i^2, \cdots, v_i^k, \cdots v_i^D\right)$, $i = 1, 2, \cdots, N$, respectively, where X_i^k and v_i^k represent the position and velocity components of the particle i on the k-dimension, respectively. Then the mass and gravity of each particle are determined through evaluating the objective function values of each particle, calculating the acceleration and updating the speed and position on this basis.

Firstly, population initialization is carried out, including the initialization of position, velocity, acceleration, and mass of particles. Then, the mass of each particle is calculated. When carrying out the t-th iteration, the inertia mass $M_i(t)$ of particle i can be updated according to its fitness value. The updated formula is as follows:

$$mass_i(t) \begin{cases} \frac{fit_i(t)-worst(t)}{best(t)-worst(t)} & if\ best(t) \neq worst(t) \\ 1 & otherwise \end{cases} \tag{10}$$

$$M_i(t) = \frac{mass_i(t)}{\sum\limits_{j=1}^{N} mass_j(t)} \tag{11}$$

where $fit_i(t)$ denotes the fitness of particle i at iteration t-th, $i = 1,2,\cdots,N$. For solving the minimum optimization problem, the optimal fitness $best(t)$ and the $worst(t)$ fitness word are expressed as follows:

$$best(t) = \min_{j \in \{1,2,\cdots,N\}} fit_j(t) \tag{12}$$

$$worst(t) = \max_{j \in \{1,2,\cdots,N\}} fit_j(t) \tag{13}$$

Conversely, it can be applied to the maximum optimization problem.

Then, calculating the gravity of each particle, when carrying out the t-th iteration, the mutual attraction of particle i and particle j in K-dimension is defined as:

$$F_{ij}^d(t) = G(t)\frac{M_i(t) \times M_j(t)}{R_{ij}(t) + \varepsilon}\left(x_j^k(t) - x_i^k(t)\right) \tag{14}$$

where M_j and M_i represent the inertia mass of particle i and j, respectively. ε is constant. $G(t)$ denotes the gravitation coefficient for t-th iterations. $R_{ij}(t)$ denotes the distance between particle i and particle j (generally taking the Euclidean distance), shown as follows:

$$G(t) = G_0 e^{-\alpha\frac{t}{T}} \tag{15}$$

$$R_{ij}(t) = \|X_i(t), X_j(t)\|_2 \tag{16}$$

where G_0 represents the universal gravitational constant of the universe at its earliest moments. α denotes the attenuation factor of the gravitational coefficient and is generally taken as a constant, and T represents the maximum number of iterations.

In GSA, the total force acting on the t-th iteration particle i in the K-dimension can be expressed as:

$$F_i^k(t) = \sum_{j=1,j\neq i}^{Kbest} rand_i \times F_{ij}^k(t) \tag{17}$$

where $rand_i$ denotes a random number of $[0,1]$. The initial value of $Kbest$ is N, which gradually decreases to 1 over time. It is defined as:

$$Kbest(t) = final\_per + \left(\frac{1-t}{T}\right) \times (100 - final\_per) \tag{18}$$

Among them, $final\_per$ represents the percentage of particles that exert forces on other particles.

Finally, the particles' position movement is calculated. According to Newton's second law, when the *t*-th iteration is performed, the acceleration of particle *i* on the *K*- dimension can be defined as:

$$a_i^k(t) = \frac{F_i^k(t)}{M_i(t)} \tag{19}$$

In each iteration of the GSA, the particle updates the velocity *v* and position *x* of particle *i* according to the following formula:

$$v_i^k(t+1) = rand_j \times v_i^k(t) + a_i^k(t) \tag{20}$$

$$x_i^k(t+1) = x_i^k(t) + v_i^k(t+1) \tag{21}$$

3.2.3. Improved Gravitational Search Algorithm

When predicting via SVM, the different choices of penalty parameter *C* and kernel parameter *g* in the Gauss kernel function have a great impact on the prediction accuracy. In order to avoid falling into the local optimum when the gravitational search algorithm optimizes the parameters of the support vector machine, the improved gravitational search algorithm (IGSA) in this paper was constructed by introducing the memory function and boundary mutation strategy of PSO (particle swarm optimization), shown as follows:

1. Introducing the memory function of PSO. When updating particles, GSA only considers the influence of the current position of particles but does not take the memory of particles into account. Therefore, the global memory function of particle swarm optimization was introduced in this paper. GSA is embedded in the memory function, that is to say, the idea of the best position that the whole population has searched so far was added to the velocity update equation, so as to improve the local development ability of GSA:

$$v_i^k(t+1) = rand_j \times v_i^k(t) + a_i^k(t) + c_1 \times rand_i \times \left(gbest - x_i^k(t)\right) \tag{22}$$

where c_1 is learning factor and *C* means the best location the whole group has found so far.

2. Introducing the boundary mutation strategy of PSO. In the process of GSA iteration, under the action of the law of gravitation and Newton's second law, if the particle's position exceeds the set range $[x_{min}, x_{max}]$, the standard GSA will force the particle to return to the boundary. If the particle gathers too much on the boundary of the feasible region, it is not conducive to the convergence of GSA. In order to improve the convergence of GSA, the boundary mutation strategy is introduced as follows:

$$\begin{aligned}&if\ x_i < x_{min}\ or\ x_i > x_{max}\\&then\ x_i = rand \times (x_{max} - x_{min}) + x_{min}\end{aligned} \tag{23}$$

After the boundary mutation, the particles do not gather on the boundary, which increases the diversity of the population and helps the algorithm to find the optimal solution faster.

3.2.4. IGSA–SVM Forecasting Model

The improved IGSA was used to optimize the parameters of SVM to ensure good prediction results. The specific steps are as follows.

Step 1 Initialization of parameter. Setting *N* as population size, x_i^k, and v_i^k represents position and velocity of *N* particles, respectively. Then determining the number of iterations *t*, the learning factor c_1, and the range of parameters *C* and *g*.

Step 2 Calculating the fitness of the particle and finding out the optimal solution *Kbest* and the position *gbest* of the optimal particle.

Step 3 Updating the inertial mass $M_i(t)$ of particles by Equations (10)–(13), the gravitational constant $G(t)$ by Equation (15), and calculating Euclidean distance $R_{ij}(t)$ among particles by Equation (16).

Step 4 Calculating the sum of forces $F_i^k(t)$ in each direction of the particle by Equation (17), and the acceleration of the particle according to Equation (19).

Step 5 Updating the position and velocity of particles by Equations (20)–(22).

Step 6 Judging whether the boundary conditions are satisfied by Equation (23).

Step 7 Return to step 2 and iterate until the maximum number of iterations or accuracy requirements are met, and output (C, g).

Step 8 The IGSA is used to optimize the parameter values of SVM, and the forecasting model is obtained.

3.3. Forecasting Process

According to the previous analysis, coal demand forecasting is affected by many factors. In order to predict coal demand scientifically and accurately, firstly, sample data of 15 influencing factors were collected. Then, according to the grey relational analysis method, the main influencing factors of coal demand were selected and used as the input value of the IGSA–SVM model. On that basis, according to the steps 1–8 above, forecasting the coal demand was based on IGSA–SVM. Repeat the above steps until the number of iterations is reached. Finally, the IGSA–SVM forecasting model was obtained to forecast the coal demand. The prediction process is shown in Figure 3.

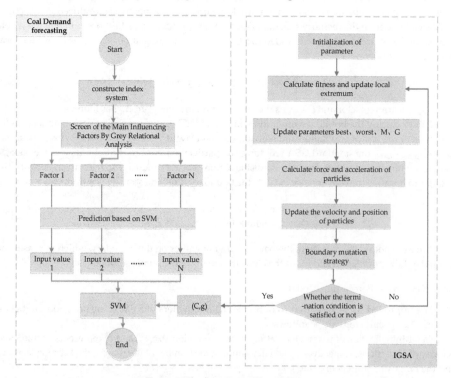

Figure 3. The flow chart of forecasting.

4. Empirical and Comparative Analysis

In this section, influencing factor screening for model input was determined first. Then the effectiveness of IGSA–SVM in predicting coal demand was further proven by comparing the errors of

BP, SVM, and GSA–SVM. Through empirical and comparative analysis, the IGSA–SVM was used to forecast China's coal demand in 2018–2024.

4.1. Influencing Factor Screening for Model Input

According to the influence factors selected in Figure 1, this paper selected the relevant data, that is, GDP, total population, CPI, value added of secondary industry, urbanization rate, energy consumption intensity, the proportion of coal consumption to energy consumption, coal reserves, inventory of coal and products, quantity of imported coal, average annual price of Shanxi premium blended coal in Qinhuangdao Port, new productivity increased in raw coal mining, national railway coal freight volume, completion of investment in industrial pollution control, and sulfur dioxide emissions, from 1990 to 2017 as the research object. Among them, in addition to the average annual price of Shanxi premium blended coal in Qinhuangdao Port and new productivity increased in raw coal mining indicators' data from the China coal industry association, the other indicators' data came from the relevant data issued by the National Bureau of Statistics and the Statistical Yearbook.

In order to determine the input index of the prediction model, the grey correlation analysis was used to analyze the correlation degree between influencing factors and coal demand, as shown in Table 1.

Table 1. The calculation results of the grey relational degrees for influencing factors.

Influencing Factor	Grey Relational Degree
GDP	0.8602
Total Population	0.943
CPI	0.9754
Value added of secondary industry	0.9517
Urbanization rate	0.9733
Energy consumption intensity	0.8982
The proportion of coal consumption to energy consumption	0.9573
Coal reserves	0.9241
Inventory of Coal and Products	0.65
Quantity of imported coal	0.6263
Average annual price of Shanxi premium blended coal in Qinhuangdao Port	0.9426
New Productivity Increased in Raw Coal Mining	0.9364
National Railway Coal Freight Volume	0.9861
Completion of Investment in Industrial Pollution Control	0.9544
Sulfur dioxide emissions	0.9285

The six factors whose grey relational degree was greater than 0.95 were selected as the coal consumption forecasting model input. They are CPI, value added of secondary industry, urbanization rate, the proportion of coal consumption to energy consumption, national railway coal freight volume, and completion of investment in industrial pollution control. The model output is the power grid investment, as shown in Table 2.

Table 2. Data of the main influencing factors.

Year	Total Coal Consumption (Ten Thousand Tons of Standard Coal)	CPI (1978 = 100)	Value Added of Secondary Industry (%)	Urbanization Rate (%)	The Proportion of Coal Consumption to Energy Consumption (%)	National Railway Coal Freight Volume (Ten Thousand Yuan)	Completion of Investment in Industrial Pollution Control (Ten Thousand Yuan)
1990	75211.686	216.4	41	26.41	76.2	62870	454465
1991	78978.863	223.8	41.5	26.37	76.1	62603	597306
1992	82641.69	238.1	43.1	27.63	75.7	64108	646661
1993	86646.771	273.1	46.2	28.14	74.7	65336	693270
1994	92052.75	339	46.2	28.62	75	65943	833313
1995	97857.296	396.9	46.8	29.04	74.6	67357	987376
1996	103794.16	429.9	47.1	29.37	74.7	72058	956135
1997	98793.695	441.9	47.1	29.92	71.5	70345	1164386
1998	92020.944	438.4	45.8	30.4	69.6	64081	1220461
1999	81862	432.2	45.4	38.89	67.1	64917	1527307
2000	100670.34	434	45.50	36.00	69.00	68545	2347895
2001	105771.96	437	44.80	38.00	68.00	76625	1745280
2002	116160.25	433.5	44.50	39.00	69.00	81852	1883663
2003	138352.27	438.7	45.60	41.00	70.00	88132	2218281
2004	161657.26	455.8	45.90	42.00	70.00	99210	3081060
2005	189231.16	464	47.00	43.00	72.00	107082	4581909
2006	207402.11	471	47.60	44.00	72.00	112034	4839485
2007	225795.45	493.6	46.90	46.00	73.00	122080.63	5523909
2008	229236.87	522.7	47.00	47.00	72.00	134325	5426404
2009	240666.22	519	46.00	48.00	72.00	132720.15	4426207
2010	249568.42	536.1	46.50	50.00	69.00	156020	3969768
2011	271704.19	565	46.50	51.00	70.00	172125.74	4443610
2012	275464.53	579.7	45.40	53.00	69.00	168515.29	5004573
2013	280999.36	594.8	44.20	54.00	67.00	167945.66	8496647
2014	279328.74	606.7	43.30	55.00	66.00	164130.57	9976511
2015	273849.49	615.2	41.10	56.00	64.00	143221.23	7736822
2016	270320	627.5	40.10	57.00	62.00	131790.73	8190041
2017	278159	637.5	40.50	59.00	60.40	149129.86	6815345

Data sources: Statistical Yearbook 1990–2016 and official website of national bureau of statistics (http://www.stats.gov.cn/).

4.2. Forecasting and Comparative Analysis

Taking 1990–2009 as the training set and 2010–2017 data as the test set to forecast, the model parameters settings are as follows: population size $N = 20$, maximum number of iterations MAX_IT = 100, the optimal range of penalty parameter C is [1,500], the optimum range of kernel parameter g is [0.001,200], $G_0 = 100$, $\alpha = 20$, and $c_1 = 2$. The forecasting results and the residuals are shown in Figure 4.

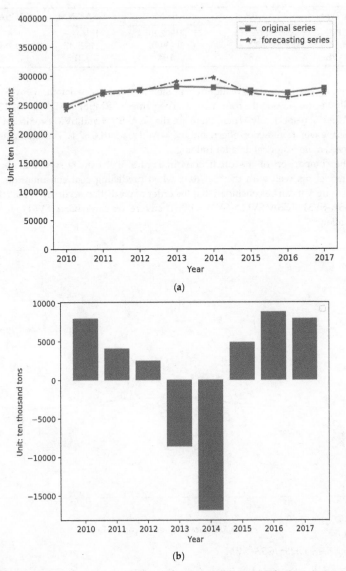

Figure 4. The prediction and residual figures by IGSA–SVM. (**a**) The prediction figure; (**b**) the residual figure.

According to Figure 3, it can be seen that IGSA–SVM has a good fitting effect on predicting coal consumption. Furthermore, the errors of IGSA–SVM, GSA–SVM, SVM, and BP in predicting coal demand in 2010–2017 were compared and analyzed, including RMSE (root mean square deviation), MAE (mean absolute error), and MAPE (mean absolute percent error). The comparison results are shown in Table 3 below.

Table 3. Error analysis and comparison.

Error Types	BP	SVM	GSA–SVM	IGSA–SVM
RMSE	19798.56	21902.20	15164.49	8721.86
MAE	18877.36	16439.78	11925.47	7694.94
MAPE	6.87	5.94	4.31	2.82

From the comparative analysis of the errors in Table 3, it can be seen that the IGSA–SVM prediction has the smallest errors among the four models. The values of RMSE, MAE, and MAPE are 8721.86, 7694.94, and 2.82%, respectively. Then, there are the GSA–SVM and SVM prediction models. The BP model has the worst prediction effect, and the MAPE reached 6.87%. The main reason is that BP prediction needs more historical data for training.

From the comparison of the predictions in Figure 5, it can be seen that the IGSA–SVM forecasting model fits well with the real data when predicting coal consumption. Combining Table 3 and Figure 5, it can be concluded that the order of prediction accuracy of the four prediction models is IGSA–SVM > GSA–SVM > SVM > BP. Therefore, we chose IGSA–SVM to forecast China's coal consumption.

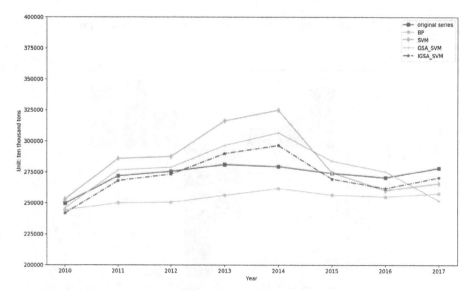

Figure 5. Forecasting results comparison.

4.3. Forecasting Based on the IGSA–SVM

According to the data of six key factors from 1990 to 2017, the data of six factors from 2018 to 2025 were predicted by SVM, as shown in Figure 6, as input variables of the IGSA–SVM forecasting model, and the forecasted results are shown in Figure 7.

As we can see from Figure 6, combined with the change trend of six key factors affecting coal demand selected in this paper, the following findings are made. CPI is on a slight upward trend. Value added of secondary industry is on a slight downward trend. Urbanization rate is on a slight upward trend. The proportion of coal consumption to energy consumption is on a steady downward trend. Finally, the national railway coal freight volume fluctuates between increase and decrease. Completion of investment in industrial pollution control shows a downward trend.

Affected by these factors, as we can see from Figure 7, coal consumption grew slowly in 2018 to 2025, peaked in 2019, and began to decline gradually. According to the predicted results, combined with the impact of the factors, value added of secondary industry, the proportion of coal consumption to energy consumption, and completion of investment in industrial pollution control are the key factors leading to the turning point of demand. This means that the decline in coal demand is related to policy constraints, such as the increase of the proportion of tertiary industry output and the reduction of the proportion of coal consumption, as well as the gradual saturation of industrial pollution control projects. If we want to further reduce the demand for coal in China, energy efficiency must be comprehensive, and the coal-based energy consumption structure needs further optimization. In terms of future energy development, we must be committed to adjusting the structure of the energy industry, developing new energy sources, improving the level of re-electrification, and realizing clean substitution and electric energy substitution.

The forecasting results are in line with the development trend of the energy revolution, especially the peak value forecasting results, and are helpful to energy strategic planning, overall planning of coal production and consumption, and ensuring the healthy and coordinated development of the coal industry. According to the prediction results of the model, the base line of coal safety supply in each year is calculated according to the coefficient of raw coal converted into standard coal. In addition, the coal production capacity can be controlled at this bottom line, providing an important reference value for the reduction of production capacity.

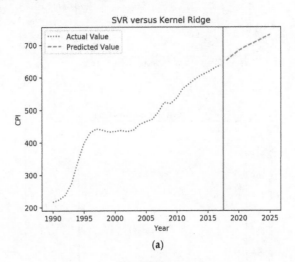

(a)

Figure 6. *Cont.*

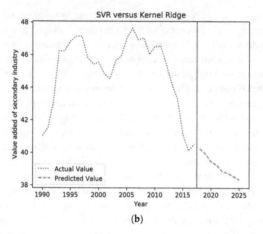

(b)

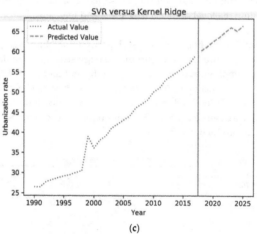

(c)

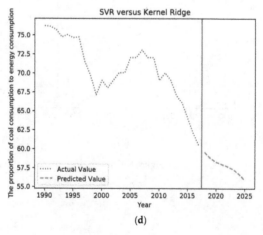

(d)

Figure 6. *Cont.*

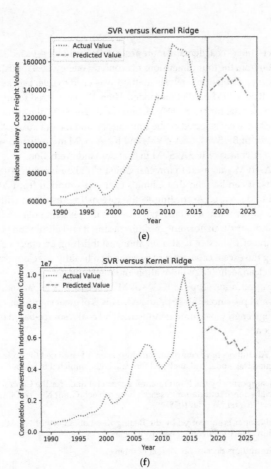

Figure 6. Coal consumption key influencing factors prediction curve. (**a**) Forecasting of consumer price index (CPI); (**b**) forecasting of value added of secondary industry; (**c**) forecasting of urbanization rate; (**d**) forecasting of the proportion of coal consumption to energy consumption; (**e**) forecasting of national railway coal freight volume; (**f**) forecasting of completion of investment in industrial pollution control.

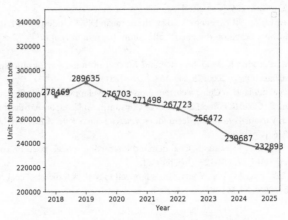

Figure 7. The forecasting results of coal consumption from 2018 to 2025 in China.

Energies **2019**, *12*, 2249

5. Conclusions

In order to predict China's coal demand more accurately, firstly, we selected 15 influence factors that affect coal consumption from the four dimensions of economy, energy, coal industry, and environmental constraints, and the grey correlation analysis method was used to select six key influence factors as input variables of the IGSA–SVM prediction model. Then, PSO was used to improve the GSA, which is used to optimize the kernel function of SVM, and the IGSA–SVM algorithm was obtained. Next, the data from 1990 to 2009 were selected as the training set, and the data from 2010 to 2017 were used as the test set. The errors of BP, SVM, GSA–SVM, and IGSA–SVM in prediction were compared, which further proved the effectiveness of IGSA–SVM in coal demand prediction.

Finally, the IGSA–SVM was used to forecast China's coal demand in 2018–2025. According to the forecasting results, it can be seen that China's coal consumption from 2018 to 2025 will have a trend from growth to decrease. According to the forecasting results, it will be better to have an active response for the government and industry, enacting the relevant policies, further optimizing the structure of the coal industry, transforming and upgrading it and eliminating backward production capacity, improving the efficiency of coal utilization, establishing an effective modern supervision system of coal, solving the excess capacity, and promoting a healthy development of the coal industry through speeding up clean technology innovation and development.

In the future, the application scope of IGSA–SVM can be further expanded, and the coal demand of different industries and provinces can be predicted. This is of great significance to the transformation and development of high energy-consuming industries and coal resource-based provinces and to adapt to the energy revolution.

Author Contributions: All authors have contributed to this paper. Y.L. initiated the project and gave guidance for the methods. Z.L. designed the model and analyzed the data and completed the paper in English.

Funding: This paper is supported by the Fundamental Research Funds for the Central Universities under **No.** 2019QN075, Beijing social science foundation research base project (Grant No. 17JDGLA009), *"Natural Science Foundation of China Project" (Grant No. 71471058)*.

Acknowledgments: This paper is supported by the Beijing Key Laboratory of New Energy and Low-Carbon Development (North China Electric Power University), Beijing.

Conflicts of Interest: The authors declare no conflict of interest.

References

1. Shi, D. Current situation and development Suggestions of coal chemical industry. *Guangzhou Chem. Ind.* **2018**, *24*, 14–16.
2. Michieka, N.M.; Fletcher, J. An investigation of the role of China's urban population on coal consumption. *Enery Policy* **2012**, *48*, 668–676. [CrossRef]
3. Chong, C.; Ma, L.; Li, Z.; Ni, W.; Song, S. Logarithmic mean Divisia index (LMDI) decomposition of coal consumption in China based on the energy allocation diagram of coal flows. *Energy* **2015**, *85*, 366–378. [CrossRef]
4. Kulshreshtha, M.; Parikh, J.K. Modeling demand for coal in India: Vector autoregressive models with ciontegrated variables. *Energy* **2000**, *25*, 149–168. [CrossRef]
5. Lin, B.; Wu, W. Coal demand in China's current economic development. *China Soc. Sci.* **2018**, *2*, 141–161.
6. Jebaraj, S.; Iniyan, S.; Goic, R. Forecasting of Coal Consumption Using an Artificial Neural Network and Comparison with Various Forecasting Techniques. *Energy Sources Part. A Recovery Utilization Environ. Eff.* **2011**, *33*, 1305–1316. [CrossRef]
7. Zhu, Q.; Zhang, Z.; Chai, J.; Wang, S. Coal demand forecasting model based on consumption structure division. *Syst. Eng.* **2014**, *32*, 112–123. (In Chinese)
8. Hao, Y.; Zhang, Z.Y.; Liao, H.; Wei, Y.M. China's farewell to coal: A forecast of coal consumption through 2020. *Energy Policy* **2015**, *86*, 444–455. [CrossRef]

9. Yawei, O.; Li, O. A forecast of coal demand based on the simultaneous equations model. In Proceedings of the 2011 IEEE International Symposium on IT in Medicine and Education, Cuangzhou, China, 9–11 December 2012.

10. Wang, Y.; Li, X. The influence of renewable energy power generation development on coal demand. *J. North. China Electr. Power Univ. (Soc. Sci. Ed.)* **2014**, *5*, 13–16.

11. Wang, Y.; Chen, X.; Han, Y. Forecast of Passenger and Freight Traffic Volume based on Elasticity Coefficient Method and Grey Model. *Proced. Soc. Behav. Sci.* **2013**, *96*, 136–147. [CrossRef]

12. Xiong, Y.; Zou, J.; Liu, L. Energy profile model based on multiple linear regression analysis. *China High.-tech Zone* **2018**, *13*, 52.

13. Tsekouras, G.J.; Dialynas, E.N.; Hatziargyriou, N.D. A non-linear multivariable regression model for midterm energy forecasting of power systems. *Electr. Power Syst. Res.* **2007**, *77*, 1560–1568. [CrossRef]

14. Huang, Y.; Zhao, G.; Wang, Y. Research on China's energy consumption combination forecasting model based on Shapley value. *Energy Eng.* **2012**, *6*, 5–9.

15. Chi, K.C.; Nuttall, W.J.; Reiner, D.M. Dynamics of the UK natural gas industry: System dynamics modelling and long-term energy policy analysis. *Technol. Forecast. Soc. Chang.* **2009**, *76*, 339–357.

16. Kou, A.; Zhou, W.; Hu, Q. GM (1,1) Model for Forecasting and Analysis of China's Energy Consumption. *China Market.* **2018**, *16*, 15–16.

17. Zhang, J.; Yao, Z.; Xu, X.; Nan, Y. Coal demand forecasting based on Grey GM (1,1) model. *J. North. China Inst. Sci. Technol.* **2016**, *13*, 106–108. (In Chinese)

18. Khotanzad, A.; Afkhami-Rohani, R.; Maratukulam, D. ANNSTLF-artificial neural network short-term load forecaster-generation three. *IEEE Trans. Power Syst.* **1998**, *13*, 1413–1422. [CrossRef]

19. De, G.; Gao, W. Forecasting China's Natural Gas Consumption Based on AdaBoost-Particle Swarm Optimization-Extreme Learning Machine Integrated Learning Method. *Energies* **2018**, *11*, 2938. [CrossRef]

20. Xie, M.; Deng, J.; Liang, J. Cooling load forecasting method based on support vector machine optimized with entropy and variable accuracy roughness set. *Power Syst. Technol.* **2017**, *41*, 210–214.

21. Dai, S.; Niu, D.; Han, Y. Forecasting of Energy-Related CO2 Emissions in China Based on GM (1,1) and Least Squares Support Vector Machine Optimized by Modified Shuffled Frog Leaping Algorithm for Sustainability. *Sustainability* **2018**, *10*, 958. [CrossRef]

22. Damrongkulkamjorn, P.; Churueang, P. Monthly energy forecasting using decomposition method with application of seasonal ARIMA. In Proceedings of the 2005 International Power Engineering Conference, Singapore, 29 November–2 December 2005.

23. Li, T.; Hu, Z.; Jia, Y.; Wu, J.; Zhou, Y. Forecasting Crude Oil Prices Using Ensemble Empirical Mode Decomposition and Sparse Bayesian Learning. *Energies* **2018**, *11*, 1882. [CrossRef]

24. Liu, X.; Mao, G.; Ren, J.; Li, R.Y.M.; Guo, J.; Zhang, L. How might China achieve its 2020 emissions target? A scenario analysis of energy consumption and CO2 emissions using the system dynamics model. *J. Clean. Prod.* **2015**, *103*, 401–410. [CrossRef]

25. Mollaiy-Berneti, S. Developing energy forecasting model using hybrid artificial intelligence method. *J. Cent. South. Univ. Engl. Ed.* **2015**, *22*, 3026–3032. [CrossRef]

26. Cao, G.; Wu, L. Support vector regression with fruit fly optimization algorithm for seasonal electricity consumption forecasting. *Energy* **2016**, *115*, 734–745. [CrossRef]

27. Kuo, P.-H.; Huang, C.-J. A High Precision Artificial Neural Networks Model for Short-Term Energy Load Forecasting. *Energies* **2018**, *11*, 213. [CrossRef]

28. Zhang, M.; Mu, H.; Li, G.; Ning, Y. Forecasting the transport energy demand based on PLSR method in China. *Energy* **2009**, *34*, 1396–1400. [CrossRef]

29. Hsu, J.F.; Chang, J.M.; Cho, M.Y.; Wu, Y.H. Development of Regression Models for Prediction of Electricity by Considering Prosperity and Climate. In Proceedings of the 2016 3rd International Conference on Green Technology and Sustainable Development (GTSD), Kaohsiung, Taiwan, 24–25 November 2016.

30. Hu, Y.C. A genetic-algorithm-based remnant grey prediction model for energy demand forecasting. *PLoS ONE* **2017**, *12*, e0185478. [CrossRef]

31. Zou, S.; Ding, Z. China's coal demand forecasting model based on ABA-ESA. *China Coal* **2018**, *44*, 9–14.

32. Kefayat, M.; Ara, A.L.; Niaki, S.A.N. A hybrid of ant colony optimization and artificial bee colony algorithm for probabilistic optimal placement and sizing of distributed energy resources. *Energy Convers. Manag.* **2015**, *92*, 149–161. [CrossRef]

33. Rahman, M.N.; Esmailipour, A.; Zhao, J. Machine Learning with Big Data An Efficient Electricity Generation Forecasting System. *Big Data Res.* **2016**, *5*, 9–15. [CrossRef]
34. Wang, R.; Li, J.; Wang, J.; Gao, C. Research and Application of a Hybrid Wind Energy Forecasting System Based on Data Processing and an Optimized Extreme Learning Machine. *Energies* **2018**, *11*, 1712. [CrossRef]
35. Dai, S.; Niu, D.; Li, Y. Forecasting of Energy Consumption in China Based on Ensemble Empirical Mode Decomposition and Least Squares Support Vector Machine Optimized by Improved Shuffled Frog Leaping Algorithm. *Appl. Sci.* **2018**, *8*, 678. [CrossRef]
36. Zhang, L.; Ge, R.; Chai, J. Prediction of China's Energy Consumption Based on Robust Principal Component Analysis and PSO-LSSVM Optimized by the Tabu Search Algorithm. *Energies* **2019**, *12*, 196. [CrossRef]
37. Han, Z.; Wan, J.; Liu, K. Prediction and model optimization of gasoline octane number based on near-infrared spectroscopy. *Anal. Lab.* **2015**, *34*, 1268–1271.
38. Liu, A.; Xue, Y.; Hu, J.; Liu, L. Super-short-term prediction of wind power based on GA optimization SVM. *Power Syst. Prot. Control.* **2015**, *2*, 90–95. (In Chinese)
39. Saini, L.M.; Aggarwal, S.K.; Kumar, A. Parameter optimisation using genetic algorithm for support vector machine-based price-forecasting model in National electricity market. *IET Gener. Trans. Distrib.* **2010**, *4*, 36–49. [CrossRef]
40. Zhu, C.; Chen, X.; Wang, Z.; Zhang, X. Defect prediction model for object oriented software based on particle swarm optimized SVM. *Comput. Appl.* **2017**, *37*, 60–64.
41. Bao, L.; Chen, H.; Guo, J.; Yuan, Y. Application of SVM based on improved PSO algorithm in methane measurement. *Chin. J. Sens. Actuators.* **2017**, *9*, 1454–1458. (In Chinese) [CrossRef]
42. Rubio, G.; Pomares, H.; Rojas, I.; Herrera, L.J. A heuristic method for parameter selection in LS-SVM: Application to time series prediction. *Int. J. Forecast.* **2010**, *3*, 725–739. [CrossRef]
43. Xu, D.; Ding, S. Research on improved GWO-optimized SVM-based short-term load prediction for cloud computing. Computer Engineering and Applications. *Comput. Eng. Appl.* **2017**, *7*, 68–73. (In Chinese)
44. Feng, T.; Zhong, Y.; Liu, X.; Yu, L. Application of least square support vector machine based on adaptive artificial swarm optimization algorithm in deformation prediction. *J. Jiangxi Univ. Sci. Technol.* **2008**, *39*, 35–39.
45. Rashedi, E.; Nezamabadi-pour, H.; Saryazdi, S. GSA: A Gravitational Search Algorithm. *Inf. Sci.* **2009**, *179*, 2232–2248. [CrossRef]
46. Wang, H.; Han, Z.; Xu, X.; Li, X. Prediction model of steam turbine exhaust enthalpy based on gray correlation analysis and gsa-lssvm. *Electr. Power Constr.* **2016**, *37*, 115–122. (In Chinese)
47. Chen, Y.; Yin, X.; Tan, R. A network security situation prediction model based on gsa-svm. *J. Air Force Eng. Univ. (Nat. Sci. Ed.)* **2018**, *19*, 82–87.
48. Cui, D.; Huang, E. Application of improved gravity search algorithm and support vector machine in monthly runoff prediction in dry season. *People's Pearl River* **2016**, *37*, 48–52.
49. Deng, J. *Grey Theory Foundation*; Huazhong University of Science and Technology Press: Wuhan, China, 2002.
50. Mei, H.; Liu, D.; He, Y. Fault Diagnosis of SVM Analog Circuits Optimized by Improved Gravity Search Algorithms. *Microelectr. Comput.* **2018**, *408*, 115–121.

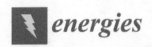

Article

Hybrid Model for Determining Dual String Gas Lift Split Factor in Oil Producers

Chew Chen Law, Mohamed Zamrud Zainal *, Kew Hong Chew * and Jang Hyun Lee

Department of Petroleum Engineering, Faculty of Engineering, Universiti Teknologi PETRONAS,
Seri Iskandar 32610, Perak Darul Ridzuan, Malaysia; chewchen87@yahoo.com.my (C.C.L.);
lee.janghyun@utp.edu.my (J.H.L.)
* Correspondence: mzamrud.zainal@utp.edu.my (M.Z.Z.); kewchew@gmail.com (K.H.C.)

Received: 22 March 2019; Accepted: 15 April 2019; Published: 14 June 2019

Abstract: Upstream oil production using dual string completion, i.e., two tubing inside a well casing, is common due to its cost advantage. High pressure gas is employed to lift the oil to the surface when there is insufficient reservoir energy to overcome the liquids static head in the tubing. However, gas lifting for this type of completion can be complicated. This is due to the operating condition where total gas is injected into the common annulus and then allowed to be distributed among the two strings without any surface control. High uncertainties often result from the methods used to determine the split factor—the ratio between the gas lift rate to one string over the total gas injected. A hybrid model which combined three platforms: the Visual Basics for Application programme, PROSPER (a nodal analysis tool) and Excel spreadsheet, is proposed for the estimation of the split factor. The model takes into consideration two important parameters, i.e., the lift gas pressure gradient along the annulus and the multiphase pressure drop inside the tubing to estimate the gas lift rate to the individual string and subsequently the split factor. The proposed model is able to predict the split factor to within 2% to 7% accuracy from the field measured data. Accurate knowledge of the amount of gas injected into each string leads to a more efficient use of lift gas, improving the energy efficiency of the oil productions facilities and contributing toward the sustainability of fossil fuel.

Keywords: dual string completion; gas lift; gas lift rate; split factor; gas robbing; gas lift optimization

1. Introduction

Gas lift is one of the common artificial lift techniques used in upstream oil production. It works by injecting high pressure gas into the annulus and tubing, reducing the fluids density and allowing the fluids to be produced to the surface at a lower bottom-hole pressure. Dual string completion refers to a single well casing housing two tubing, see Figure 1. The tubings are fitted into the casing side by side with all of the required accessories. Typically, the tubing will be of a different length, with one shorter than the other, to allow production from different zones. The shorter tubing is named Short String (SS), and the longer tubing is named Long String (LS) [1]. Gas lift is particularly suitable for this type of completion due to the space limitation.

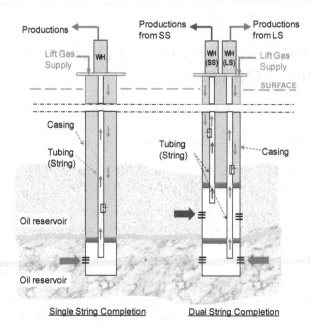

Productions ← | Productions ← from SS | Productions → from LS

Lift Gas Supply | WH | WH (SS) | WH (LS) | Lift Gas Supply

SURFACE

Casing

Tubing (String) | Tubing (String) | Casing

Oil reservoir

Oil reservoir

Single String Completion | Dual String Completion

Figure 1. Single string and dual string completions.

The design of gas lifting for the dual string completion has proven to be a challenge. The total gas injection rate for both strings are measured at the surface, but the amount of the gas going into each tubing is difficult to determine [2]. This may lead to one of the strings getting too much gas while the other string is starved. This phenomenon is known as gas robbing. Excessive or inadequate injected gas can lead to lower well fluids production and flow instability. At the same time, it is desirable to inject the gas at the optimum level in each string to operate the gas lift system efficiently. Widianoko et al. [3] suggested placing the orifices for both LS and SS at the same depth to prevent such phenomena. This, however, may not suit the lifting requirement. There is also industry practice that simply assumes an equal split of the gas lift rate (GLIR) between LS and SS, which would hardly be the case, noting that the production behaviour may vary significantly between the two.

While the gas allocation to optimize the oil production from single string completions has been widely discussed by many researchers, the literature on gas lift optimization for dual string completion is scarce. Widianoko et al. [3] proposed the use of a trial and error technique in the field to maximize the production from the dual string well, but this is easier said than done operationally. Nishikiori et al. [4] suggested the use of a non-linear optimization method to generate the most optimum gas lift operating condition for a set of multiple single string wells. This is only applicable to a single string well where the determination of GLIR is straightforward. The optimization of the gas lift in a dual string completion would require the computation of the amount of gas going into the individual string, and the consideration of the interaction between the strings as well as with the common annulus. The split factor is defined as the ratio between the gas lift rate (GLIR) for SS and LS over the total gas lift injected. A higher split factor for a particular string means that more gas will be entering that string. This parameter is important for dual string gas lift optimization, ensuring an efficient usage of energy from the produced gas for oil production.

Eikrem et al. [5] discussed gas lift instability in a single point gas lifted dual string well using both a mathematical model and laboratory measurement. A single point gas injection for both strings does not represent a realistic field operation. In field operations, the SS gas lift injection valve was always placed shallower than that of the LS. Conejeros [6] proposed to optimize the dual string production by producing water through LS and oil through SS; his work does not consider the gas lift.

Kamis et al. [2] proposed to use the average of the GLIR for LS and SS obtained from individual well modelling. The method assumed that well tests were carried out on one string at a time while the other string is on production with the stable gas lift. The second string is to be tested immediately or as soon as possible after the first test. This is to maintain similar test conditions for both of the strings, which may not be operationally practical. The calculated gas injection rate from the single well model is matched against the surface measured test rate based on the first string's parameters. The gasses injection rate for the second string is obtained through a simple deduction from the total gas injected. A similar approach is carried out for the second string. Consequently, each individual string will have its gas lift rate estimated, and the average of the values is then used to estimate the split factor.

Petroleum Experts (PETEX) proposed two methods to estimate the amount of lift gas going into the LS and SS using their nodal analysis tool, PROSPER [7]. Note that PROSPER does not have the capability of modelling the dual string completion and treat the dual string completion as two separate single string completions. The first method used the measured injection or casing head pressure (CHP) to find the corresponding GLIR. The GLIR is changed, repeatedly, until the CHP estimated by the software matched the measured injection pressure. The workflow required the two strings to be solved independently. The second method uses the measured total GLIR and allocates the lift gas between the two strings assuming the same casing pressure for both of the single wells. This method reiterates the total GLIR allocation ratio as a variable, and a solution is reached when the calculated CHP is the same for both strings. The second method ensures the sum of the GLIR to the LS and SS is equal to the input of the total GLIR. Figure 2 summarises the workflow for the two methods.

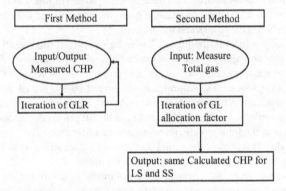

Figure 2. Workflow comparison between two methods proposed by PETEX.

Both of the two methods proposed by PETEX relied on the availability of field measured data. The solution is obtained by an iteration technique. As the software can only model a single well, the solution is obtained without considering the impact of one string on the other. The wells or the strings are considered to flow independently. The interaction of the annulus and the two strings is also ignored. The calculated casing pressure by the software is likely to be the average of the CHP at the surface, as if the LS and SS are flowing as a separate single well. The calculated CHP is almost always lower than the actual measured CHP. The estimation of the gas lift rates to the SS and LS is a mathematical exercise using the iteration technique to match a measured value. No consideration is given to the gas flow phenomenon inside the annulus and tubing.

A study by Chia et al. [8] noted that for the dual completion gas lift, as both strings share the same common annulus, it is difficult to determine the exact individual GLIR of each string. The researchers suggested to use a pre-known fix value of the gas-oil ratio (GOR) to compute the individual GLIR from the total gas measures at the surface. The assumption of a constant GOR is fallacious as the GOR value changes from time to time. It also required the well to be tested individually, which may not be practical due to operational constraints.

A few researchers have, in more recent years, proposed the use of an artificial intelligence (AI) approach to tackle gas lift optimization. Patterns of changes in operating conditions were identified through the measurement of parameters, and AI algorithms were then applied to correct the conditions. Abbasov et al. proposed a method using machine learning to attain the minimum field measured tubing head pressure oscillation, so as to achieve an optimum production. It relied mainly on field testing and no well modelling is required [9]. Xiao et al. suggested the use of a calibrated well model to establish the operating envelope for a stable production. An automated workflow is created to monitor, troubleshoot and correct the well for any instabilities during the production [10]. Both methods are only applicable to single string wells where the wells behave independently of each other.

To obtain an accurate estimate of the gas lift rates for dual string completions, the industry has to resort to a field measurement via a well tracer survey. The tracer method allows the measurement, when both strings are flowing, to account for the interaction between them. However, it is costly and operationally impractical as it requires the mobilization of expensive equipment and personnel. In addition, there is a potential production deferment related to the well work. The well tracer for the dual string can be challenging, as discussed by Abu Bakar et al. [11]. An average of six hours per survey is required with a 25% rerun for dual strings due to too many points of gas returns.

Hermank et al. proposed the use of distributed thermal sensing (DTS) and distributed acoustic sensing (DAS) for the gas lift surveillance [12]. The method is less invasive and can provide a precise injection point based on changing temperature and sound signals. However, it would require high capital investment to equip the wells with downhole optical fibers. This method will work well with the single string well but not necessarily with the dual string well. The interpretation of DTS and DAS for the dual string well is a lot more complicated given the pressure and flow dynamics within the annulus as well as between the strings and the annulus.

For an optimum production, the lift gas has to be injected at the correct rate. Excessive or inadequate injected gas can lead to a suboptimal production and flow instability [13]. So far, the various methods proposed by the researchers do not directly compute the gas lift rates to individual strings for the dual string completion. The solutions were derived by the use of an iterative approach, with simplification (such as an injection at the same depth), averaging, or equating the dual string as two separate single strings. The interactions between the lift gas inside the common annulus and the production fluids in the tubing were not taken into consideration. The results obtained can be misleading as they do not represent the actual operation of a dual string well. Little insight can be derived for gas lift troubleshooting to optimize the lift gas usage.

The objective of this study is to develop a hybrid model to determine the gas lift split factor considering the fluids flow phenomena both in the annulus and the tubing. The hybrid model will combine three platforms: the Visual Basic for Application (VBA) programme, PROSPER (a nodal analysis tool) and Excel spreadsheet to compute the gas lift rate into an individual string and subsequently the split factor. The model will be validated using a set of field measured data. This hybrid model allows for an accurate estimation of the lift gas rate for the dual string completion, enabling an efficient use of lift gas and yielding an opportunity for improving the energy efficiency of upstream production facilities.

2. Methods

The study is carried out in 3 stages:

- Development of the hybrid model for the split factor in dual string gas lift.
- Validation of the hybrid using a set of available field data
- Case study.

2.1. Modelling of Split Factor

A flow chart for modelling the split factor is given in Figure 3.

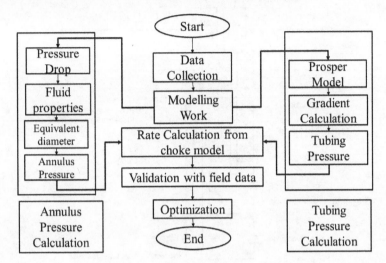

Figure 3. Process flow of split factor modelling.

The proposed model used the casing injection pressure (CHP) and total gas lift rate (GLIR) measured at the surface as the input to estimate the GLIR to both the SS and LS. Based on these individual GLIRs (estimated) and the total GLIR (measured), the split factor can be determined.

The tubing experiences pressure from the annulus (from the gas lift injection) and from the well fluid flowing from the reservoir through it. To estimate the gas lift rate, three pressure drops need to be considered (see Figure 4):

- Pressure drop in the annulus
- Pressure drop across the orifice
- Pressure drop in the tubing

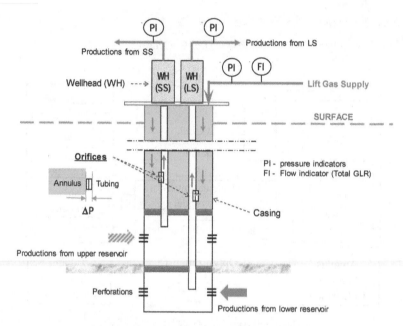

p/s: Only the gas lift valve (at the deepest depth) is shown, as an orifice, on each of the string. The other gas lift valves (at the shallower depths) for unloading purposes during start-up are not included.

Figure 4. Relationship between casing, tubing and orifice pressure drop.

The lift gas is introduced to the casing/annulus, and the pressure (CHP) is measured at the wellhead. The lift gas flowing in the annulus imposes a pressure upstream of the orifice, while the production fluid flowing inside the tubing will exert a pressure downstream of the orifice. The pressure differential across the orifice determines the amount of lift gas flowing into the tubing.

The estimation of the annulus pressure will incorporate the concept of the equivalent hydraulic diameter, which is commonly applied in a heat exchanger design to account for a non-regular flow path. PROSPER [14], a multiphase flow simulator, will be used to estimate the pressure along the tubing from the reservoir to the wellhead as the multiphase production fluids are expected to flow inside the production string from the reservoir to the wellhead. The orifice flow correlation will be used to compute the GLIR as a function of the annulus and tubing pressure difference.

2.1.1. Annulus Pressure Calculations

The energy balance equation expressed in the pressure gradient form, dP/dL [15]:

$$\frac{dP}{dL} = (-\frac{g}{gc}\rho sin\varnothing - \frac{24fV^2]\rho}{gcD} - \frac{\rho V}{\alpha gc}\frac{dV}{dL})$$

(1)

There are three pressure loss components, i.e., elevation, frictional and acceleration. ρ is the density, v is the velocity, gc is the gravity constant, f is the friction factor and D is the flow diameter. Equation (1) is applicable for the single-phase gas flow. The considered cross-sectional flow area will be adjusted accordingly with the equivalent hydraulic diameter, to account for two concentric cylinders inside the casing (see Figure 5). The Hydraulic diameter is defined as the cross-sectional area of the channel divided by the wetted perimeter. It uses the perimeter and area to provide the diameter such that conservation of momentum is maintained. By using this term, one can handle any dimension of a flow path as one would for a round tube. The uniform flow path is expected within the annulus to the point of injection, no significant change in velocity is expected and hence the acceleration component may be ignored.

Figure 5. Cross-section of the dual strings well.

The fluid velocity V, in Equation (1), can be derived from the flowrate Q:

$$Q = \frac{\pi}{4}D^2V \tag{2}$$

The equivalent hydraulic diameter, De, can be calculated using the generic equation:

$$De = \frac{4 * Cross\ Sectional\ Area}{Wetted\ Perimeters} \tag{3}$$

From Figure 4, it can be shown that:

$$De = \frac{4 * \pi(r_c^2 - 2r_c^2)}{\pi(2r_c + 4r_t)} \tag{4}$$

where r_c and r_t are the casing radius and tubing radius, respectively.

The Colebrook Equation, which is applicable for turbulent flow, is used to calculate the friction factor, f. The software Visual Basics for Application (VBA) is employed to handle the complicated iterations needed to solve for the friction factor, and ε is the roughness:

$$\frac{1}{\sqrt{f}} = 2\log_{10}(\frac{\varepsilon/D}{3.7} + \frac{2.51}{Re * \sqrt{f}}) \tag{5}$$

Re is the Reynold Numbers:

$$Re = \frac{\rho * V * d}{v} \tag{6}$$

The gas properties, critical properties, Z factor and viscosity are estimated using the well-known empirical correlations. The critical temperature and pressure are required to complete the calculation for the Z factor. The Z factor is required for the calculation of the density. The viscosity is required in the calculation of the Reynolds Numbers, which is part of the solution for the friction factor calculation. The selection of the correlations used is based on accuracy and practicality.

The critical pressure Pc, and critical temperature Tc are estimated using the Sutton correlation [16] due to its simplicity for the single gas phase, requiring only the viscosity v as the input. Deckle et al. [17] recommended the use of the correlation for gas gravity under 0.75, which is compatible with the lift gas used for this study.

$$T_{c,} = 164.3 + 357.7 * v - 67.7 * v^2 \tag{7}$$

$$P_{c,} = 744 - 125.4 * v + 5.9 * v^2 \tag{8}$$

The Z factor is estimated using the Hall Yarborough correlation. This correlation has the highest accuracy according to the work done by Lateef in comparison to the other correlation [18]. After the Lateef modification is not significant, an original equation is used as variance in the accuracy.

$$t = \frac{1}{T_{pr}} \tag{9}$$

$$A = 0.06125te^{-1.2*(10t)^2} \tag{10}$$

$$B = 14.76t - 9.76t^2 + 4.58t^2 \tag{11}$$

$$C = 90.7t - 242.2t^2 + 4.58t^3 \tag{12}$$

$$D = 2.18 + 2.82t \tag{13}$$

$$- AP_{pr} + \frac{y + y^2 + y^3 - y^4}{(1 - y)^3} - By^2 + Cy^D = 0 \tag{14}$$

$$T_{pr\ (pseudo-reduced)} = \frac{T}{T_c} \tag{15}$$

$$P_{pr(\ pseudo-reduced)} = \frac{P}{P_c} \tag{16}$$

$$z = \frac{AP_{Pr}}{y} \tag{17}$$

The Lee, Gonzalez and Eakin [19] correlation is chosen for the viscosity estimate due to its simplicity. The accuracy does not vary much from the other correlation reported by Al-Nasser et al. [20]. The original correlation was used in comparison with one optimized by Al-Nasser et al., as the difference in accuracy is not significant.

$$v_g = 10^{-4} K e^{[x(T)\rho_g^{Y(T)}]} \tag{18}$$

$$v_{gsc} = K = \frac{(9.4 + 0.02M)T^{1.5}}{209 + 19M + T} \tag{19}$$

$$X(T) = 3.5 + \frac{986}{T} + 0.01 \tag{20}$$

$$Y(T) = 2.4 - 0.2X \tag{21}$$

The density is estimated as follows:

$$\rho = \frac{P * M_{gas}}{Z * R * T} \tag{22}$$

where the molecular weight of the gas, M_{gas}, is:

$$M_{gas} = M_{air} * SG_{gas} \tag{23}$$

The annulus pressure along the casing length is estimated by dividing it into n, the number of node sections from the casing head to the injection point. A loop was created with VBA to iterate the pressure at the inflow, until it equals the outflow from the previous section (see Figure 6). The process continues until the pressure reaches the specified point of interest, which is the orifice valve location, the injection point. This is done to improve the resolution of the calculation and accuracy of the computation.

$$P_{out,1} = P_{in,1} + \frac{dP}{dz} * n \tag{24}$$

$$P_{out,\,1} = P_{in,2} \tag{25}$$

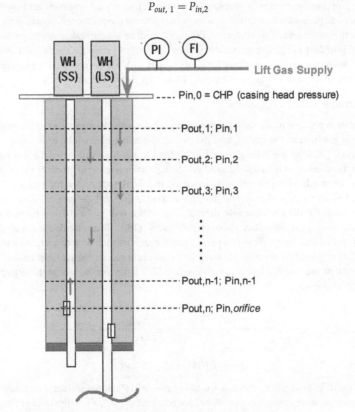

Figure 6. Division of pipe (annulus) section into smaller nodes in VBA.

The composite flow path with a different inclination is created to account for the non-vertical flow direction. This is to improve the accuracy of the model. A total of 5 sections with different angles is created to mimic the flow path of the upper well section. The number of sections can be added easily for a better resolution with VBA. However, this is usually not required, as the angle would typically be low at the gas lift valve depth. The gas lift valves need to be placed at a shallow wireline accessible depth for the change-out during the well intervention, and normally this is below 60 degrees.

The temperature is assumed to follow the geothermal gradient, as a continuous gas lift under a steady state condition is expected to have achieved a thermal equilibrium. The interpolation is done to estimate the pressure at the depth, to simplify the calculation. The temperature input and output at the 5 segments is specified. In VBA, it is further broken down into n, the number of nodes, as below, similarly to the pressure for the calculation between the segments:

$$T_{out} = T_{in} + \frac{\Delta T}{n} * L \tag{26}$$

$$\Delta T = T_{reservoir} - T_{surface} \tag{27}$$

2.1.2. Tubing Pressure Calculations

The pressure drop required to lift the reservoir fluid to the surface at a certain rate is controlled by the wellhead choke. The pressure along the tubing is a function of the mechanical configuration, fluid properties and production rate. Several empirical and analytical correlations have been developed to estimate the pressure drop in multiphase flow depending on the reservoir and well conditions. The selection of an optimum correlation is essential to estimate the pressure drop along the tubing [21]. In this study, the multiphase flow simulator, PROSPER, will be used to calculate the pressure gradient along the inside of the tubing. The software is widely recognized for well modelling within the industry. Among the advantages are that it allows for a calibration with previous test data and that it provides a wide selection of Vertical Lift Performance (VLP) and PVT correlations.

2.1.3. Gas Lift Rate (GLIR) Calculations

The common practice in the industry is to estimate the pressure upstream of the orifice from the known lift gas gradient in the casing, and the pressure downstream of the orifice from the flowing gradient survey (FGS) in the tubing, to calculate the lift gas flow rate from the throughput chart. However, in the absence of actual fluid gradient data, usually a general gradient value (for example 0.2 psi/ft.) is assumed. This practice is more suited to a field application for a quick check. It will calculate the LS and SS individually, and does not reconcile the total gas rate or injection pressure.

In this study, the lift gas flow rate through the orifice will be calculated using the Thornhill Craver equations [22] for the orifice valve (see Equation (28)). This equation is for a compressible, one-dimensional and isentropic flow of a perfect gas through restriction; a correction factor (discharge coefficient, Cd) is added to account for deviations encountered in the real gas case [23]. The gas rate is corrected to the real condition using Equation (29), to reflect the rate at the actual down hole temperature and pressure:

$$Q_{sc} = \frac{155.5 * Cd(A * P1 \sqrt{2(g) * (\frac{k}{k-1})[(F_{du})^{\frac{2}{k}} - (F_{du})^{\frac{k+1}{k}}]}}{\sqrt{SG * (T1)}} \tag{28}$$

$$Q_{inj} = 0.0544 * (SG * T)^{0.5} * Q_{sc} \tag{29}$$

where Cd is the discharge coefficient, A is the flow area (in^2), P1 is the upstream pressure (psia), P2 is the downstream pressure (psia), k is the ratio of specific heat (=1.27), F_{du} is the ratio of P1/P2, SG is the gas specific gravity, T is the temperature (°R), Q_{sc} is the gas rate at the standard condition (Mscf/D) and Q_{inj} is the gas rate at the actual condition (Mcf/D).

An orifice requires a more than 40% pressure drop across it to achieve a critical flow. Although a slight variation in the tubing flow regime will result in an unsteady injection rate, an orifice can accommodate a wide operation range, and it remains the most commonly used operating gas lift valve in the industry. The upstream and downstream pressure are put into Equations (28) and (29) to calculate the injection gas rate. The discharge coefficient, Cd, is taken from experimental work done by Nieberding et al. [24] (see Figure 7).

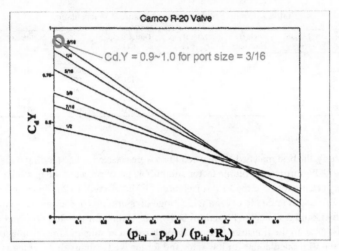

Figure 7. Cd estimation for the Camco R-20 valve [24].

Cd is the discharged coefficient. Y, the expansion factor, equals 1 for an ideal gas at a high temperature and pressure, representative of the down-hole condition. From Figure 7, CdY is between 0.9 to 1, for the Camco R-20 with a port size equal to 3/16.

The determination of the gas lift injection assumes a steady state, continuous injection and single point injection at the SS and LS, respectively. This is representative of the gas lifted oil production whereby only the deepest gas lift valve is open during the normal operation after unloading. The valves that are installed at a shallower depth for unloading purposes are excluded, as unloading is a transient operation during the well start-up, which is to say an infrequent event.

2.2. Model Validation

The actual field data from a brown field is used to validate the proposed model. The field has a long history of gas lift application. The gas lift allocation for the dual string completion was a big problem for the field.

The field data were measured using the Well Tracer method. The method measured the concentration and travel time to allocate the amount and location of gas injected [25]. Although costly and strenuous in operation, this method provides the most accurate measured data.

2.3. Case Study

A case study was carried out to demonstrate the application of the proposed hybrid model in determining the split factor for a dual completion oil produce well "X".

3. Results and Discussions

3.1. Annulus Pressure Calculation

The pressure calculation is conducted for the annulus via a breakdown into 5 sections, as shown in Table 1. The diameter calculated here is the equivalent hydraulic diameter. The inlet in the first section is the injection pressure, and the outlet pressure in the last section represents the choke upstream pressure. It is observed that the pressure increases from the point of the inlet down to the orifice depth, for the condition assessed, and there is an estimated increase of 42 psi that accounts for both the hydrostatic and frictional pressure differential. The number of sections is deemed sufficient as the well is vertical in the top sections, and the deviation is only observed in the last two sections.

Table 1. Example of the annulus pressure VBA calculation.

Sec	Diameter (in)	Length (ft)	Angle from Horizontal	Inlet Temp. (°F)	Outlet Temp. (°F)	No. Seg.	Inlet Press. (psia)	Outlet Press. (psia)
0								
1	6.62	500	90	86	96.5	100	641	649
2	6.62	500	90	96.5	101	100	649	657
3	6.62	500	90	101	108.5	100	657	665
4	6.62	500	87	108.5	116	100	665	672
5	6.62	795	67	116	127.4	100	672	684

3.2. Tubing Pressure Calculation

For this study, the best matched PVT correlations were selected. The Standing correlation is used to estimate the GOR, formation volume factor and bubble point pressure. Begg's correlation is used for the viscosity calculation. For the VLP correlation, PE2 is selected. PE2 is an empirical correlation developed by PETEX, intended to cover a wide range of operating conditions. This correlation uses the flow map by Gould & et al., the Hagedorn Brown correlation for the slug flow and the Duns and Ros for the mist flow. In the transition regime, a combination of slug and mist results is used. It also has improved the VLP calculations for low rates and for the well stability. It provides a more accurate prediction of the minimum load-up rates [14].

PE2 appeared to have the best match with 3 previous well tests, as shown in Figure 8. The calculated rate and flowing bottom hole pressure (FBHP) was well within the ±10% of the measured test data. It also has the lowest average absolute error compared to other vertical lift performance correlations. The inflow Performance Relationship (IPR) Model is generated with the Productivity Index (PI) model.

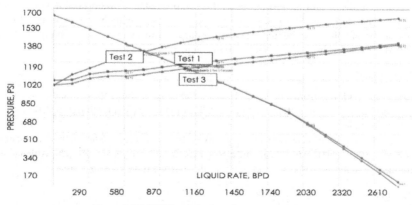

Figure 8. PROSPER matching to the previous well test.

The measured parameters from the well test: the flowing tubing head pressure (FTHP), GOR, and water cut (WC), are the required input for the pressure gradient calculation. The generated pressure profile from the wellhead to the perforation depth can be seen in Figure 6. The pressure at the orifice is singled out from the depth-pressure profile to represent the downstream orifice pressure (highlighted in Figure 9).

Point	Label	Bottom Measured Depth	True Vertical Depth	Pressure	Temperature	Gradient
		(feet)	(feet)	(psig)	(deg F)	(psi/ft)
1		44.0	44.0	180.00	111.61	
2		245.0	245.0	207.09	114.29	0.13479
3		446.0	446.0	234.73	116.93	0.13754
4		446.0	446.0	234.91	116.93	0.01
5		473.0	473.0	238.66	117.28	0.139
6		500.0	500.0	242.42	117.63	0.13934
7		750.0	750.0	277.67	120.84	0.141
8		1000.0	1000.0	313.62	123.97	0.14378
9		1250.0	1250.0	350.20	127.00	0.14632
10		1500.0	1500.0	387.37	129.93	0.14869
11		1750.0	1750.0	425.09	132.74	0.15089
12		2000.0	2000.0	463.40	135.42	0.15323
13		2250.0	2246.1	503.09	137.94	0.15876
14		2500.0	2492.3	544.72	140.30	0.16651
15		2647.5	2627.6	568.98	141.60	0.16448
16		2795.0	2762.8	593.84	142.85	0.16853
17		2897.5	2856.8	614.00	143.68	0.19674
18		3000.0	2950.8	634.51	144.48	0.20012
19		3250.0	3161.7	681.88	146.30	0.18947
20		3500.0	3372.6	730.93	147.92	0.1962
21		3750.0	3559.2	775.84	149.35	0.17962
22		4000.0	3745.8	821.99	150.58	0.18461
23		4250.0	3932.2	869.31	151.60	0.18926
24		4500.0	4118.6	917.81	152.36	0.19403
25		4638.5	4222.5	945.36	152.66	0.19896
26		4777.0	4326.5	973.28	152.86	0.20156

Figure 9. Pressure gradient calculation with PROSPER.

3.3. Gas Lift Rate (GLIR) Calculation

An example of the gas lift rate calculation is given in Table 2. The GLIR is obtained via subtracting the SS injection rate from the total gas injected, assuming that all the injections from the surface go into the tubing as per the basis used by both Kamis et al. [2] and Chia et al [8]. This also provides a reconciliation to the surface measured gas rate.

Table 2. Example of the gas lift rate calculation.

Discharge Coefficient, Cd	0.98
Flow area, A (in^2)	0.028
Upstream pressure, $P1$ (psia)	684
Downstream pressure, $P2$ (psia)	528
Gravity constant, g (ft^2/s)	32.17
Ratio of specific heat, k	1.27
Upstream temperature, $T1$, (°R)	582
Ratio P1/P2, F_{du}	0.77
Gas gravity, SG	0.65
Gas rate at standard conditions, Qsc (Mscf/D)	485.5
Correction factor	1.06
Gas injection at depth, Qg (Mscf/D)	514.0

3.4. Model Validaton

The split factor is calculated for the two wells, namely, wells "X" and "Y". Both wells are dual string completion type wells, with $3\frac{1}{2}$ inch production tubing and a low angle deviation angle. A wide range is observed in the measured GOR and WC for the wells. "X" SS and LS and "Y" LS are producing from the deeper reservoir, with a water breakthrough and high GOR due to depletion. "Y" SS produced from the shallower reservoir with a low water cut and is still above its bubble point, which explains the noticeable lower GOR.

A comparison between the calculated values from the model and measured date from the field is given in Table 3.

Table 3. Comparison of model results and measured data.

	Well "X"		Well "Y"	
	SS	LS	SS	LS
Tubing size (in)	3.5	3.5	3.5	3.5
Deviation (degree)		23		18
Total Liquid (Bpd)	1350	2655	540	1510
Water Cut (%)	92	82	15	80
Total GOR	3430	1778	504	1500
Injection Pressure, (psia)	641	641	600	600
Tubing Pressure, (psia)	134	140	120	108
Injection Depth (ft.)	2795	2865	2350	3560
	Gas Lift Rate, (Mscf/D)			
Field measured data	525	275	495	405
Model results	515	285	530	370
	Split Factor			
Field measured data	0.66	0.34	0.55	0.45
Model results	0.64	0.36	0.59	0.41
% Difference (average relative error)		2%		7%

The model demonstrated that it can estimate the lift gas distributions for both well "X" and well "Y", over a considerable range of operating conditions, to an average relative error of between 2% and 7 %, as shown in Table 3. The higher discrepancy in well "Y" is possibly due to poor matching between the well test data and the calculated PROSPER results. This could be due to inaccuracies in the measurement data collected during the well test, discrepancies in the PROSPER modelling of the actual well conditions, or both. The collected well test data were within the acceptable accuracies. For well "Y", during the tracer survey, it was observed that multi-pointing had occurred at the LS. The PROSPER model has assumed a single injection point for LS (well "Y"). It could not mimic the multi-pointing actual conditions, hence the poor match between the measured well test data and the simulated results.

3.5. Case Studies

The proposed hybrid model is used to generate the lift gas distribution and compared to the available field well test data for well "X", see Table 4. The field well test data were based on PETEX methods, assuming the dual strings as a single well model, as described previously in Section 1.

Table 4. Comparison of model results and field well test data.

Well "X"	Gas Lift Rate (Mscf/D) Well Test Data 1		Gas Lift Rate (Mscf/D) Well Test Data 2	
	SS	LS	SS	LS
Field well test data	450	550	200	1000
Model results	530	470	485	715

There is a tendency to under-allocate the lift gas for SS, overlooking gas robbing phenomena. The well test data 1 was off by 80 Mscf/D, whilst the well test data 2 was off by almost 185 Mscf/D. The calculated split ratios from the proposed hybrid model in both tests are closer to the Well Tracer measured data, indicating a higher consistency and better accuracy. This is in line with the understanding that the single well models, such as those proposed by PETEX, are limited for application due to the interaction of the gas distribution between the 2 strings.

Having an accurate GLIR estimation is fundamental to the optimization of the gas lift to improve the production and hence the profitability of the well. The potential impact of an accurate GLIR estimate on the productions from well "X" can be demonstrated by Figure 10, and Tables 5 and 6.

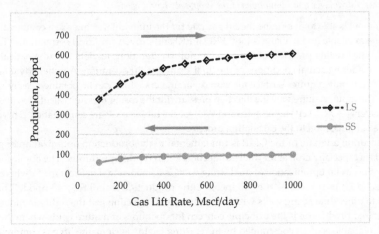

Figure 10. Gas lift performance curve for LS and SS.

Table 5. GLIR sensitivity against production.

GLIR (Mscf/D)	SS (Bopd)	LS (Bopd)
300	82	501
400	85	533
500	87	555

Table 6. GLIR and orifice sizes.

Well "X"	GLIR (Mscf/D)	
Orifice Size (/64 in.) for SS	SS	LS
8	228	572
10	356	444
12	490	310

Figure 10 showed that the production for LS is more sensitive to gas lift rate changes. The red arrow indicates the directional changes on the GLIR redistribution for SS and LS to achieve a production optimization. Reducing the GLIR for SS by 200 Mscf/D merely cut back the production by 5 Bopd, whilst increasing the same amount will increase the production by almost 50 Bopd for LS. This is because LS is producing from a better productivity index (PI) reservoir and the water cut is also higher at 80%, resulting in the requirement for more gas to lift the total liquid. On the other hand, SS, due to a low water cut and relatively lower liquid rate, is nearing it optimum operating condition. There is an opportunity to reduce the gas lift rate for the SS and redistribute to LS to improve the production as a whole.

The variation of the production with the lift gas rate is tabulated in Table 5.

The GLIR can be varied by changing the orifice sizes. The proposed hybrid model is used to generate the split ratio between the LS and SS for the selected 3 different orifice sizes, as shown in Table 6.

The existing orifice for SS is 12/64th. By reducing the orifice size for the SS to 10/64 in. and 8/64 in., the GLIR is cut back by 128 and 262 Mscf/D respectively. This saving in the GLIR from the SS may be re-allocated to LS or other wells in the field to enhance production. An estimated 48 Bopd gain can be achieved by reducing the orifice size for the SS from 12/64 in. to 8/64 in. based on the gas lift performance curve of LS and SS.

3.6. Features, Advantages and Limitations of the Proposed Hybrid Model

The model is able to determine the lift gas rate for the individual string, via a combination usage of VBA, the available nodal analysis tool (PROSPER) and proven empirical correlations. This sets it apart from the existing practices of conducting multiple iterations to yield a mathematical solution without considering actual conditions of the dual string gas lift operation. Another advantage of the model is that it only requires surface measured parameters which can be obtained easily with high accuracy. The inputs required are the injection pressure for the casing pressure calculation and the well test parameters (GOR, FTHP, liquid rates and WC) for the tubing pressure computation. This approach provides a simplification to the calculation process.

An accurate estimate of the GLIR is fundamental to the production optimization of gas-lifted wells. This is especially crucial for dual string completion type oil producers. The application of the model for the gas lift optimization of dual string wells is demonstrated in Section 3.5., where a saving of 262 Mscf/d GLIR from SS is used to increase the production from LS by 48 Bpod. This can be extended to other dual string wells within the same field. Injecting just the right amount of gas lift for the desired production is the principle concern for operators of mature fields where the amount of gas lift available is often constrained by the existing facility system, the lift gas compressors [26]. The opportunities to enhance the production can also be realized through the optimizing of parameters such as the casing head pressure, injection depth and orifice size.

The model is developed for steady state, continuous injection conditions, which are commonly fulfilled in oil production following the unloading operations, when and where only the deepest gas lift valve is operating. It does not cater to transient operations such as those during well unloading or an unstable injection resulting from the flow instability of the production conditions. Nor does the model apply to multi-pointing, conditions whereby the lift gas enters the tubing through more than one point along the tubing. This could be due to poor design or unintentional changes in the operating condition. Under such circumstances, a transient analysis tool or field measurement may be more suitable for diagnostic purposes.

The proposed model offers a better method to quantify the gas lift split factor, thereby enabling the efficient use of gas lift, paving a way for the improvement on gas lift optimization in dual string completion and for opportunities to understand the overall produced gas usage. This would inevitably improve the energy efficiency of upstream oil production facilities.

The model can be adapted easily into the current industry gas lift optimization work flow. Lastly, the proposed hybrid model provides a potential substitute to the Well Tracer measurement method, averting the costs and risks of well intervention.

4. Conclusions

By considering the pressure drops of production fluids in tubing and the pressure gradients of the lift gas along the annulus, this proposed hybrid model enables a more accurate estimate of the gas lift split factor for dual string completion wells. In this study, the model was able to predict the split factors for well "X" and well "Y" within an accuracy of 2% and 7% from the actual measured data. This hybrid model provides far better results than the current methods, which are based on an approximation of dual strings as two separate single strings, disregarding any interactions between the strings within the same casing. An accurate knowledge of the amount of gas injected into each string leads to a more efficient use of lift gas, improving the energy efficiency in oil productions facilities and contributing toward the sustainability of fossil fuel.

Author Contributions: The idea of this research was conceptualized by all authors. K.H.C., M.Z.Z. and J.H.L. provided the guidance and supervisions; C.C.L. implemented the research, performed the analysis and wrote the paper. All authors contributed significantly to this work.

Funding: This research was funded by Yayasan Universiti Teknologi PETRONAS, YUTP-FRG research grant 015LC0-009.

Conflicts of Interest: The authors declare no conflict of interest.

Abbreviations

The following abbreviations are used in this manuscript.

CHP	Casing Head Pressure
FGOR	Free Gas Oil Ratio
FTHP	Flowing Tubing Head Pressure
GLIR	Gas Lift Rate
GOR	Gas Oil Ratio
LS	Long String
PFD	Process Flow Diagram
PI	Productivity Index
PVT	Pressure, Temperature and Volume
SS	Short String
SGOR	Solution Gas Oil Ratio
VBA	Visual Basic Application
WC	Water Cut
$\varnothing$	Angle from the horizon, degree
Z	Compressibility factor
T_c	Critical Temperature, °R
P_c	Critical Pressure, psia
P	Density, lb/ft^3
D	Diameter, inches
C_d	Discharge Coefficient
P2	Downstream pressure, psia
De	Equivalent hydraulic diameter, inches
A	Flow area, in^2
f	Friction factor
gc	Gravity constant
Q_{sc}	Gas rate at standard condition, Mscf/D
Q_{inj}	Gas Rate at actual condition, Mscf/D

L	Length, ft
M	Molecular weight, lbmol
N	Number of nodes
P	Pressure, psia
T_{pr}	Pseudo reduced Temperature
P_{pr}	Pseudo reduced Pressure
r_c	Radius of tubing, in
r_t	Radius of casing, in
K	Ratio of specific heat= 1.27
F_{du}	Ratio of P1/P2
Re	Reynolds numbers
ε	Roughness, inches
SG	Specific gravity
T	Temperature, °F
P1	Upstream pressure, psia
V	Velocity, ft/s
V	Viscosity, cP

References

1. Bellarby, J. Well Completion Design. In *Development in Petroleum Science*, 1st ed.; Elsevier: Oxford, UK, 2009; Volume 56, pp. 303–370, 662–663. ISBN 0-444-81743-3.
2. Kamis, A.; Zulkifli, S.; AbdulRani, M.; Alvarez, J.; Tello, C.; Chuliwanlee, C.; Iskandar, O.; Barbarino, S.; Konakom, K.; Khan, M. Real Time Production Surveillance Overcomes challenges in Offshore Dual String Gas Lift Wells in Baram Field, Malaysia. In Proceedings of the Offshore Technology Conference, Kuala Lumpur, Malaysia, 25–28 March 2014. [CrossRef]
3. Widianoko, G.; Subekti, H.; Jatmiko, W. Casing head Pressure Fine Tuning in Dual String Gas Lift Well Injection Pressure Operated Valve. In Proceedings of the ASME Workshop, Kuala Lumpur, Malaysia, 5–8 May 2013.
4. Nishikiori, N.; Redner, R.; Doty, D.; Schmidt, Z. An Improve Method for Gas Lift Allocation Optimization. In Proceedings of the SPE Annual Technical Conference and Exhibition, San Antonio, TX, USA, 8–11 October 1989. [CrossRef]
5. Eikrem, G.; Aamo, O.; Foss, B. Stabilization of Gas Distribution Instability in Single Point Dual Gas Lifted Wells. *SPE J. Prod. Oper.* **2012**, *21*, 252–259. [CrossRef]
6. Conejeros, R.; Lenoach, B. Model-based Optimal Control of Dual Completion Wells. *J. Pet. Sci. Eng.* **2003**, *42*, 1–14. [CrossRef]
7. Petroleum Expert Ltd. DOF Dual String Gas Lift. DOF User Group Meeting, June 2015. Available online: http://www.petex.com/media/2278/dof-brochure.pdf (accessed on 1 March 2018).
8. Chia, Y.C.; Hussain, S. Gas Lift Optimization Efforts and Challenges. In Proceedings of the SPE Asia Pacific Improved Oil Recovery Conference, Kuala Lumpur, Malaysia, 25–28 October 1999. [CrossRef]
9. Abbasov, A.; Suleymanov, A.; Abbasov, E. Gas Lifted Wells Optimization Based on Tubing Head Pressure Fluctuation. In Proceedings of the SPE Annual Caspian Technical Conference and Exhibition, Baku, Azerbaijan, 13 November 2017. [CrossRef]
10. Xiao, F.; Dani, N.; Long, T.; Gantt, J.; Borden, Z.; El-Bakry, A. A Robust Surveillance and Optimization Workflow for Offshore Gas Lifted Wells. In Proceedings of the Abu Dhabi International Petroleum Exhibition & Conference, Abu Dhabi, UAE, 13–16 November 2017. [CrossRef]
11. Abu Bakar, A.; Ali Jabris, M.; Abd Rahman, H.; Abdullaev, B.; Idris, K.; Kamis, A.; Yusop, Z.; Kok, J.; Kamaludin, M.; Zakaria, M.; et al. CO_2 Tracer Application to Supplement Gas Lift Optimization Effort in Offshore Field Sarawak. In Proceedings of the SPE Asia Pacific Oil & Gas Conference and Exhibition, Brisbane, Australia, 23–23 October 2018. [CrossRef]
12. Hemink, G.; Horst, J. On the Use of Distributed Temperature Sensing and Distributed Acoustic Sensing for the Application of Gas Lift Surveillance. *SPE J. Prod. Oper.* **2018**. [CrossRef]
13. Saepudin, D.; Soewono, E.; Sidarto, K.; Gunawan, A.; Siregar, S.; Sukarno, P. An Investigation on Gas Lift Performance Curve in an Oil Producing Well. *Int. J. Math. Math. Sci.* **2007**, *1*. [CrossRef]

14. Petroleum Expert Ltd. *Prosper Single Well Systems Analysis in User Guide, Version 8*; Petroleum Expert Ltd.: Edinburgh, UK, 2003; Chapter 9; pp. 3–28.

15. White, F. Compressible Flow. In *Fluid Mechanics*, 2nd ed.; McGraw-Hill Book Company: New York, NY, USA, 1986; pp. 571–657.

16. Sutton, R. Fundamental PVT Calculation for Associated and Gas Condensate Natural Gas System. *J. SPE Reserv. Eval. Eng.* **2005**, *10*, 270–283. [CrossRef]

17. Decker, K.; Sutton, R. Gas Lift Annulus Pressure. In Proceedings of the SPE Artificial Lift Conferences and Exhibition-America, Woodlands, TX, USA, 28–30 August 2018. [CrossRef]

18. Kareem, L.A. Z Factor Implicit Correlation, Convergence Problem, and Pseudo-Reduced Compressibility. In Proceedings of the SPE Nigeria Annual International Conference and Exhibition, Lagos, Nigeria, 5–7 August 2014. [CrossRef]

19. Lee, A.; Gonzalez, M.; Eakin, B. The Viscosity of Natural gases. *J. Pet. Technol.* **1965**, *18*, 997–1000. [CrossRef]

20. Al-Nasser, K.; Al-Marhoun, M. Development of New Viscosity Correlation. In Proceedings of the SPE International Production and Operations Conference & Exhibition, Doha, Qatar, 14–16 May 2012. [CrossRef]

21. Azin, R.; Sedaghati, H.; Fatehi, R.; Osfouri, S.; Sakhael, Z. Production Assessment of Low Production Rate of Well in Supergiant Gas Condensate Reservoir: Application of an integrated Strategy. *J. Pet. Explor. Prod. Technol.* **2017**, 1–18. [CrossRef]

22. Cook, H.; Dotterweich, F. *Report on Calibration of Positive Flow Beans Manufactured*; Thornhill-Craver Company Inc.: Houston, TX, USA, 1946.

23. Almeda, A. Practical Equations Calculate Gas Flow Rates through Venturi Valves. *Oil Gas J.* **2010**. Available online: www.ogj.com/articles/print/volume-108/issue-5/technology/practical-equations.html (accessed on 3 May 2017).

24. Nieberding, N.; Schimdt, Z.; Blais, R.; Doty, D. Normalization of Nitrogen-Loaded Gas Lift Valve Performance Data. In Proceedings of the SPE Annual Technical Conference and Exhibition, New Orleans, LA, USA, 23–26 September 1993. [CrossRef]

25. Smith, S. *Gas Lift WinGLUE Training Course*; AppSmiths Technology: Houston, TX, USA, 2013.

26. Guyaguler, B.; Byer, T.J. A New Rate-Allocation-Optimization Framework. *SPE J. Prod. Oper.* **2008**, *23*, 448–457. [CrossRef]

Article

Laboratory Study on Changes in the Pore Structures and Gas Desorption Properties of Intact and Tectonic Coals after Supercritical CO_2 Treatment: Implications for Coalbed Methane Recovery

Erlei Su [1], Yunpei Liang [1,*], Lei Li [1,2,3], Quanle Zou [1,*] and Fanfan Niu [4]

[1] State Key Laboratory of Coal Mine Disaster Dynamics and Control, Chongqing University, Chongqing 400044, China; suerlei1992@163.com (E.S.); leili80mky@163.com (L.L.)
[2] State Key Laboratory of Gas Disaster Monitoring and Emergency Technology, Chongqing 400037, China
[3] China Coal Technology and Engineering Group Chongqing Research Institute, Chongqing 400037, China
[4] Zhengzhou Engineering Co., Ltd. of China Railway Seventh Group, Zhengzhou 450052, China; niufanfan1991@163.com
* Correspondence: liangyunpei@cqu.edu.cn (Y.L.); quanlezou2016@cqu.edu.cn (Q.Z.)

Received: 31 October 2018; Accepted: 3 December 2018; Published: 6 December 2018

Abstract: Tectonic coals in coal seams may affect the process of enhanced coalbed methane recovery with CO_2 sequestration (CO_2-ECBM). The main objective of this study was to investigate the differences between supercritical CO_2 ($ScCO_2$) and intact and tectonic coals to determine how the $ScCO_2$ changes the coal's properties. More specifically, the changes in the tectonic coal's pore structures and its gas desorption behavior were of particular interest. In this work, mercury intrusion porosimetry, N_2 (77 K) adsorption, and methane desorption experiments were used to identify the difference in pore structures and gas desorption properties between and intact and tectonic coals after $ScCO_2$ treatment. The experimental results indicate that the total pore volume, specific surface area, and pore connectivity of tectonic coal increased more than intact coal after $ScCO_2$ treatment, indicating that $ScCO_2$ had the greatest influence on the pore structure of the tectonic coal. Additionally, $ScCO_2$ treatment enhanced the diffusivity of tectonic coal more than that of intact coal. This verified the pore structure experimental results. A simplified illustration of the methane migration before and after $ScCO_2$ treatment was proposed to analyze the influence of $ScCO_2$ on the tectonic coal reservoir's CBM. Hence, the results of this study may provide new insights into CO_2-ECBM in tectonic coal reservoirs.

Keywords: supercritical CO_2; tectonic coal; pore structure; methane desorption

1. Introduction

As a major greenhouse gas, CO_2 causes global warming and initiated a series of negative effects on the balance of the natural ecosystem and the sustainable development of human society [1–7]. To mitigate CO_2 emissions into the atmosphere, enhanced coalbed methane recovery with CO_2 sequestration (CO_2-ECBM) is considered to be a promising technology, and it receives widespread attention [8–12]. This technology can not only provide CO_2 storage, but can also enhance the production of coalbed methane (CBM). To date, many scholars proved the feasibility of injecting CO_2 into methane-bearing coal seams using various experimental methods and field tests. Fulton et al. observed that the recovery of CH_4 after CO_2 injection was 57% higher than that of natural emissions [13]. Reznik et al., Jessen et al., and Dutka et al. also verified that CO_2 injection could contribute to the improvement of CH_4 recovery [14–16]. In 1996, the San Juan Basin was selected as the first demonstration site for CO_2-ECBM. After CO_2 injection, CBM production was increased by five times, and the recovery of CH_4 reached 77–95%, lasting for over six years [17].

The coal seams most suitable for CO_2 sequestration are commonly more than 800 m deep, and the temperature and pressure in coal seams at the most suitable depths are, in many cases, above the critical point for CO_2 (critical temperature and pressure: 31.05 °C and 7.39 MPa) [18]. Numerous studies were undertaken to better understand the interaction between supercritical CO_2 (ScCO_2) and intact coal. When coal is exposed to ScCO_2, the ScCO_2 is incorporated into the coal and it rearranges the coal's structure [19]. It is generally agreed that ScCO_2 can change the pore morphology of coal irreversibly, including pore volume, pore size distribution, surface area, and pore connectivity [20,21]. Additionally, ScCO_2 is an organic solvent and can mobilize some of the organic matter in the coal. If some of the CO_2 is dissolved in the reservoir water, the pH decreases and the CO_2-enriched solution can dissolve some of the minerals present [22–24]. Studies also showed that CO_2 adsorption can cause the coal's matrix to swell [25]. However, the above studies were all focused on the effects of ScCO_2 on intact coal, and there was little research on tectonic coal–ScCO_2 interactions.

Tectonic coal is deformed, sheared coal. The deformation and shearing was caused by tectonism that occurred long after coal formation and diagenesis of the coal seams [26]. Tectonic coal is present all over the world, and there is substantial tectonized coal in China [27,28]. The existence of tectonic coal in some coalfields complicates safe and efficient CO_2 sequestration. This is because the physical and chemical properties of tectonic coal differ from those of intact coal. Additionally, the permeability of tectonic coal is much lower than that of intact coal. This leads to the formation of CBM-rich coals; thus, tectonic coals have huge CBM exploitation potential. However, a CBM extraction well drilled in an area hosting significant tectonic coal may seriously restrict CBM production because the tectonic coal's low permeability restricts gas desorption and migration [29,30]. Therefore, in such an area, CO_2-ECBM is likely to be the technique implemented to solve this problem.

The coal's pore structure affects gas adsorption, diffusion, and seepage. Previous studies [31,32] showed that the pore structure in intact coal and tectonic coal are different and affect how gas desorbs from these coals. These differences are important when the subject of long-term CO_2 storage in coal seams is considered. However, limited research focused on this topic, especially the comprehensive and systematic analysis of the differences of pore structures and gas desorption properties between intact coal and tectonic coal during CO_2-ECBM. Therefore, a better understanding of the pore morphologies and desorption in intact and tectonic coals after ScCO_2 treatment is of significance for CO_2-ECBM in tectonic coal reservoirs.

In this study, the effects of ScCO_2 on the pore structures of intact and tectonic coals was studied using mercury intrusion porosimetry (MIP) and N_2 (77 K) adsorption experiments. Additionally, gas desorption of intact and tectonic coals after ScCO_2 treatment was analyzed using methane desorption experiments. The main focus of this study was the analysis of the pore structures and gas desorption properties of intact and tectonic coals before and after ScCO_2 treatment, and results were compared to analyze the influence of tectonic coal on the CO_2-ECBM process.

2. Materials and Methods

2.1. Coal Samples

Both the intact coal and tectonic coal samples used in the experiments were collected from the Shanxi Formation #3 coal seam in the Sihe coal mine. This mine is in the southern Qinshui basin in Shanxi Province, China (Figure 1). The southern Qinshui basin is not only a major CBM production area, but also the area in which China first carried out a pilot CO_2-ECBM project [33]. The coal samples for this study were collected from fresh working faces in the mine. The samples were then immediately sealed, packed, and sent to the laboratory with minimal delay to prevent oxidation. At the laboratory, standard crushing and screening equipment was used to crush and screen the coal samples to different particle sizes for the different experiments. For each experiment, the crushed and sieved intact coal and the tectonic coal sub-samples were divided into two fractions for comparative analyses: one fraction was treated with ScCO$_2$, and the other was not treated. Selected properties of the intact coal and the tectonic coal are listed in Table 1. Compared with intact coal, the mineral content and Langmuir volume of the tectonic coal is greater, indicating that tectonic coal contains more minerals and can absorb more methane. The moisture, ash, and $R_{o, max}$ values of the tectonic coal are slightly higher than those of the intact coal.

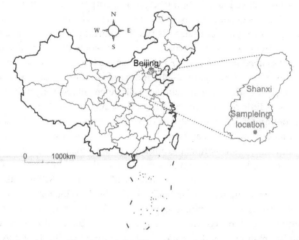

Figure 1. Map showing the location of the Sihe coal mine, the mine from which the samples were collected.

Table 1. Maximum vitrinite reflectances, Langmuir volume and pressure, macerals, and partial proximate analyses for the coal samples used in this study.

Samples	Type	$R_{o, max}$ (%)	Langmuir Volume	Langmuir Pressure	Proximate Analysis (wt.%)			Macerals (vol.%)		
					M_{ad} (%)	A_{ad} (%)	V_{daf} (%)	V (%)	I (%)	M (%)
Intact coal	Anthracite	3.13	45.83	0.81	1.35	14.26	8.46	79.70	17.22	3.08
Tectonic coal		3.26	48.77	0.89	1.58	17.09	8.24	76.42	16.72	6.86

$R_{o, max}$, maximum vitrinite reflectance; M_{ad}, moisture content; A_{ad}, ash yield; V_{daf}, volatile matter. The subscript "ad" stands for air-dried basis. V, vitrinite; I, inertinite; M, mineral.

2.2. Experimental Methods

For this study, MIP, N$_2$ (77 K) adsorption, and methane desorption experiments on both intact and tectonic coal specimens with and without ScCO$_2$ treatment were conducted.

A schematic of the set-up used for the ScCO$_2$ treatment is shown in Figure 2. According to the pressure gradient and temperature gradient in the southern Qinshui basin, the temperature of the reservoir is 35 °C and the pressure is 8 MPa at the depth of 800 m. Therefore, 35 °C and 8 MPa were chosen as the treatment conditions to replicate the in situ conditions. Before the ScCO$_2$ treatment, coal samples were first degassed in a vacuum chamber at 50 °C and 4 Pa for 24 h. Then, the constant temperature bath was set to 35 °C and the sample tank was filled with CO$_2$ using an ISCO pump. Subsequently, the pressure was maintained at 8 MPa for 72 h, and the constant temperature bath was held at 35 °C.

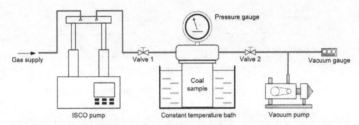

Figure 2. Schematic diagram showing the set-up used for the supercritical CO_2 treatments.

The pore structure of both $ScCO_2$-treated and untreated coal samples was analyzed using a PM33-GT-12 mercury porosimeter. This instrument can measure pore diameters between 0.007 and 1000 µm over a pressure range of 1.5–231,000 kPa. The porosimetry data were modeled using the Washburn Equation [34]:

$$r = \frac{2\sigma\cos\theta}{p_c},$$ (1)

where r is the pore radius of the porous material (nm), σ is the surface tension of mercury (dyn/cm^2), θ is the contact angle between mercury and the porous material's surface (°), and p_c is the capillary pressure (MPa).

The pore size distribution (PSD) was determined using the N_2 (77 K) adsorption method employing a Quadrasorb SI instrument. The N_2 adsorption–desorption isotherms of $ScCO_2$-treated and untreated coal samples were obtained at a temperature of −196 °C with a relative pressure (p/p_0) range of 0.01–0.99. The Barrett-Joyner-Halenda (BJH) method was used to calculate total pore volume (TPV), density functional theory (DFT) was used to calculate PSD, and the Brunauer-Emmett-Teller (BET) method was used to calculate the specific surface area (SSA) [35–37].

Methane desorption data from both treated and untreated intact and tectonic coals were collected during the gas desorption experiments. The experiments were run according to China National Standards AQ/T 1065-2008 and GB 474-2008; the experimental set-up is shown in Figure 3. Coal particles from 1 to 3 mm in diameter were placed in a container and degassed at 50 °C and 4 Pa for 24 h. After degassing, the container, in a 30 °C water bath, was quickly filled with 99.9% pure methane to achieve the methane adsorption equilibrium (2 MPa and 4 MPa). When adsorption equilibrium was reached, valve 4 was opened to vent all the free methane from the container. The container was then connected to the gas desorption measuring cylinder for the gas desorption experiment. During the experiment, the volume of methane desorbed was recorded at specific times; the desorption segment of each experiment lasted two hours.

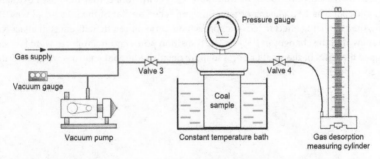

Figure 3. Schematic diagram showing the set-up used for the gas desorption experiments.

The whole experimental process is shown in Figure 4.

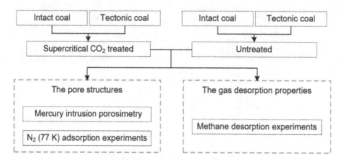

Figure 4. The whole experimental process.

3. Results and Discussion

3.1. Pore Structure Analysis

3.1.1. Pore Size Distribution

To analyze the MIP data from treated and untreated intact and tectonic coal samples quantitatively, the Hodot classification for coal pore sizes was used. This classification scheme divides the pores into macropores (>1000 nm), mesopores (100–1000 nm), transition pores (10–100 nm), and micropores (<10 nm) [38].

Figure 5 shows the PSDs for both $ScCO_2$-treated and untreated coal samples. It is clear from the figure that the macropores and mesopores in the tectonic coal were more developed. This effect was reported by a number of previous studies [39,40]. It is interesting to note that the macropores and mesopores in the tectonic coal were obviously larger after $ScCO_2$ treatment, and the effect of the $ScCO_2$ treatment on the tectonic coal's pores was greater than its effect on the intact coal's pores. According to the data in Table 2, the proportion of macropores and mesopores in the tectonic coal was 33.74% before $ScCO_2$ treatment, but 36.02% after treatment, greater than the pore percentages in the intact coal (15.91% before and 21.37% after). It can be deduced that $ScCO_2$ treatment may observably promote the development of mesopores and macropores of tectonic coal. The data also show that the porosity of the treated intact and tectonic coal samples increased by 18.90% (from 4.18% to 4.97%) and 23.14% (from 5.23% to 6.44%), respectively. Previous studies [22,23,41] showed that $ScCO_2$ can mobilize some of the polycyclic aromatic hydrocarbons and aliphatic hydrocarbons in the coal. Additionally, CO_2 will form carbonic acid when it is dissolved in the water in the coal seams, and this acid can dissolve some of the inorganic minerals in the coal, such as calcites, dolomites, and magnesites [42]. This may result in increased pore sizes. Additionally, Cao et al. found that the mean extraction yield of the tectonic coal was 1.45% using organic solvent, whereas that of the intact coal was 0.44% [43]. Apparently, the mobilizing effect in tectonic coal is more pronounced than its effect in intact coal. These MIP data show that the changes in PSD for the transition pores and micropores were less significant. To investigate the PSDs of the smaller pores in more detail, additional analyses were performed using N_2 (77 K) adsorption.

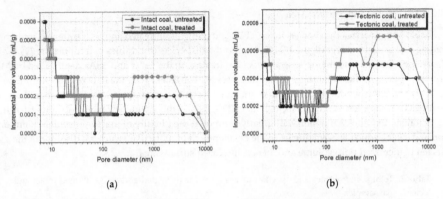

(a)   (b)

Figure 5. Pore size distributions determined by mercury intrusion porosimetry (MIP) for samples of (**a**) intact coal and (**b**) tectonic coal with and without supercritical CO_2 (ScCO_2) treatment.

Table 2. Mercury intrusion porosimetry porosities and pore volume distributions for supercritical CO_2 (ScCO_2)-treated and untreated coal samples.

Sample	Porosity (%)	Pore Volume Distribution (%)			
		V_1/V_t	V_2/V_t	V_3/V_t	V_4/V_t
Intact coal, untreated	4.18	27.73	56.36	12.27	3.64
Intact coal, treated	4.97	20.23	58.40	16.03	5.34
Tectonic coal, untreated	5.23	19.63	46.63	24.85	8.90
Tectonic coal, treated	6.44	19.43	44.55	24.88	11.14

V_1 = pore volume of micropores; V_2 = pore volume of transition pores; V_3 = pore volume of mesopores; V_4 = pore volume of macropores; V_t = total pore volume.

Using the N_2 (77 K) manometric adsorption technique can avoid the destruction of pores under high pressure; thus, it is commonly used for PSD measurements of smaller pores [26]. As shown in Figure 6, after ScCO_2 treatment, the PSDs for small pores were not significantly affected, although a slight tendency for pore sizes to increase could be discerned. This is consistent with the MIP results. The PSDs of both the treated and the untreated intact and tectonic coals had two peaks in the portion of the isotherm, representing pores with diameters smaller than 10 nm, indicating that the micropore structure of these coal samples was relatively developed.

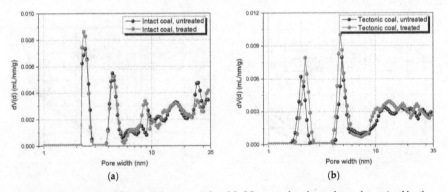

(a)   (b)

Figure 6. Pore size distributions of the untreated and ScCO_2-treated coal samples as determined by the density functional theory (DFT) analyses of N_2 (77 K) adsorption isotherms for (**a**) intact coal and (**b**) tectonic coal.

3.1.2. Pore Volume and Surface Area

Table 3 summarizes the changes in the TPVs and SSAs for untreated and ScCO$_2$-treated intact and tectonic coal samples. Overall, the BJH-TPVs and the BET-SSAs (the values determined by N$_2$ (77 K) adsorption) showed a slight increase after ScCO$_2$ treatment; the TPVs and SSAs determined by MIP showed a more significant increase. These increases are consistent with the PSD results. Additionally, the TPVs and SSAs of both the treated and untreated tectonic coal samples were higher than those for the intact coal samples. A study by Qu et al. [44] explained this point very well, and suggested that the tectonism that the coal underwent greatly promoted the fracture of macromolecular chains and aromatic layers, which was conducive to the development of molecular structural disorder in the coal. This led to increased pore volume and, hence, increased surface area.

Table 3. Total pore volumes and specific surface areas for untreated and ScCO$_2$-treated intact and tectonic coal samples.

Sample	BJH-TPV (mL/g)	BET-SSA (m^2/g)	MIP-TPV (mL/g)	MIP-SSA (m^2/g)
Intact coal, untreated	0.015	3.421	0.0220	0.165
Intact coal, treated	0.016	3.862	0.0262	0.190
Tectonic coal, untreated	0.019	3.964	0.0326	0.214
Tectonic coal, treated	0.021	4.402	0.0422	0.256

BET, Brunauer-Emmett-Teller; BJH, Barrett-Joyner-Halenda; SSA, specific surface area; TPV, total pore volume. The BJH-TPV and BET-SSA values are from N$_2$ (77 K) adsorption determinations.

3.1.3. Pore Connectivity

The pores in coal can be divided into four types according to their shapes: cross-linked, passing, dead end, and closed pores. The first three types are called open pores [45]. Some useful information about the pores can be extracted from the MIP injection/ejection curves and the hysteresis loops. As shown in Figure 7, after ScCO$_2$ treatment, the mercury injection volumes for the intact coal and the tectonic coal increased by 19.09% (from 0.0220 to 0.0262 mL/g) and 29.45% (from 0.0326 to 0.0422 mL/g), respectively. The mercury withdrawal volume for these two coals showed the same trend, increasing by 9.14% (from 0.0197 to 0.0215 mL/g) and 17.42% (from 0.0264 to 0.0310 mL/g), indicating that the open pore volume increased after ScCO$_2$ treatment [46]. Moreover, the hysteresis loops of all treated coals increased, indicating that the pore network was more complex with more bottleneck pores, and there was an increase in the number of open pores [47]. Thus, the connectivity of the treated tectonic coal sample was excellent, which may affect the methane desorption, diffusion, and seepage.

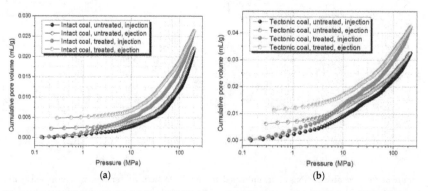

Figure 7. Mercury intrusion porosimetry injection and ejection curves for untreated and treated coal samples: (**a**) intact coal; (**b**) tectonic coal.

3.2. Desorption Analysis

Carbon dioxide sequestration in coal seams enhances CBM production. Simplistically, CO_2 sequestration can be thought of as just gas in/gas out: carbon dioxide seepage, diffusion, and adsorption in, coupled with methane desorption, diffusion, and seepage out. The preceding pore structure analysis showed that $ScCO_2$ can change the pore structure in intact and tectonic coals, and this will undoubtedly affect the desorption, diffusion, and seepage of the methane. Previous studies showed that the diffusion of methane through the coal's matrix has an important influence on CBM production [48,49]. However, research on how $ScCO_2$ affects methane diffusion in tectonic coal is still limited. Therefore, the methane desorption from $ScCO_2$-treated intact and tectonic coals was analyzed.

3.2.1. Gas Desorption Curves

Methane desorption curves for untreated and treated coal samples are shown in Figure 8. The cumulative desorption volume increased as the desorption time increased. The slopes of the desorption curves were steepest in the early stages of the desorption experiments, and the slopes gradually decreased with time. Compared with intact coal, the slopes of the desorption curves of tectonic coal were greater. This is because the mesopores and macropores of tectonic coal were more developed. When the adsorption equilibrium pressure was 4 MPa, the volumes of gas desorbed from the coal samples were higher than the volumes of gas desorbed when the absorption equilibrium pressure was 2 MPa. This is because, during adsorption, the coal matrix could absorb more methane at higher equilibrium pressures [50].

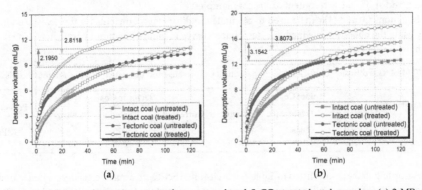

Figure 8. Methane desorption curves for untreated and $ScCO_2$-treated coal samples: (**a**) 2 MPa; (**b**) 4 MPa.

In general, $ScCO_2$ treatment has a considerable effect on the coal sample's gas desorption. As shown in Figure 7, when the equilibrium pressure was 4 MPa and 2 MPa, the final desorption volume of treated tectonic coal increased by 2.8118 mL/g and 3.8073 mL/g, respectively. However, under the same conditions, the final volume of gas desorbed from treated intact coal was slightly greater than volume of gas desorbed from untreated tectonic coal (2.1950 mL/g and 3.1542 mL/g). The final desorption volume for treated tectonic coal was 1.33 to 1.40 times that of the desorption volume for intact coal. These results indicate that more CBM can be extracted from a tectonic coal CBM reservoir after the coal in the reservoir is injected with $ScCO_2$.

To further analyze the changes in gas desorption from $ScCO_2$-treated intact and tectonic coal, their gas diffusion coefficients were compared in the subsequent section.

3.2.2. Gas Desorption Diffusion Coefficients

Based on Fick's diffusion laws, an analytic solution for a diffusion equation was presented by Crank [51] under Dirichlet boundary conditions (Equation (2)). However, this analytical solution is difficult to apply to practical engineering problems because it is in the form of an infinite series. Yang [52] simplified Equation (1) to a more practical form (Equation (3)), and this equation is widely used in engineering and can represent coal's gas desorption very well. The two equations are as follows:

$$\frac{Q_t}{Q_\infty} = 1 - \frac{6}{\pi^2} \sum_{n=1}^{\infty} \frac{1}{n^2} e^{-\frac{Dn^2\pi^2 t}{r_c^2}}, \tag{2}$$

$$\frac{Q_t}{Q_\infty} = \sqrt{1 - e^{-B_1 t}}, \tag{3}$$

where D is a diffusion coefficient, t is the desorption time, r_c is the average particle diameter, Q_t is the cumulative desorption amount at time t, Q_∞ is the ultimate desorption amount, and B_1 is a fitting parameter related to the diffusion coefficient D and the average diameter r_c, which can be calculated from $B_1 = K((4\pi^2 D)/r_c^2)$. K is a correction parameter commonly taken to be equal to 1.

The variable Q_∞ is, in most cases, calculated using Equation (4) [53].

$$Q_\infty = \left(\frac{V_L P_{eq}}{P_L + P_{eq}} - \frac{V_L P_a}{P_L + P_a} \right) (1 - M_{ad} - A_{ad}), \tag{4}$$

where P_{eq} is the definite equilibrium pressure, P_a is the atmospheric pressure, M_{ad} is the moisture content, and A_{ad} is the ash content of the coal samples.

Equations (3) and (4) can be used to calculate the diffusion coefficients for the coals used in this study using the data from the methane desorption curves in Figure 7 as input parameters. The results are shown in Table 4.

Table 4. Gas diffusion coefficients for supercritical CO_2-treated and untreated tectonic and intact coals.

Samples	Intact Coal			Tectonic Coal			Pressure
	Untreated	Treated	Change	Untreated	Treated	Change	
$D \, (\times 10^{-12}$	2.6360	3.8480	45.98%	3.9991	7.5020	87.59%	2 MPa
$m^2/s)$	4.2155	6.0091	42.55%	5.2912	9.8848	86.82%	4 MPa

According to the results listed in Table 4, it is clear that ScCO$_2$ treatment considerably increased the tectonic coal's diffusion coefficient, and the treatment increased the tectonic coal's diffusion coefficient more than it increased the coefficient for intact coal. For example, when the equilibrium pressure was 2 MPa, ScCO$_2$ treatment increased the diffusion coefficient for treated tectonic coal by 87.59%, but the treatment only increased intact coal's diffusion coefficient by 45.98%. It can also be seen in Table 4 that, under the same conditions, tectonic coal's diffusion coefficients were higher than those for intact coal. This is because desorption largely depends on the properties of the pores and the pore structure in the coal. As indicated by the MIP test results, the proportion of macropores and mesopores increased after ScCO$_2$ treatment, especially for tectonic coal, and larger pores can provide better channels for gas migration. In short, tectonic coal has stronger diffusivity capacity.

3.2.3. Implication for CO$_2$-ECBM in Tectonic Coal Reservoirs

Liu et al. surveyed the distribution of tectonic coal reservoirs in China and indicated that the CBM resource in these reservoirs was as much as 5.60 trillion cubic meters. This amounts to 39.20% of China's CBM resources [54]. Although the tectonic coal reservoirs have great potential for CBM development, the permeability of this type of gas reservoir is very low. This means that developing the resources in these CMB reservoirs would be very difficult.

A simplified illustration of the migration of methane in a coal seam is show in Figure 9. In Figure 9, methane migration is divided into three stages: desorption, diffusion, and seepage. Red circles represent adsorbed methane in the coal matrix, green circles represent free methane, yellow squares represent the coal skeleton, and purple squares represent pores in the matrix. In tectonic coal reservoirs, gas migration to the fractures is slow because of the tectonic coal's low permeability. Therefore, only a small amount of methane can flow to CBM wells and the production of CBM is low (Figure 9, untreated). The results of the methane desorption experiments described in previous sections show that both the desorption capacities and diffusion coefficients of treated coals were higher than in untreated coals. Therefore, more adsorbed methane was desorbed from the coal's matrix and diffused to the fractures after CO_2 was injected into the tectonic coal reservoir (Figure 9, treated). The pressure gradient must increase when more free methane is present in the fractures, as shown in Figure 9, treated (more green circles exist in the fracture). According to Darcy's law, an increasing pressure gradient must lead to an increase in gas flow per unit time; thus, more methane will flow to CBM wells. Additionally, the pore structure analysis described in Section 3.1 indicated that the number of the macropores and mesopores of the coal increased after $ScCO_2$ treatment, which provides more space for gas migration and improves the absolute permeability of coal seams to some extent. Furthermore, Liu et al. (2018) demonstrated that CBM production was mainly controlled by diffusion after six days of extraction when the diffusion coefficient was 1×10^{-12} m^2/s [49]. Therefore, a higher diffusion coefficient has a positive effect on CBM production. In short, CO_2-ECBM could overcome the negative effects of low permeability in tectonic coal reservoirs to some extent, thereby promoting CBM development.

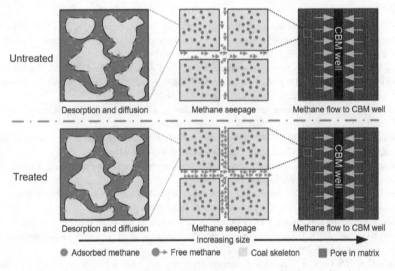

Figure 9. Sketch showing methane migration in untreated and treated coal.

4. Conclusions

This study investigated changes in the pore structures of intact coal and tectonic coal after $ScCO_2$ treatment using mercury intrusion porosimetry and N_2 (77 K) adsorption, and determined the effects of these changes on the coal's gas desorption and diffusion properties. The changes in desorption and diffusion were used to study the implications for enhanced coalbed methane recovery with CO_2 sequestration in tectonic coal reservoirs. The experimental results suggest the following conclusions:

(1) Compared with intact coal, the macropores and mesopores in tectonic coal were obviously larger after $ScCO_2$ treatment. Additional, the TPV, SSA, and pore connectivity of treated tectonic coal were significantly improved. Pore structure analysis showed that tectonic coal was significantly

Energies 2018, 11, 3419

affected by ScCO$_2$ treatment. This was because tectonic coal contained more minerals and the mobilizing effect in tectonic coal was more pronounced.

(2) The results of the methane desorption experiment showed that the desorption capacity of intact coal and tectonic coal was improved to a certain extent by ScCO$_2$ treatment; however, the diffusion coefficient of the treated tectonic coal increased twice as much as that of intact coal. This change was consistent with the pore structure experimental results. The enhancement of the tectonic coal's diffusion capacity after ScCO$_2$ treatment can partially overcome the limitation imposed on tectonic coal reservoir CBM development by the coal's inherent low permeability. The results of this study may provide new insights into CO$_2$-ECBM in tectonic coal reservoirs.

Author Contributions: E.S. and Y.L. conceived and designed the experiments. E.S. and L.L. performed the experiments. E.S. and Q.Z. analyzed the data. E.S. and F.N. wrote the paper.

Funding: This work was financially supported by the National Science and Technology Major Project of China (Grant No. 2016ZX05043005 and 2016ZX05045004), the State Key Research Development Program of China (Grant No. 2016YFC0801404 and 2016YFC0801402), and the National Natural Science Foundation of China (51674050 and 51704046), which are gratefully acknowledged.

Acknowledgments: The authors thank the editor and anonymous reviewers for their valuable advice. The authors thank David Frishman, PhD, from Liwen Bianji, Edanz Group China (www.liwenbianji.cn/ac), for editing the English text of a draft of this manuscript.

Conflicts of Interest: The authors declare no conflicts of interest.

References

1. Zou, Q.; Lin, B.; Zheng, C.; Hao, Z.; Zhai, C.; Liu, T.; Liang, J.; Yan, F.; Yang, W.; Zhu, C. Novel integrated techniques of drilling-slotting-separation-sealing for enhanced coal bed methane recovery in underground coal mines. *J. Nat. Gas Sci. Eng.* **2015**, *26*, 960–973. [CrossRef]
2. Gale, W. A review of energy associated with coal bursts. *Int. J. Min. Sci. Technol.* **2018**, *28*, 755–761. [CrossRef]
3. Yu, G.; Zhai, C.; Qin, L.; Tang, Z.; Wu, S.; Xu, J. Changes to coal pores by ultrasonic wave excitation of different powers. *J. China Univ. Min. Technol.* **2018**, *47*, 264–270.
4. Liu, C.; Li, S.; Yang, S. Gas emission quantity prediction and drainage technology of steeply inclined and extremely thick coal seams. *Int. J. Min. Sci. Technol.* **2018**, *28*, 415–422.
5. Liang, B.; Jia, L.; Sun, W.; Jiang, Y. Experimental on the law of coal deformation and permeability under desorption and seepage. *J. China Univ. Min. Technol.* **2018**, *47*, 935–941.
6. Chang, K.; Tian, H. Technical scheme and application of pressure-relief gas extraction in multi-coal seam mining region. *Int. J. Min. Sci. Technol.* **2018**, *28*, 483–489.
7. Zou, Q.; Lin, B. Fluid−solid coupling characteristics of gas-bearing coal subjected to hydraulic slotting: An experimental investigation. *Energy Fuels* **2018**, *32*, 1047–1060. [CrossRef]
8. White, C.M.; Smith, D.H.; Jones, K.L.; Goodman, A.L.; Jikich, S.A.; LaCount, R.B.; DuBose, S.B.; Ozdemir, E.; Morsi, B.I.; Schroeder, K.T. Sequestration of carbon dioxide in coal with enhanced coalbed methane recovery: A review. *Energy Fuels* **2005**, *19*, 659–724. [CrossRef]
9. Haszeldine, R.S. Carbon capture and storage: How green can black be? *Science* **2009**, *325*, 1647–1652. [CrossRef]
10. Zhang, X.; Ranjith, P.G.; Perera, M.S.A. Gas transportation and enhanced coalbed methane recovery processes in deep coal seams: A review. *Energy Fuels* **2016**, *30*, 8832–8849. [CrossRef]
11. Gale, J. Geological storage of CO$_2$: What do we know, where are the gaps and what more needs to be done? *Energy* **2004**, *29*, 1329–1338. [CrossRef]
12. Perera, M. A Comprehensive Overview of CO$_2$ Flow Behaviour in Deep Coal Seams. *Energies* **2018**, *11*, 906. [CrossRef]
13. Fulton, P.F.; Parente, C.A.; Rogers, B.A.; Shah, N.; Reznik, A. A laboratory investigation of enhanced recovery of methane from coal by carbon dioxide injection. In *SPE Unconventional Gas Recovery Symposium*; Society of Petroleum Engineers: Pittsburgh, PA, USA, 1980.
14. Reznik, A.A.; Singh, P.K.; Foley, W.L. An analysis of the effect of CO$_2$ injection on the recovery of in-situ methane from bituminous coal: An experimental simulation. *Soc. Petrol. Eng. J.* **1984**, *24*, 521–528. [CrossRef]

15. Jessen, K.; Tang, G.-Q.; Kovscek, A.R. Laboratory and simulation investigation of enhanced coalbed methane recovery by gas injection. *Transp. Porous Media* **2008**, *73*, 141–159. [CrossRef]

16. Dutka, B.; Kudasik, M.; Pokryszka, Z.; Skoczylas, N.; Topolnicki, J.; Wierzbicki, M. Balance of CO_2/CH_4 exchange sorption in a coal briquette. *Fuel Process. Technol.* **2013**, *106*, 95–101. [CrossRef]

17. Stevens, S.H.; Kuuskraa, V.A.; Spector, D.; Riemer, P. CO_2 sequestration in deep coal seams: Pilot results and worldwide potential. In Proceedings of the Greenhouse Gas Control Technology, Saskatoon, SK, Canada, 4–6 October 1999; pp. 175–180.

18. Span, R.; Wagner, W. A new equation of state for carbon dioxide covering the fluid region from the triple-point temperature to 1100 K at pressures up to 800 MPa. *J. Phys. Chem. Ref. Data* **1996**, *25*, 1509–1596. [CrossRef]

19. Larsen, J.W. The effects of dissolved CO_2 on coal structure and properties. *Int. J. Coal Geol.* **2004**, *57*, 63–70. [CrossRef]

20. Zhang, K.; Cheng, Y.; Jin, K.; Guo, H.; Liu, Q.; Dong, J.; Li, W. Effects of Supercritical CO_2 Fluids on Pore Morphology of Coal: Implications for CO_2 Geological Sequestration. *Energy Fuels* **2017**, *31*, 4731–4741. [CrossRef]

21. Chen, R.; Qin, Y.; Wei, C.; Wang, L.; Wang, Y.; Zhang, P. Changes in pore structure of coal associated with Sc-CO_2 extraction during CO_2-ECBM. *Appl. Sci.* **2017**, *7*, 931. [CrossRef]

22. Liu, C.J.; Wang, G.X.; Sang, S.X.; Rudolph, V. Changes in pore structure of anthracite coal associated with CO_2 sequestration process. *Fuel* **2010**, *89*, 2665–2672. [CrossRef]

23. Kolak, J.J.; Hackley, P.C.; Ruppert, L.F.; Warwick, P.D.; Burruss, R.C. Using ground and intact coal samples to evaluate hydrocarbon fate during supercritical CO_2 injection into coal beds: Effects of particle size and coal moisture. *Energy Fuels* **2015**, *29*, 5187–5203. [CrossRef]

24. Zhang, D.; Gu, L.; Li, S.; Lian, P.; Tao, J. Interactions of supercritical CO_2 with coal. *Energy Fuels* **2013**, *27*, 387–393. [CrossRef]

25. Busch, A.; Gensterblum, Y. CBM and CO_2-ECBM related sorption processes in coal: A review. *Int. J. Coal Geol.* **2011**, *87*, 49–71. [CrossRef]

26. Cai, Y.; Liu, D.; Pan, Z.; Yao, Y.; Li, J.; Qiu, Y. Pore structure and its impact on CH_4 adsorption capacity and flow capability of bituminous and subbituminous coals from Northeast China. *Fuel* **2013**, *103*, 258–268. [CrossRef]

27. Lu, S.Q.; Cheng, Y.P.; Li, W.; Wang, L. Pore structure and its impact on CH_4 adsorption capability and diffusion characteristics of normal and deformed coals from Qinshui Basin. *Int. J. Oil Gas Coal Technol.* **2015**, *10*, 76–78. [CrossRef]

28. Ju, Y.W.; Jiang, B.; Hou, Q.L.; Wang, G.L. The new structure-genetic classification system in tectonically deformed coals and its geological significance. *J. China Coal Soc.* **2004**, *29*, 513–517.

29. Shi, J.Q.; Durucan, S. A bidisperse pore diffusion model for methane displacement desorption in coal by CO injection. *Fuel* **2003**, *82*, 1219–1229. [CrossRef]

30. An, F.H.; Cheng, Y.P. An explanation of large-scale coal and gas outbursts in underground coal mines: The effect of low-permeability zones on abnormally abundant gas. *Nat. Hazard Earth Syst.* **2014**, *14*, 4751–4775. [CrossRef]

31. Clarkson, C.R.; Bustin, R.M. The effect of pore structure and gas pressure upon the transport properties of coal: A laboratory and modeling study. 1. Isotherms and pore volume distributions. *Fuel* **1999**, *78*, 1333–1344. [CrossRef]

32. Pan, J.; Zhao, Y.; Hou, Q.; Jin, Y. Nanoscale pores in coal related to coal rank and deformation structures. *Transp. Porous Media* **2015**, *107*, 543–554. [CrossRef]

33. Wong, S.; Law, D.; Deng, X.; Robinson, J.; Kadatz, B.; Gunter, W.D.; Ye, J.; Feng, S.; Fan, Z. Enhanced coalbed methane and CO_2 storage in anthracitic coals—Micro-pilot test at South Qinshui, Shanxi, China. *Int. J. Greenhouse Gas Control* **2007**, *1*, 215–2228. [CrossRef]

34. Washburn, E.W. The dynamics of capillary flow. *Phys. Rev. Ser.* **1921**, *17*, 273–283. [CrossRef]

35. Barrett, E.P.; Joyner, L.G.; Halenda, P.P. The determination of pore volume and area distributions in porous substances. I. Computations from nitrogen isotherms. *J. Am. Chem. Soc.* **1951**, *73*, 373–380. [CrossRef]

36. Brunauer, S.; Emmett, P.H.; Teller, E. Adsorption of gases in multimolecular layers. *J. Am. Chem. Soc.* **1938**, *60*, 309–319. [CrossRef]

37. Landers, J.; Gor, G.Y.; Neimark, A.V. Density functional theory methods for characterization of porous materials. *Colloids Surf. A* **2013**, *437*, 3–32. [CrossRef]

38. Hodot, B.B. *Outburst of Coal and Coalbed Gas*; China Industry Press: Beijing, China, 1966.
39. Hou, Q.; Li, H.; Fan, J.; Ju, Y.; Wang, T.; Li, X.; Wu, Y. Structure and coalbed methane occurrence in tectonically deformed coals. *Sci. China Earth Sci.* **2012**, *55*, 1755–1763. [CrossRef]
40. Li, M.; Jiang, B.; Lin, S.; Lan, F.; Wang, J. Structural controls on coalbed methane reservoirs in faer coal mine, southwest china. *J. Earth Sci.-China* **2013**, *24*, 437–448. [CrossRef]
41. Perera, M.S.A. Influences of CO_2 injection into deep coal seams: A review. *Energy Fuels* **2017**, *31*, 10324–10334. [CrossRef]
42. Wellman, T.P.; Grigg, R.B.; Mcpherson, B.J.; Svec, R.K.; Lichtner, P.C. Evaluation of CO_2–brine–reservoir rock interaction with laboratory flow tests and reactive transport modeling. In *International Symposium on Oilfield Chemistry*; Society of Petroleum Engineers: Pittsburgh, PA, USA, 2003.
43. Cao, Y.; Davis, A.; Liu, R.; Liu, X.; Zhang, Y. The influence of tectonic deformation on some geochemical properties of coals—A possible indicator of outburst potential. *Int. J. Coal Geol.* **2003**, *53*, 69–79. [CrossRef]
44. Qu, Z.; Wang, G.G.X.; Jiang, B.; Rudolph, V.; Dou, X.; Li, M. Experimental study on the porous structure and compressibility of tectonized coals. *Energy Fuels* **2010**, *24*, 2964–2973. [CrossRef]
45. Rouquerol, F.; Rouquerol, J.; Sing, K. *Adsorption by Powders and Porous Solids*; Academic Press: Cambridge, MA, USA, 2014.
46. Wang, H.; Fu, X.; Jian, K.; Li, T.; Luo, P. Changes in coal pore structure and permeability during N_2 injection. *J. Nat. Gas Sci. Eng.* **2015**, *27*, 1234–1241. [CrossRef]
47. Guo, H.; Cheng, Y.; Yuan, L.; Wang, L.; Zhou, H. Unsteady-state diffusion of gas in coals and its relationship with coal pore structure. *Energy Fuels* **2016**, *30*, 7014–7024. [CrossRef]
48. Pan, Z.; Connell, L.D.; Camilleri, M.; Connelly, L. Effects of matrix moisture on gas diffusion and flow in coal. *Fuel* **2010**, *89*, 3207–3217. [CrossRef]
49. Liu, Z.; Cheng, Y.; Dong, J.; Jiang, J.; Wang, L.; Li, W. Master role conversion between diffusion and seepage on coalbed methane production: Implications for adjusting suction pressure on extraction borehole. *Fuel* **2018**, *223*, 373–384. [CrossRef]
50. Jiang, H.; Cheng, Y.; Yuan, L. A Langmuir-like desorption model for reflecting the inhomogeneous pore structure of coal and its experimental verification. *RSC. Adv.* **2014**, *5*, 2434–2440. [CrossRef]
51. Crank, J. Mathematics of Diffusion. In *Handbook of Chemistry Physic*; Oxford University Press: Oxford, UK, 1975.
52. Yang, Q.; Wang, Y. Theory of methane diffusion from coal cuttings and its application. *J. China Coal Soc.* **1986**, *11*, 87–94.
53. Shouqing, L.; Yuanping, C.; Hongxing, Z.; Haijun, G. Gas desorption characteristics of the high-rank intact coal and fractured coal. *Int. J. Min. Sci. Technol.* **2015**, *25*, 819–825.
54. Liu, C.; Zhou, F.; Yang, K.; Xiao, X.; Liu, Y. Failure analysis of borehole liners in soft coal seam for gas drainage. *Eng. Failure Anal.* **2014**, *42*, 274–283. [CrossRef]

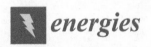

Article

An Integrally Embedded Discrete Fracture Model with a Semi-Analytic Transmissibility Calculation Method

Renjie Shao and Yuan Di *

College of Engineering, Peking University, Beijing 100871, China; shaorenjie@pku.edu.cn
* Correspondence: diyuan@mech.pku.edu.cn; Tel.: +86-135-5270-9129

Received: 12 November 2018; Accepted: 12 December 2018; Published: 14 December 2018

Abstract: The embedded discrete fracture model (EDFM) combines the advantages of previous numerical models for fractured reservoirs, achieving a good balance between calculation cost and simulation accuracy. In this work, an integrally embedded discrete fracture model (iEDFM) is introduced to further improve the simulation accuracy and expand the application of the model. The iEDFM has a new gridding method that can arbitrarily grid the fractures according to the requirements rather than finely subdividing fracture elements. Then, with a more precise pressure distribution assumption inside the matrix blocks, we are able to obtain a semi-analytic calculation method of matrix-fracture transmissibility applied to iEDFM. Several case studies were conducted to demonstrate the advantage of iEDFM and its applicability for intersecting and nonplanar fractured reservoirs, and a 3D case with a modified dataset from a reported seismic survey could be used to demonstrate the potential application of the iEDFM in real field studies.

Keywords: embedded discrete fracture model; fractured reservoir simulation; matrix-fracture transmissibility

---

1. Introduction

Fractured reservoirs are commonly found all over the world. In many geoscience applications, such as petroleum extraction, the target formations are fractured [1,2]. In these formations, matrix rock is crossed by several fractures at multiple length scales, behaving as hydraulic conductors.

In order to evaluate the economic feasibility and to manage production, numerical simulation tools have to be used. However, when modeling flow in fractured reservoirs, the high heterogeneity caused by complex fractures and complicated matrix-fracture fluid exchange will cause inefficiency and inaccuracy [3–5].

Research in this area has been advanced significantly over the past several decades. The dual porosity model (DPM) [6–8] is one of the earliest methods for modelling fractured systems, and is still widely used in the petrol industry because of its simplicity and practicability. Since then, many other methods based on a similar concept have been developed to expand the application of DPM, such as the dual-porosity dual-permeability (DPDK) model [9,10], multiple interacting continua (MINC) model [11,12], subdomain model [13], triple-porosity dual-permeability (TPDK) model [14,15], and multi-porosity model [16]. These multi-continuum methods provide an efficient approach to simulate micro scale fractures. However, the assumption of fracture uniformity is limited due to losing detailed information (such as geometry and location) of the discovered macro-scale factures.

A more accurate and physics-based approach was proposed to discretize the discovered macro-scale fractures explicitly, which is called the discrete fracture model (DFM). Unstructured grids are commonly used in DFM [17–22]. Based on DFM, some complex fractures, such as intersecting fractures and nonplanar fractures, can be represented through appropriate gridding [22–25]. However, the use of unstructured fine grids in real field studies is still limited because of its complexity in gridding and computational cost [26,27].

The embedded discrete fracture model (EDFM) has been developed as a compromise. EDFM borrows the dual-porosity concept from DPM, using traditional Cartesian gridding for the matrix to keep the efficiency, but also incorporates the effect of each fracture explicitly as with DFM to account for the complexity and heterogeneity of reservoirs. The concept was first proposed by Lee [28]. The fracture element within its parent matrix block (element) is represented by a control volume and is connected to the parent matrix block and other fracture element. This concept was implemented by Li and Lee [29] to vertical fracture cases and implemented by Moinfar [26] to non-vertical fracture cases, in which the fractures have arbitrary dip and strike angles. Xu [30] and Yu [31] implemented EDFM to some more complex fracture-networks, such as nonplanar shape and variable aperture. Li [32] combined EDFM with DPDK for reservoirs with different scale fractures.

However, all these EDFMs are based on a simplified matrix-fracture fluid exchange assumption, which leads to inaccuracy in some cases and also limits the application of the model, as it is necessary to finely subdivide fracture elements. Matei [33] proposed a projection-based EDFM (pEDFM) where the fracture and matrix grids are independently defined. The pEDFM is proposed to deal with highly conductive fractures and flow barriers, thus it makes little improvement on computational efficiency, but still provides a useful method to improve gridding and transmissibility calculation upon the classic EDFM.

In this work, an integrally embedded discrete fracture model (iEDFM) is introduced to improve simulation accuracy and expand the applications of the model. The iEDFM has a new gridding method that can arbitrarily grid the fractures according to the requirements, and then embed them integrally in matrix blocks. A semi-analytic matrix-fracture transmissibility calculation method is applied to iEDFM with a more precise pressure distribution assumption inside the matrix blocks.

This paper is organized as follows. First of all, the basic mathematical method is introduced. Second, the improvements of iEDFM upon EDFM are described, where the gridding method and the semi-analytic calculation method of transmissibility are the most important. Subsequently, we demonstrate the applicability and advantages of iEDFM through a single-phase case and two flooding tests. Finally, by using a modified 3D case with a reported dataset from the seismic survey, we are able to demonstrate the potential applications of the iEDFM in real field studies.

2. Methodology

2.1. Basic Mathematical Method

Models in this paper (including the fine-grid model, EDFM and iEDFM) have been implemented into a multiphase multidimensional black-oil reservoir simulator, named MSFLOW [34]. The basic mathematical model of this multiphase multidimensional black-oil method is introduced as follows.

In an isothermal system containing three mass components, three mass-balance equations are needed to describe flow in fracture elements and matrix blocks. For the flow of phase β ($\beta = g$ for gas, $\beta = w$ for water, and $\beta = o$ for oil), the mass-balance equation is given by:

$$\frac{\partial}{\partial t}(\phi \, S_\beta \, \rho_\beta) = -\nabla \bullet (\rho_\beta \mathbf{v}_\beta) + q_\beta \qquad (1)$$

where the velocity of phase β is defined by Darcy's law:

$$\mathbf{v}_\beta = -\frac{k k_{r\,\beta}}{\mu_\beta}(\nabla P_\beta - \rho_\beta g \nabla D) \qquad (2)$$

where S_β is the saturation of phase β; k is the absolute permeability of the formation; $k_{r\,\beta}$ is relative permeability to phase β; μ_β is the viscosity of phase β; P_β is the pressure of phase β; ρ_β is the density of phase β under reservoir conditions; g is gravitational acceleration; ϕ is the effective porosity; q_β is the sink/source term of phase β; and D is depth from a reference datum.

As implemented numerically, Equation (1) is discretized in space with an integral finite-difference or control-volume scheme for the fracture elements and matrix blocks. And in time, it's discretized with a backward, first-order, finite-difference scheme. The discrete equations are as follows:

$$\left\{ (\phi \, S_\beta \rho_\beta)_i^{n+1} - (\phi \, S_\beta \rho_\beta)_i^{n} \right\} \frac{V_i}{\Delta t} = \sum_{j \in \eta_i} F_{\beta,i\,j}^{n+1} + Q_{\beta i}^{n+1} \tag{3}$$

where n is the previous time level; $n+1$ is the current time level; t is time step size; V_i is the volume of element i (matrix or fracture element); η_i contains the set of neighboring elements j (matrix or fracture element) to which element i is directly connected; $F_{\beta,i\,j}$ is the flow term for phase β between element i and j; and $Q_{\beta i}$ is the sink/source term at element i of Phase β.

The flow term $F_{\beta,i\,j}$ in Equation (3) is described by a discrete version of Darcy's law, given by:

$$F_{\beta,i\,j} = \lambda_{\beta,ij+1/2} T_{i\,j} \left(\psi_{\beta j} - \psi_{\beta i} \right) \tag{4}$$

where $\lambda_{\beta,ij+1/2}$ is the mobility term to phase β, defined as:

$$\lambda_{\beta,ij+1/2} = \left(\frac{\rho_\beta k_{r\beta}}{\mu_\beta} \right)_{ij+1/2} \tag{5}$$

where $ij+1/2$ is a proper averaging or weighting of properties at the interface between element i and j. The flow potential term is defined as:

$$\psi_{\beta i} = P_{\beta i} - \rho_{\beta,i\,j+1/2}\, g\, D_i \tag{6}$$

where D_i is the depth from a reference datum to the center of element i, and T_{ij} is transmissibility. If the integral finite-difference scheme [14] is used, the transmissibility will be calculated as:

$$T_{ij} = \frac{A_{ij}\, k_{ij+1/2}}{d_i + d_j} \tag{7}$$

where A_{ij} is the common interface area between elements i and j; d_i is the distance from the center of element i to the interface; and $k_{i\,j+1/2}$ is an averaged (such as harmonically weighted) absolute permeability along the connection between elements i and j.

2.2. Embedded Discrete Fracture Model

EDFM creates fracture elements connected with corresponding matrix blocks (each represents a matrix element) to account for the mass transfer between matrix and fractures. Once a fracture penetrates a matrix block, an additional element is created to represent the fracture segment in the physical domain (Figure 1a). Each individual fracture is discretized into several fracture elements by the fracture intersections (Figure 1b) and the matrix block boundaries (Figure 1c).

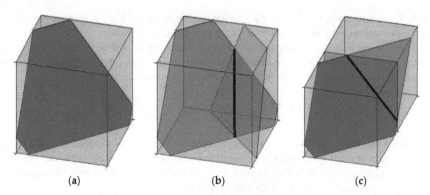

Figure 1. Explanation of the embedded discrete fracture model (EDFM) gridding [26]: (**a**) a fracture segment is embedded in a matrix block; (**b**) two fracture planes intersect in a matrix block; and (**c**) two fracture segments embedded in neighboring matrix blocks.

Thus, there exist three kinds of connections: matrix–matrix (M–M), fracture–fracture (F–F), and matrix–fracture (M–F). The transmissibility of each connection can be calculated referring to Equation (7).

The parameters for the M–M connection give clear physical meanings, and so the transmissibility can be easily obtained. For the F–F connection, a simplified approximation from Karimi–Fard [23] is used. The two-point flux approximation scheme is:

$$T_{F_1-F_2} = \frac{T_1 T_2}{T_1 + T_2} \tag{8}$$

$$T_1 = \frac{k_F A_c}{d_{F_1}} \quad T_2 = \frac{k_F A_c}{d_{F_2}} \tag{9}$$

where k_F is the absolute permeability of fracture, A_c is the common interface area for these two fracture elements, and d_{F_1} and d_{F_2} are the average distances from fracture elements 1 and 2 to the common interface.

Transmissibility of *M–F* depends on the matrix permeability and fracture geometry. When a fracture segment fully penetrates a matrix block, EDFM assumes a uniform pressure gradient in the matrix element, and that pressure gradient is normal to the fracture plane, creating a linear pressure distribution assumption. Then, the *M–F* transmissibility referring to the equation:

$$T_{M-F} = \frac{2 A_F k_M}{d_{M-F}} \tag{10}$$

where A_F is the area of the fracture element on one side, k_M is the absolute permeability of matrix (when using the harmonically weighted average permeability, the huge fracture permeability can be ignored), and d_{M-F} is the average normal distance from matrix to fracture, which is calculated as:

$$d_{M-F} = \frac{\int\limits_V x_n dV}{V} \tag{11}$$

where V is the volume of the matrix element, dV is the volume element of matrix, and x_n is the distance from the volume element to the fracture plane. If the fracture does not fully penetrate the matrix element, most of the EDFMs make the same assumption as Li [29] that the transmissibility is proportional to the area of the fracture element inside the matrix element, which actually further simplifies the previous linear pressure distribution assumption.

2.3. The Improvement of iEDFM

The integrally embedded discrete fracture model (iEDFM) is implemented on EDFM with a new gridding method, a semi-analytic matrix-fracture transmissibility equation from a more realistic pressure distribution assumption inside the matrix element, which would improve simulation accuracy and expand the scope of application. An iEDFM preprocessor was developed with the inputs of reservoir features and fracture geometries. The gridding and transmissibility calculation processes are conducted in this preprocessor.

In Figure 2, we illustrate the procedure to add fracture elements with a 2D case with 4 matrix blocks and 2 fractures. Figure 2a shows that in EDFM, 6 fracture elements have to be added with 14 F–F connections and 6 F–M connections because of the matrix element boundaries and fracture intersections. However, in iEDFM, the fractures can be embedded integrally—either discretizing by matrix element boundaries (Figure 2b) or taking the intersecting fracture group as one element (Figure 2c) is permitted. As a result, the number of fracture elements can be reduced to 3 or 1, and the number of F–F and F–M connections can be reduced to 2 or 0, 3 or 3.

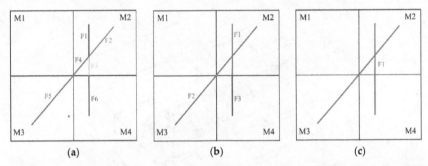

Figure 2. Explanation of the integrally embedded discrete fracture model (iEDFM) gridding: (**a**) EDFM gridding method; (**b**) iEDFM gridding method I—discretizing by matrix element boundaries; and (**c**) iEDFM gridding method II—taking the intersecting fracture group as one element.

With the fracture added, we determined the calculation method of transmissibility in iEDFM. For M–M and F–F connections, Equations (7) and (8) are applicable. However, other than using Equation (10), iEDFM has a new semi-analytic transmissibility equation for M–F connections.

When calculating M–F transmissibility in iEDFM, the analytic solution of pressure distribution around the fractures and superposition principle of potential are applied. The detailed method will be explained later.

In general, EDFM has four weaknesses which have been overcome by iEDFM:

1. The pressure difference between adjacent fracture pieces is relatively small due to the high conductivity in the fracture. Therefore, such fine gridding for fracture in EDFM may bring unnecessary calculation costs and difficulty in convergence;
2. When the matrix block is coarse or the embedded fractures are more complicated, the linear assumption in EDFM is too rough (showed in Case 1);
3. If there are more than one fracture pieces inside one matrix element or the fracture has complex geometries, the pressure distribution inside the matrix element will no longer be available, which will limit the application of EDFM (shown in Case 1(c) and Case 3);
4. Only the fracture piece and its background matrix block are used for the transmissibility calculation, while the global fracture network is not taken into consideration. For example, when calculating the transmissibility of F5–M3 in Figure 2a, according to the linear pressure distribution assumption in EDFM, the pressure drop at the red point is only influenced by F5.

This is inaccurate because there is also F6 nearby, which will also cause a pressure drop at the red point (showed in Case 1(b)).

2.4. Calculation of M–F Transmissibility in iEDFM

The M–F transmissibility calculation is based on the assumption of pressure distribution near the fracture. The linear pressure distribution assumed in EDFM is rough, especially when the embedded fractures are non-penetrating or complicated. In iEDFM, fractures can be considered as a bunch of point sinks. Around each point sink, an analytic pressure distribution formula exists. Then, the superposition principle of potential allows us to obtain the semi-analytic pressure distribution near the fracture. With this pressure distribution, the M–F transmissibility equation is obtained.

Given a 2D gridded fractured reservoir as an example (Figure 3a), we could take these two intersecting fractures as one element (iEDFM gridding method II) and consider the incompressible steady state single-phase flow in this reservoir.

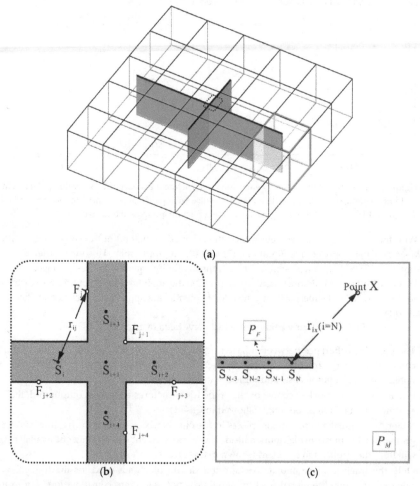

Figure 3. Illustration showing the calculation of M–F transmissibility in iEDFM: (**a**) a 2D example of the fractured reservoir; (**b**) the top view of the dotted frame part; and (**c**) the top view of the red matrix block.

2.4.1. Point Sinks Imitate the Fracture

The pressure of the fracture element is P_F (for the iEDFM gridding method I, a fracture piece in each matrix block has a different pressure, but the formula derivation is also similar, as follows).

We use point sink S_i ($i = 1$ to N) to replace whole intersecting fractures. If the point sinks are dense enough (e.g., more than 50 sinks inside one matrix block), the pressure distribution near the fracture could be imitated near these point sinks. Point F_j ($j = 1$ to N) is on the surface of the fracture, thus having the same fracture pressure P_F. The top view of the dotted frame part in Figure 3a is presented in Figure 3b.

We define a potential:

$$\Phi = \frac{k}{\mu}P \tag{12}$$

when the flow reaches the steady state, we have the analytic potential distribution formula of a single point sink from the integral of the plane radial flow equation:

$$\Phi = \frac{q}{2\pi h}\ln r + c \tag{13}$$

Then, the N-dimensional linear equations are obtained from Equation (13) and the superposition principle of potential:

$$\Phi_F = \sum_{i=1}^{N}\frac{q_{S_i}}{2\pi h}\ln r_{ij} + C \ (j = 1 \text{ to } N) \tag{14}$$

where q_{S_i} is the flow rate of the point sink S_i, h is the height of the reservoir, r_{ij} is the distance between the point sink S_i and the point F_j, and C and c are constant numbers. We define:

$$\xi_i = \frac{q_{S_i}}{\Phi_F - C} \tag{15}$$

where ξ_i can be solved out from Equation (14).

2.4.2. The Semi-Analytic Calculation Method

Here, we consider the transmissibility between a specific matrix element (for example, the red matrix block in Figure 3a, the top view shown in Figure 3c and the fracture element. The average pressure of the whole matrix element (block) is P_M.

Similar to Equation (14), the potential of point X near the fracture inside this specific matrix element can be determined by:

$$\Phi_X = \frac{k_M}{\mu}P_X = \frac{\sum\limits_{i=1}^{N}\xi_i \ln r_{ix}}{2\pi h}(\Phi_F - C) + C \tag{16}$$

Thus, we obtain:

$$P_M = \frac{\iint\limits_{M}P_X \cdot dV_x}{V_M} \tag{17}$$

Combining Equations (16) and (17) and the definition of transmissibility (Equation (4)), the M–F transmissibility and the pressure anywhere inside the matrix block can be calculated by:

$$T_{M-F} = \frac{k_M \sum\limits_{S_i \in M}\xi_i}{\varepsilon - 1} \tag{18}$$

$$P_X = \left(\frac{\sum\limits_{i=1}^{N} \xi_i \ln r_{ix}}{2\pi h} - 1 \right) \cdot \frac{P_M - P_F}{\varepsilon - 1} + P_F \tag{19}$$

where:

$$\varepsilon = \frac{\iint\limits_{M} \frac{\sum\limits_{i=1}^{N} \xi_i \ln r_{ix}}{2\pi h} dV_x}{V_M} \tag{20}$$

where V_M is the volume of the specific matrix element, $\in M$ means that the point is inside this matrix block, dV_x is an element volume of this matrix element, r_{ix} is the distance between the point sink S_i and dV_x. T_{M-F} is only related to the properties of the reservoir and can be determined at the step of pre-processing before the simulation starts.

For the 3D situation, referring to Equations (6) and (12), the analytic potential distribution formula of a single point sink is:

$$\Phi^{3D} = \frac{k}{\mu}\psi = \frac{q}{2\pi r} + c \tag{21}$$

After a similar derivation, the—F transmissibility and the pressure distribution can be written as:

$$T_{M-F}^{3D} = \frac{k_M \sum\limits_{S_i \in M} \xi_i}{\varepsilon^{3D} - 1} \tag{22}$$

$$\Phi_X^{3D} = \left(\frac{\sum\limits_{i=1}^{N} \xi_i}{2\pi r_{ix}} - 1 \right) \cdot \frac{\Phi_M^{3D} - \Phi_F^{3D}}{\varepsilon^{3D} - 1} + \Phi_F^{3D} \tag{23}$$

where,

$$\varepsilon^{3D} = \frac{\iint\limits_{dV_x \in M} \frac{\sum\limits_{i=1}^{N} \xi_i}{2\pi r_{ix}} dV_x}{V_M} \tag{24}$$

3. Verifications and Applications

In the following simulation studies, we present four cases to demonstrate the applicability of iEDFM.

First, a single-phase case is considered to demonstrate the improvement of iEDFM upon EDFM. Then, in the second case, we demonstrate the accuracy of our model by comparing the flow rates and saturation profiles with the fine-grid model through a flooding test. Both gridding methods I and II are used in this case. Then, a nonplanar fractures case is presented to show the applicability of iEDFM for a complex geometry situation. At last, a 3D field case demonstrates the potential application of iEDFM in real field studies.

Most of the reservoir properties and operation parameters were kept the same in all the cases, as shown in Table 1.

Table 1. Basic reservoir and fluid parameters in simulations for case 1~4.

Parameter	Value	Unit
Reservoir permeability	1×10^{-14}	m^2
Fracture permeability	1×10^{-10}	m^2
Reservoir porosity	30%	-
Oil density	800	kg/m^3
Oil viscosity	2.0	Pa·s

3.1. Case 1: Constant Pressure Pumping from Fractures

The calculation method of M–F transmissibility is based on the assumption of pressure distribution pattern in the vicinity of fractures. EDFM assumes a uniform pressure gradient in the matrix element, while iEDFM uses a semi-analytic method to calculate the pressure distribution.

In this case, we consider an ideal reservoir with two intersecting fractures (Figure 4), using simulation results from the fine-grid model to verify the effectiveness of iEDFM over EDFM. EDFM or iEDFM simulation is not conducted in this case. Only the pressure distribution and M–F transmissibility for red block (a)/(b)/(c) are calculated through the pre-processing procedure of iEDFM and EDFM, comparing with the simulation result of the whole reservoir of the fine-grid model.

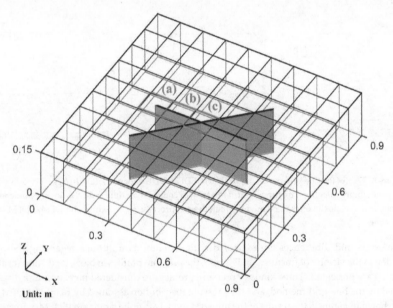

Figure 4. The fractured reservoir in Case 1. The 9 × 9 grids showed a possible gridding choice for iEDFM or EDFM.

Single-phase fluid (water) is pumping out from the constant pressure vertical fractures simulated in the fine-grid model discretized by 900 × 900 fine gridblocks horizontally. The element dimensions are non-uniform in the x and y directions to accommodate refinement around fractures. The widths of the fracture elements and their adjacent matrix elements were equal to the fracture aperture. Table 2 supplements some parameters in addition to Table 1.

Table 2. Some parameters in simulations for Case 1.

Parameter	Value	Unit
Initial reservoir pressure	2×10^6	Pa
Pumping pressure (Fracture pressure)	1×10^6	Pa
Simulation time	1.0	s

This pumping pressure represents the pressure of the fracture element, and the average pressure of the matrix inside the red blocks represents the pressure of the matrix element. The fluid exchange can be obtained from simulation results of the fine-grid model. Therefore, the equivalent M–F transmissibility by the fine-grid model can be calculated through Equation (4).

Figure 5b shows that the transmissibility calculation methods in EDFM and iEDFM are more accurate when the fracture fully penetrates the matrix block. It can be seen from Figure 5a that when the fracture does not fully penetrate, the EDFM method will cause some error which cannot be ignored, which is in agreement with the estimation by Xu [30], while the iEDFM method can still be relatively accurate in this situation. In Figure 5c, we can see that the EDFM method produces more obvious errors in the complex situation with fracture intersection, whereas the iEDFM method is still relatively accurate.

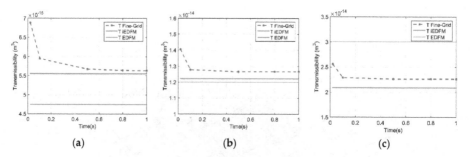

(a)   (b)   (c)

Figure 5. $T_{M\text{-}F}$ of the fine-grid model (equivalent)/iEDFM/EDFM for block (a)/(b)/(c). The equivalent transmissibility calculated by the fine-grid model will change over time until the flow approaches a steady state, while the ones calculated through the pre-processing procedure of iEDFM/EDFM stay the same.

The errors and differences are mainly due to different assumptions regarding the pressure distribution in the vicinity of fractures. In Figure 6, the pressure profiles in block (a) (b) and (c) after 0.1s of pumping are presented. Three kinds of pressure profiles are considered here: pressure distribution calculated by the fine-grid method, and the pressure distribution assumed by EDFM/iEDFM.

As shown in Figure 6, the results of the fine-grid model are quite similar to iEDFM, with negligible difference. However, the results of EDFM indicate that when the matrix block is not fully penetrated by the fracture (Figure 6a) or the embedded fractures are complicated (Figure 6c), the linear distribution of EDFM's assumption can be rough. Even if the fracture penetrates the block (b), the linear distribution still does not exactly reflect the real situation, because of the effect of the fracture segment outside this matrix block, which EDFM does not take into consideration but iEDFM does.

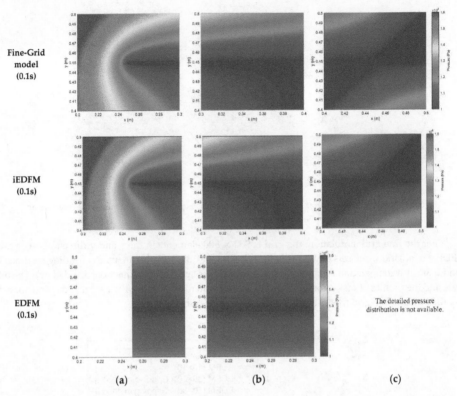

Figure 6. Pressure distribution profiles after 0.1 s pumping of the fine-grid model/iEDFM (assumed)/EDFM (assumed) for block (**a**)/(**b**)/(**c**).

In summary, iEDFM has a more accurate M–F transmissibility calculation method based on a more realistic pressure distribution assumption. In addition, iEDFM can assume a specific pressure field for any complex embedding situation. If pressure-related physical properties, such as adsorption analysis and diffusion effects, need to be considered, iEDFM is able to show a greater applicability.

3.2. Case 2: Intersecting-Fractured Reservoir Flooding Test

Figure 7 shows a 2D fractured reservoir containing three intersecting vertical fractures. This case is a displacement of oil by water, applied in the fine-grid model and the iEDFM model. Table 3 supplements some parameters of this case in addition to Table 1, which will also be used in the following cases.

Table 3. Some parameters in simulations for Cases 2–4.

Parameter	Value	Unit
Initial reservoir pressure	1×10^6	Pa
Producer pressure	1×10^6	Pa
Initial water saturation	20%	-

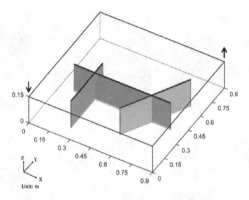

Figure 7. The fractured reservoir in Case 2. A water injector is placed in one corner of the reservoir and a producer is located in the opposite corner.

For the fine-grid simulation, the grid is 900 × 900 elements in the x and y directions. For the iEDFM simulation, two sets of uniform matrix grids (30 × 30/10 × 10) are used. Gridding methods I and II are also compared in this case (gridding method I is applied if not mentioned, as below). The oil rate and the profiles of oil saturation after 115 days of water injection (8.64×10^{-5} m^3/day) calculated by iEDFM and fine-grid model are presented in Figures 8 and 9.

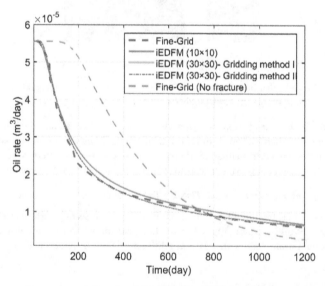

Figure 8. Comparison of oil production rate by the iEDFM/fine-grid model in Case 2.

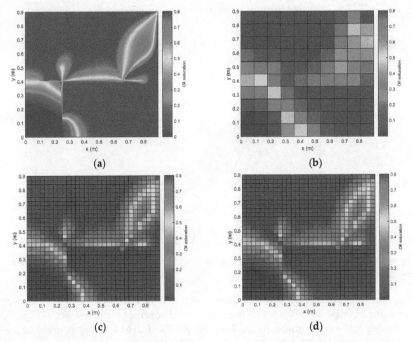

Figure 9. Profiles of oil saturation after 115 days' injection for Case 2 in: (**a**) the fine-grid model; (**b**) iEDFM (10 × 10, gridding method I); (**c**) iEDFM (30 × 30, gridding method I); and (**d**) iEDFM (30 × 30, gridding method II).

Figure 8 compares the oil production rates, confirming the accuracy of the iEDFM approach. A good agreement exists in both 30 × 30/10 × 10. The curves of gridding methods I and II almost coincide, indicating that the results of these two methods are not much different when the permeability of fracture is much higher than that of the matrix, respectively. A curve of the same reservoir without any fracture is also present on this figure to show the influence of the existence of fractures. The effect of phase behavior on oil saturations is also pronounced in this case, as Figure 9 shows.

The computational times for iEDFM (30 × 30), iEDFM (10 × 10) and the fine-grid model are 9.1 s, 4.3 s and 6.7 h, respectively, which indicates a high efficiency of iEDFM.

3.3. Case 3: Nonplanar-Fractured Reservoir Flooding Test

Recent advances in fracture-diagnostic tools and fracture-propagation models make it necessary to model fractures with complex geometries in reservoir-simulation studies. A nonplanar shape is one of the most common complex geometries [30].

Fractures tend to grow in the direction perpendicular to the minimum horizontal stress. In some cases, the preferred direction of fracture may not keep the same, which often leads to a nonplanar shape.

The methodology introduced in this study can be directly applied in a nonplanar fractures case. As Figure 10 shows, the location of the assumed points S_i and F_i will reflect the geometry of the fracture. As a result, the semi-analytic method can bring iEDFM enough flexibility in modeling nonplanar fractures.

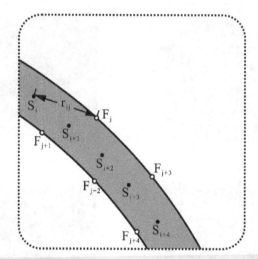

Figure 10. Schematic of equivalent point sinks and other parameters used in the nonplanar fracture case corresponding to Figure 3b.

We present an ideal case as shown in Figure 11 in which the fracture is formed as two connected arcs seen in the top view and is vertical in the z direction. The same parameters as Case 2 are applied here, and a fine-grid model similar to Case 2 is built. For the iEDFM simulation, gridding method I and a uniform matrix grid (30 × 30 with one layer in z direction) are used.

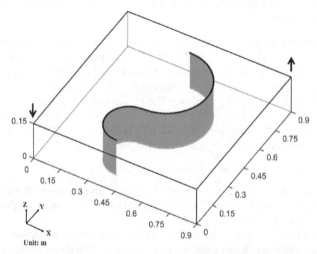

Figure 11. The nonplanar-fractured reservoir in Case 3. A water injector (8.64×10^{-5} m³/day) is located in one corner of the reservoir, and a producer is placed in the opposite corner.

The oil rate curves are presented in Figure 12 and the oil saturation profiles after 115 days of injection are shown in Figure 13, where a good agreement demonstrates the accuracy of the iEDFM in modeling the nonplanar fractures reservoir. The curves of the same reservoir without any fracture and with a planar fracture with the same starting and ending location are also present on this figure to show the influence of the existence of nonplanar fractures.

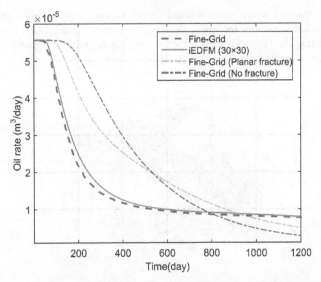

Figure 12. Comparison of oil production rate by the iEDFM/fine-grid model in Case 3.

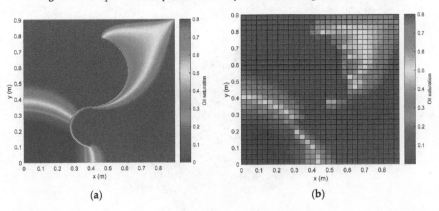

(a)   (b)

Figure 13. Profiles of oil saturation after 115 days' injection for Case 3 in (**a**) the fine-grid model and (**b**) iEDFM (30 × 30).

Actually, as Figures 3b and 10 show, different locations of point S_i and F_i are able to reflect any geometry of the fractures, such as fractures with variable apertures and vuggy-fractures, which means that iEDFM is naturally suitable for any other complex geometry cases beside nonplanar fractures.

3.4. Case 4: 3D Case with a Modified Dataset of a Real Field

As mentioned previously, iEDFM can be used as a general procedure in both 2D and 3D cases. In real field applications, the reservoir may have multiple layers, and the height of the fractures can be smaller than the reservoir height. Therefore, a 3D simulation example is presented to show the application of iEDFM in a typical field study.

In this case, the geological model is modified from a reported dataset which is interpreted from a 3D seismic survey [35,36]. Some irregular, sparsely distributed large-scale fractures are present as main fractures (black planes in Figure 14a). Some stochastic medium-scale fractures (blue planes in Figure 14a) are added to test iEDFM's applicability in a complex fracture-network situation. The main fractures and the wells are assumed to extend throughout the entire depth of the reservoir, while the stochastic medium-scale fractures are of different heights. Figure 14b shows the matrix grids (25 × 25 × 5) and the position of fractures and wells. An injection well (4×10^4 m^3/day) is placed in the center, while producers are placed in four corners of the reservoir. Again, the reservoir without any fracture is also simulated as a comparison.

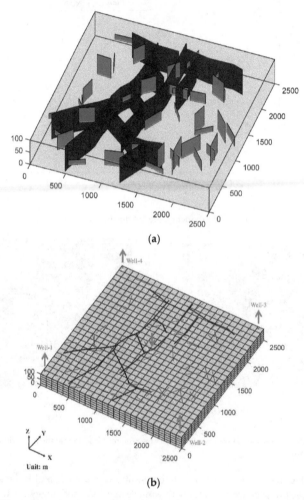

(a)

(b)

Figure 14. (a) Three-dimensional view of main fractures and medium-scale fractures; (b) The matrix grids of the reservoir and the position of each fracture and each well.

The oil rate curves are presented in Figure 15. The curves of wells 1–4 in the reservoir without fractures are coincident because of the symmetry. The pressure profiles of the top and bottom layers after 7300 days of injection are shown in Figure 16. As we can see, the reservoir shows strong heterogeneity due to the existence of fractures, and the water's breakthrough is advanced, which reduces the recovery efficiency. As medium-scale fractures are added, these phenomena become more pronounced. Because of gravity, the average oil saturation of the first layer is higher than that of the bottom layer (0.6952 of Figure 16c than 0.5671 of Figure 16d, respectively).

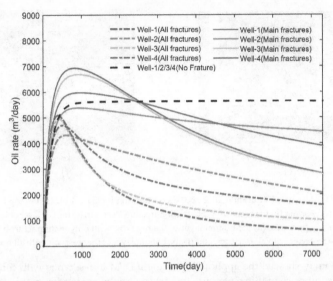

Figure 15. Comparison of oil production rate for Case 4.

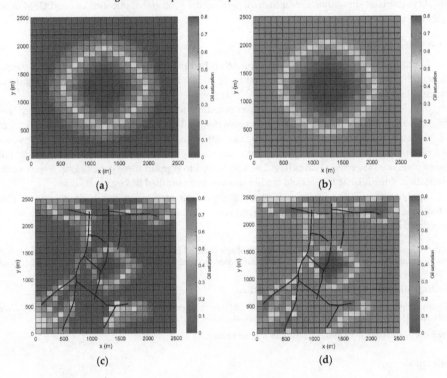

Figure 16. *Cont.*

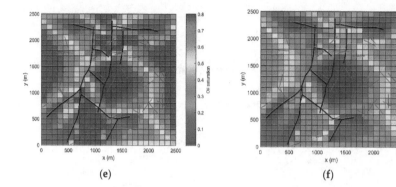

(e)   (f)

Figure 16. Profiles of oil saturation after 7300 days' injection for Case 4 of: (**a**) the top layer of the non-fracture reservoir; (**b**) the bottom layer of the non-fracture reservoir; (**c**) the top layer of the reservoir with only main fractures; (**d**) the bottom layer of the reservoir with only main fractures; (**e**) the top layer of the reservoir with all fractures; and (**f**) the bottom layer of the reservoir with all fracture.

The case study showed the applicability of the iEDFM in reservoirs with complex fracture networks. The influence of different scales of fractures can be modeled appropriately by iEDFM. The 3D multiphase simulation example demonstrates the potential application of the iEDFM in real field studies.

4. Conclusions and Future Work

In this study, we developed a new approach called the integrally embedded discrete fracture model (iEDFM). This approach, for the first time, avoids the limitations of the need to subdivide fracture elements in EDFM.

In iEDFM, we can arbitrarily grid fractures according to the requirements, and then embed them integrally in matrix blocks. As a precise pressure distribution assumption inside the matrix blocks is introduced, we can obtain a semi-analytic calculation method of matrix-fracture transmissibility. As a result, the simulation accuracy is improved and the application is also expanded to fractures with complex geometries. Several cases have been presented to support these conclusions. The potential application of the iEDFM in real field studies has also been testified through a 3D case.

Applying the iEDFM to real field study and guiding production is our ultimate goal. Thus, heterogeneous and more complex actual reservoir examples with actual production data will be considered in our on-going project.

Author Contributions: R.S. and Y.D. conceived of the idea. R.S. developed the theory and conducted the computations. Y.D. guided the data analysis and supervised the work. This paper was written by R.S. and modified by Y.D.

Funding: This research received no external funding.

Acknowledgments: This work was supported by National Science and Technology Major Project of China (2016ZX05014) and National Natural Science Foundation of China (51674010).

Conflicts of Interest: The authors declare no conflict of interest.

References

1.   Saller, S.P.; Ronayne, M.J.; Long, A.J. Comparison of a Karst Groundwater Model with and without Discrete Conduit Flow. *Hydrogeol. J.* **2013**, *21*, 1555–1566. [CrossRef]
2.   Gurpinar, O.M.; Kossack, C.A. Realistic Numerical Models for Fractured Reservoirs. *SPE J.* **2000**, *5*, 485–491. [CrossRef]

3. Basquet, R.; Cohen, C.E.; Bourbiaux, B. Fracture Flow Property Identification: An Optimized Implementation of Discrete Fracture Network Models. In Proceedings of the SPE Middle East Oil and Gas Show and Conference, Manama, Kingdom of Bahrain, 12–15 March 2005.

4. Sarda, S.; Jeannin, L.; Basquet, R.; Bourbiaux, B. Hydraulic Characterization of Fractured Reservoirs: Simulation on Discrete Fracture Models. In Proceedings of the SPE Reservoir Simulation Symposium, Houston, TX, USA, 11–14 February 2001.

5. Yan, B.; Mi, L.; Chai, Z.; Wang, Y.; Killough, J.E. An Enhanced Discrete Fracture Network Model for Multiphase Flow in Fractured Reservoirs. *J. Pet. Sci. Eng.* **2018**, *161*, 667–682. [CrossRef]

6. Warren, J.E.; Root, P.J. The Behavior of Naturally Fractured Reservoirs. *SPE J.* **1963**, *3*, 245–255. [CrossRef]

7. Kazemi, H. Pressure Transient Analysis of Naturally Fractured Reservoirs with Uniform Fracture Distribution. *SPE J.* **1969**, *9*, 451–462. [CrossRef]

8. Swaan, A.D. Theory of Waterflooding in Fractured Reservoirs. *SPE J.* **1978**, *18*, 117–122. [CrossRef]

9. Gilman, J.R. Efficient Finite-Difference Method for Simulating Phase Segregation in the Matrix Blocks in Double-Porosity Reservoirs. *SPE Reserv. Eng.* **1986**, *1*, 403–413. [CrossRef]

10. Gilman, J.R.; Kazemi, H. Improved Calculations for Viscous and Gravity Displacement in Matrix Blocks in Dual-Porosity Simulators. *J. Pet. Technol.* **1988**, *40*, 60–70. [CrossRef]

11. Pruess, K.; Narasimhan, T.N. Practical Method for Modeling Fluid and Heat Flow in Fractured Porous Media. *SPE J.* **1985**, *25*, 14–26. [CrossRef]

12. Wu, Y.S.; Pruess, K. Multiple-porosity Method for Simulation of Naturally Fractured Petroleum Reservoirs. *SPE Reserv. Eng.* **1988**, *3*, 335–350. [CrossRef]

13. Fung, L.S.K. Simulation of Block-to-Block Processes in Naturally Fractured Reservoirs. *SPE Reserv. Eng.* **1991**, *6*, 477–484. [CrossRef]

14. Wu, Y.; Di, Y.; Kang, Z.; Fakcharoenphol, P. A Multiple-Continuum Model for Simulating Single-phase and Multiphase Flow in Naturally Fractured Vuggy Reservoirs. *J. Pet. Sci. Eng.* **2011**, *78*, 13–22. [CrossRef]

15. Sun, H.; Chawathe, A.; Hoteit, H.; Shi, X.; Li, L. Understanding Shale Gas Flow Behavior Using Numerical Simulation. *SPE J.* **2015**, *20*, 142–154. [CrossRef]

16. Yan, B.; Alfi, M.; An, C.; Cao, Y.; Wang, Y.; Killough, J.E. General Multi-porosity Simulation for Fractured Reservoir Modeling. *J. Nat. Gas Sci. Eng.* **2016**, *33*, 777–791. [CrossRef]

17. Noorishad, J.; Mehran, M. An Upstream Finite-Element Method for Solution of Transient Transport-Equation in Fractured Porous-Media. *Water Resour. Res.* **1982**, *18*, 588–596. [CrossRef]

18. Karimi-Fard, M.; Firoozabadi, A. Numerical Simulation of Water Injection in Fractured Media Using the Discrete-Fracture Model and the Galerkin Method. *SPE Reserv. Eval. Eng.* **2003**, *6*, 117–126. [CrossRef]

19. Monteagudo, J.; Firoozabadi, A. Control-Volume Method for Numerical Simulation of Two-phase Immiscible Flow in Two- and Three-Dimensional Discrete-Fractured Media. *Water Resour. Res.* **2004**, *40*. [CrossRef]

20. Hoteit, H.; Firoozabadi, A. Compositional Modeling of Discrete-Fractured Media without Transfer Functions by the Discontinuous Galerkin and Mixed Methods. *SPE J.* **2006**, *11*, 341–352. [CrossRef]

21. Matthai, S.K.; Mezentsev, A.; Belayneh, M. Finite Element-Node-Centered Finite-Volume Two-Phase-Flow Experiments with Fractured Rock Represented by Unstructured Hybrid-Element Meshes. *SPE Reserv. Eval. Eng.* **2007**, *10*, 740–756. [CrossRef]

22. Sandve, T.H.; Berre, I.; Nordbotten, J.M. An Efficient Multi-Point Flux Approximation Method for Discrete Fracture-Matrix Simulations. *J. Comput. Phys.* **2012**, *231*, 3784–3800. [CrossRef]

23. Karimi-Fard, M.; Durlofsky, L.J.; Aziz, K. An Efficient Discrete-Fracture Model Applicable for General-Purpose Reservoir Simulators. *SPE J.* **2004**, *9*, 227–236. [CrossRef]

24. Olorode, O.M.; Freeman, C.M.; Moridis, G.J.; Blasingame, T.A. High-Resolution Numerical Modeling of Complex and Irregular Fracture Patterns in Shale-Gas Reservoirs and Tight Gas Reservoirs. *SPE Reserv. Eval. Eng.* **2013**, *16*, 443–455. [CrossRef]

25. Al-Hinai, O.; Signh, G.; Pencheva, G.; Almani, T.; Wheeler, M.F. Modeling Multiphase Flow with Nonplanar Fractures. In Proceedings of the SPE Reservoir Simulation Symposium, The Woodlands, TX, USA, 18–20 February 2013.

26. Moinfar, A.; Varavei, A.; Sepehrnoori, K.; Johns, R.T. Development of an Efficient Embedded Discrete Fracture Model for 3d Compositional Reservoir Simulation in Fractured Reservoirs. *SPE J.* **2014**, *19*, 289–303. [CrossRef]

27. Boyle, E.J.; Sams, W.N. Nfflow: A Reservoir Simulator Incorporating Explicit Fractures. In Proceedings of the SPE Western Regional Meeting, Bakersfield, CA, USA, 21–23 March 2012.

28. Lee, S.H.; Lough, M.F.; Jensen, C.L. Hierarchical Modeling of Flow in Naturally Fractured Formations with Multiple Length Scales. *Water Resour. Res.* **2001**, *37*, 443–455. [CrossRef]

29. Li, L.; Lee, S.H. Efficient Field-Scale Simulation of Black Oil in a Naturally Fractured Reservoir Through Discrete Fracture Networks and Homogenized Media. *SPE Reserv. Eval. Eng.* **2008**, *11*, 750–758. [CrossRef]

30. Xu, Y.; Cavalcante Filho, J.S.A.; Yu, W.; Sepehrnoori, K. Discrete-Fracture Modeling of Complex Hydraulic-Fracture Geometries in Reservoir Simulators. *SPE Reserv. Eval. Eng.* **2017**, *20*, 403–422. [CrossRef]

31. Yu, W.; Xu, Y.; Liu, M.; Wu, K.; Sepehrnoori, K. Simulation of Shale Gas Transport and Production with Complex Fractures Using Embedded Discrete Fracture Model. *AIChE J.* **2018**, *64*, 2251–2264. [CrossRef]

32. Li, W.; Dong, Z.; Lei, G. Integrating Embedded Discrete Fracture and Dual-Porosity, Dual-Permeability Methods to Simulate Fluid Flow in Shale Oil Reservoirs. *Energies* **2017**, *10*. [CrossRef]

33. Tene, M.; Bosma, S.B.M.; Al Kobaisi, M.S.; Hajibeygi, H. Projection-Based Embedded Discrete Fracture Model (pEDFM). *Adv. Water Resour.* **2017**, *105*, 205–216. [CrossRef]

34. Wu, Y.S. A Virtual Node Method for Handling Well Bore Boundary Conditions in Modeling Multiphase Flow in Porous and Fractured Media. *Water Resour. Res.* **2000**, *36*, 807–814. [CrossRef]

35. Shuck, E.L.; Davis, T.L.; Benson, R.D. Multicomponent 3-D Characterization of a Coalbed Methane Reservoir. *Geophysics* **1996**, *61*, 315–330. [CrossRef]

36. Zhang, Y.; Gong, B.; Li, J.; Li, H. Discrete Fracture Modeling of 3D Heterogeneous Enhanced Coalbed Methane Recovery with Prismatic Meshing. *Energies* **2015**, *8*, 6153–6176. [CrossRef]

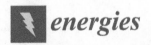

Article

An Experimental Study of Combustion of a Methane Hydrate Layer Using Thermal Imaging and Particle Tracking Velocimetry Methods

Misyura S. Y. [1,2,*], Voytkov I. S. [1], Morozov V. S. [2], Manakov A. Y. [3], Yashutina O. S. [1] and Ildyakov A. V. [3]

[1] National Research Tomsk Polytechnic University, Tomsk 634050, Russia; vojtkov12@mail.ru (V.I.S.); yashutina1993@mail.ru (Y.O.S.)
[2] Institute of Thermophysics Siberian Branch, Russian Academy of Sciences, 1 Acad. Lavrentiev Ave., Novosibirsk 630090, Russia; morozov.vova.88@mail.ru
[3] Nikolaev Institute of Inorganic Chemistry, Siberian Branch, Russian Academy of Sciences, 3 Acad. Lavrentiev Ave., Novosibirsk 630090, Russia; manakov@niic.nsc.ru (M.A.Y.); avildyakov@gmail.com (I.A.V.)
* Correspondence: misura@itp.nsc.ru

Received: 9 November 2018; Accepted: 13 December 2018; Published: 17 December 2018

Abstract: In this paper, the combustion of methane hydrate over a powder layer is experimentally studied using thermal imaging and Particle Tracking Velocimetry (PTV) methods. The experiments are carried out at different velocities of the external laminar air-flow from zero to 0.6 m/s. Usually, simulation of methane hydrate combustion is carried out without taking into account free convection. A standard laminar boundary layer is often considered for simplification, and the temperature measurements are carried out only on the axis of the powder tank. Measurements of the powder temperature field have shown that there is a highly uneven temperature field on the layer surface, and inside the layer the transverse temperature profiles are nonlinear. The maximum temperature always corresponds to the powder near the side-walls, which is more than 10 °C higher than the average volumetric temperature in the layer. Thermal imager measurements have shown the inhomogeneous nature of combustion over the powder surface and the highly variable velocity of methane above the surface layer. The novelty of the research follows from the measurement of the velocity field using the PTV method and the measurement of methane velocity, which show that the nature of velocity at combustion is determined by the gas buoyancy rather than by the forced convection. The maximum gas velocity in the combustion region exceeds 3 m/s, and the excess of the oxidizer over the fuel leads to more than tenfold violation of the stoichiometric ratio. Despite that, the velocity profile in the combustion region is formed mainly due to free convection, it is also necessary to take into account the external flow of the forced gas U_0. Even at low velocities U_0, the velocity direction lines significantly deviate under the forced air-flow.

Keywords: combustion; methane hydrate; hydrate dissociation; PTV method

1. Introduction

1.1. The Key Parameters Controlling the Dissociation of Gas Hydrates

Huge deposits of natural gas in the form of methane hydrates are found in the marine sediments [1,2]. Attention to alternative clean energy sources is increasing every year. Today, scientific interest is focused on the methods of extraction of alternative raw materials, technologies for their storage, transportation, as well as methods of efficient combustion of natural fuel [1–3]. The complexity of the study of methane hydrates combustion is associated with the presence of three interrelated processes: The dissociation of solid particles of methane hydrate; gas filtration through a multicomponent medium (methane–water–water vapor); and diffusion combustion of methane in the mixing layer (methane–air–water vapor), which is usually implemented over the powder layer. Therefore, it is necessary to consider the features of these processes.

The gas hydrate dissociation rate is determined by the following driving forces: Deviation of temperature and pressure from the equilibrium states, size of particles, and structural characteristics of ice crust [4,5]. Dissociation at positive temperatures has a thin film of water formed on the surface of solid particles. The dissociation rate is quasi-constant for most of the dissociation time. The case of negative temperatures is significantly more difficult to describe due to the emergence of the phenomenon of self-preservation [6]. In the temperature range of 230–268 K, the dissociation rate of methane hydrate decreases by several orders of magnitude and strongly depends on temperature. This annealing temperature region was called the self-preservation region. The lowest dissociation rate of methane hydrate is achieved at the temperature of 265–267 K [6]. A sharp decrease in the dissociation rate is bound with the appearance of a strong fine-grained ice shell without pores on the surface of the granules. Thus, in addition to the expression describing the system deviation from the equilibrium, it is necessary to take into account the structural properties of the ice crust. At atmospheric air pressure, the equilibrium temperature of methane hydrate is approximately equal to 188 K. The mechanism of self-preservation was studied in References [7–19]. The influence of structural parameters of the ice crust on the dissociation process was studied by scanning electron microscopy (cryo-FE-SEM) [6,15–17]. The effect of temperature and pressure on self-preservation was considered in Reference [14]. When the methane hydrate temperature changes, the texture changes, which are clearly visible by scanning electron microscopy, occur on the surface of the ice crust [16,17]. The behavior of mixed methane hydrates at an abnormally low dissociation rate was considered in Reference [19]. Kinetics of methane hydrate dissociation depends on the external heat flux, curvature of granules, and pore characteristics [20,21]. Collective effects of crystallohydrate growth on the free surface of the liquid were studied using thermal imaging technology [22].

When modeling filtration, it is necessary to take into account diffusion and heat transfer for both a porous methane hydrate particle and a porous medium (solid particle powder–methane–water–water vapor) [20,23].

1.2. Combustion of Gas Hydrates

Methods of organization of combustion may vary. Thus, natural hydrate of sedimentary rocks is burned in large reactors. In the tank there is sand with methane hydrate, over which a layer of water is formed during combustion. In this case, the combustion above the layer surface is uneven and unstable, and separate flame tips are formed [24]. Combustion of synthetic hydrate without impurities (sand and clay) is more stable. In this case, the flame height for pure methane hydrate is higher than for natural sedimentary rocks. However, even in this case, periodic flame tips appear on the surface, and their location may change [24]. Combustion of methane hydrate and double hydrate (methane-alcohol) at free gas convection above the surface of the powder layer is also considered in References [25–27]. Combustion is shown to lead to a multiple increase in the dissociation rate. In addition, in the presence of combustion, the methane velocity above the layer surface is 10–20 times higher than in the case without combustion [25].

The above combustion option is implemented without the organization of forced air-low. To organize a more stable combustion, a forced flow of oxidizer (air) is used. The methane flow moves perpendicular to the powder surface. Methane is formed by the methane hydrate dissociation. Above the layer surface, there is a laminar air-flow [28–31]. The complexity of the research is that it is difficult to determine the value of methane velocity, as it depends on the temperature of the powder. The temperature in the experiments can vary from zero to -80 °C in the transverse direction of the layer. In this range, there are several characteristic temperature zones [10–14]: From −80 to −45–50 °C there is a multiple increase in the dissociation rate with temperature growth; from −40 to −5 °C there is self-preservation when the dissociation rate falls by several orders of magnitude and the rate varies greatly with temperature; at about 0 °C and above, there is a very high dissociation rate and a high density of methane flow over the powder. Thus, it is obvious that it is extremely difficult to analyze the methane combustion in the entire specified range of powder temperature and that there may be conflicting results. For the development of calculation models, there is a need in conditions when the temperature of the powder during combustion is quasi-constant over time, which is difficult to organize in practice.

Often the analysis is carried out based on the powder temperature near the upper surface of the methane hydrate layer T_s. It has been experimentally established that depending on temperature T_s there may be both the low speed of flame spreading and the high speed.

The effect of external air velocity on the speed of flame spreading is considered in Reference [29]. The velocity of the external forced air-flow U_0 varied in the range 0.1–1.5 m/s. The initial temperature of the powder surface before combustion (in the middle of the tank) was the same for all experiments and was equal to −80 °C. It has been established that the tenfold growth of speed led only to the 20–30% growth of U_s, and the high speed of flame spreading mode was absent. Possible causes of such flame behavior will be discussed further in the analysis of experimental data. In another experimental work [31], the ignition was carried out at a certain temperature T_s, which varied for different experiments in the range from −80 °C to −10 °C. The external air velocity was constant $U_0 = 0.4$ m/s. With an increase in the initial temperature T_s, a change in combustion modes was observed: (1) The low speed flame spreading ($T_s < -50$ °C); (2) the high speed of flame spreading ($T_s = -40$–50 °C); (3) the low speed of flame spreading ($T_s > -50$ °C). Surprisingly, the increase in methane velocity V_{CH4} by only 20–40% (from 0.5 mm/s to 0.6–0.7 mm/s) led to about 100 times increase in the speed of flame spreading U_s (from 10–15 mm/s to 1000 mm/s). The authors of the article in Reference [31] attribute such a strong increase in speed to the fact that in the first case the stoichiometric line was absent (lack of fuel for the reaction), and in the second case (the high speed of flame spreading), the stoichiometric line was present and was located above the dark region (the area with no combustion). However, more experimental and theoretical studies are needed to justify such strong U_s growth.

Let us consider another method of mixing fuel (methane) and oxidizer (air) [32]. In this paper, the motion of methane and air is organized in opposite directions in a narrow slit, which allows for a good mixing of the oxidant and fuel and provides an approximate stoichiometric ratio. The air velocity and the slit height were selected to perform the stoichiometric ratio. The methane velocity was estimated using the gravimetric method (the fall of the mass of methane hydrate powder over time was recorded using weights). This method proved to be successful, and the maximum flame temperature was about 1700 K. In this case, the calculation has shown that the temperature of the combustion is possible at the mass fraction of water of 0.6. The injection of pure methane (without water vapor) led to a combustion temperature value of about 1950 K. Thus, high concentration of water leads to an underestimation of fuel combustion temperature by 200–250 K.

An experimental study of combustion of a single sphere of methane hydrate was performed in Reference [33]. The rate of change in the sphere diameter decreases with time and depends on the sphere temperature. Theoretical calculation of combustion of a separate sphere taking into account dissociation, formation of a film of water, steam bubbles, and water vapor was considered in Reference [34]. The simulation was performed for a sphere of small diameter of 0.1 mm and for positive temperatures. The curve of change of a square of sphere radius over time had quasi-linear character for the most time of dissociation. Thus, this calculation differs significantly from the experimental conditions [33]. The maximum combustion temperature in the flame in Reference [34] corresponds to approximately 1700–1750 K, which closely corresponds to the data [32]. The mass fraction of vapor near the sphere surface is 0.65–0.7 and decreases rapidly at a distance from the surface, which also closely corresponds to the calculation in Reference [32] (0.65). It is obvious that for a narrow slit [32] and for a large number of methane hydrate particles, water saturation occurs and water concentration in the flame is the same as on the surface of the powder layer (0.65).

The influence of the blowing parameter on the combustion efficiency in the methane-air mixing layer is considered in Reference [35]. Since the considered air flow rates are low and there is a noticeable excess or lack of oxidant, the combustion stability in the diffusion layer will depend on the Richardson number $Ri = Gr/Re^2$ (Gr is Grashof number, and Re is Reynolds number) [29]. The process of mixing the fuel and oxidizer will depend on both buoyancy and forced convection. The modes of non-premixed combustion were considered in Reference [36] depending on the D_a numbers. When burning methane hydrate, depending on the temperature of the powder, the thickness of the dynamic layer and the mixing layer, the blowing parameter, and the rate of chemical reaction can vary significantly. In addition, low air velocities and high temperature gradient in the mixing layer lead to the formation of a vortex flow. In this case, knowledge of the mean and local flow parameters is needed to accurately model the transfer processes. This information can be obtained using modern optical non-contact methods of Shadow Photography (SP) [37], Planar Laser Induced Fluorescence (PLIF) [38], Particle Tracking Velocimetry (PTV) [39,40], and Particle Image Velocimetry (PIV) [38]. These optical methods allow for a deeper clarification of the influence of free convection on the evaporation rate and heat transfer and are already are widely used in the study of droplets, films, sprays, micro-channel, heterogenic flows, two-phase (vapor-liquid) flows and emulsions, boundary layers, and crystallization processes in multicomponent solutions.

From the above, it follows that, to increase the efficiency of combustion of methane hydrate, the following problems must be surmounted: (1) Maintaining the powder temperature quasi-constant for the entire volume and during the entire combustion time; (2) the temperature of the powder should provide a high rate of dissociation of the methane hydrate, i.e., to exclude the phenomenon of self-preservation and at the same time to ensure the maximum deviation of temperature and pressure from the equilibrium curve (the driving forces of dissociation); (3) it is necessary to ensure the kinetic combustion condition, as the excess of oxidant and fuel lead to low reaction rates during combustion. To do this, it is necessary to accurately calculate the diffusion mixing layer and to select the optimal conditions for mixing the fuel and oxidizer; and (4) the concentration of steam that enters the combustion region should be reduced.

Analysis of the existing literature has shown that for correct modeling and further development of methane hydrate combustion technologies, additional experimental data on the instantaneous characteristics of the velocity field in the mixing layer are needed. The existing data for the case of pure methane injection through a porous wall are not suitable for the considered problem (methane hydrate combustion). In addition, there are no experimental data on the rate of dissociation during combustion in the organization of forced air-flow.

The aim of this work is to determine the dependences for the dissociation rate when the velocity of the external air flow changes in a wide range and to visualize the velocity field during combustion using the thermal imaging method and the optical contactless Particle Tracking Velocimetry (PTV) method. The obtained experimental data will help to improve and test the existing computational methods.

2. The Connection of Dissociation Rate and Heat Transfer

It is well known that the internal kinetics of the process plays an important role only at initial stages of crystal growth/decomposition. In real processes, taking into account the observed crystallization time (relatively long times), the point is not the formation and development of the crystallization nucleus with the size r_{cr}, but the growth of crystals (crystallohydrates) of rather large size $r \gg r_{cr}$. In this case, the rate of growth or dissociation will be determined by heat transfer and diffusion, since the characteristic kinetic time will be several orders of magnitude less. It has been previously shown that only for extremely low pore density (or very small pore radii) it is necessary to take into account the diffusion of methane through the pores and the value of the kinetic constant [20]. These conditions correspond to abnormally low dissociation rates when the values of the dissociation rate j fall by three to four orders of magnitude [6]. In the present work, there are no such low rates, despite the fact that the powder temperature passes through the annealing temperature window from $-50\,^\circ\text{C}$ to $-3\,^\circ\text{C}$. This contradiction is associated with a strong transverse and longitudinal temperature gradient. Experimental data on temperature profiles will be presented further. As a result of uneven temperature, in different places of the powder there will be areas of both high and very low values of j. The total flow j will have a relatively high value.

As mentioned above, the main factor regulating the dissociation dynamics is heat transfer. Thus, it is important to investigate the effect of heat flux q on the dissociation rate. Earlier it has been shown that at free convection the growth of heat flux leads to an increase in j. Combustion led to about 55–65 times increase in q (compared to the dissociation without combustion). The dissociation rate increased only eight–nine times [21,25]. Based on experimental data [21,25], as well as the data of this article, it is possible to associate the heat flux q (W/m^2) with the dissociation rate j (kg/s) in expression (1).

$$j = a_1(q)^{n_1}, \text{ where } a_1 = 0.012 \cdot 10^{-6}, n_1 = 0.7 \tag{1}$$

This expression relates the processes of diffusion and heat transfer both for the solid phase (methane hydrate granules) and for the gas mixture (air–methane–water vapor).

It is rather difficult to analytically connect the methane transfer with heat transfer, since they are described by different equations. However, it is possible to use a triple analogy and similarity between the transfer of heat, momentum, and diffusion [41–44]. In fact, this similarity is not performed even when there is gas injection through the wall [42,44]. In our case, blowing is associated with dissociation, which depends on the heat flux and temperature, i.e., the velocity of the injected air cannot be constant. In addition, the presence of free convection will also lead to a significant dissimilarity of dimensionless temperature and concentration profiles in the mixing layer. Therefore, we can use a triple analogy for qualitative evaluation, since the qualitative behavior of j and q is similar in many ways. For example, with an increase in q, j grows as well. For such a comparison, the stationarity condition must also be satisfied, i.e., the change of dC/dt and dT/dt *over* time must be much smaller than that for the convective terms $V\rho C_p dT/dy$, $U\rho C_p dT/dx$ and $V\rho dC/dy$, $U\rho dC/dx$ (where C and T are the methane concentration and temperature in the near-wall boundary layer; x, y are the longitudinal and transverse coordinates; and U, V are the longitudinal and transverse velocities. In addition, the three-dimensionality should also be insignificant, i.e., a quasi-plane approximation is considered. Even when burning, the total period of methane hydrate dissociation is large enough and has an order of 50–100 s. Therefore, the process of dissociation can be approximately considered as quasi-stationary.

It is important to note that in this problem there is a transfer of methane both in the powder layer and in the gas mixture above the powder surface. Thus, we have three characteristic processes of diffusion and three characteristic diffusion relaxation times t_d: (1) Diffusion through the solid particle (methane transport through the pores in the outer crust of ice) (t_{d1}); (2) diffusion through the porous space within the powder layer (t_{d2}); and (3) diffusion of methane in the mixing layer, developing over the powder surface (t_{d3}). Under the condition $\Delta C_1/\Delta T_1 = \text{const}$ (where ΔC_1 and ΔT_1 are dimensionless values; the dimensionalization can be done based on initial differences), $j \sim q \sim (Re)^{0.5} \sim (U_0)^{0.5}$. In accordance with expression (1), a fourfold increase in the heat flux leads to approximately a double increase in the dissociation rate, i.e., to a two-fold increase in the average density of the methane flux. A noticeable lag in the growth of j is probably due to the fact that the growth of q leads to an increase in the temperature gradient in the boundary layer ΔT_1. At that, the growth of ΔC_1 for the same period of time is significantly slower. It can be reasonably assumed that $t_{d1} \gg t_{d2} \gg t_{d3}$. Thus, the main diffusion resistance is associated with the transfer of methane through the pores. As mentioned above, in the powder volume there are local areas of partial "self-preservation", which lead to the formation of closed pores and the appearance of a high-strength ice shell. As a result, the total methane flow falls significantly and is determined by the flow of methane through a region without self-preservation (the area of the powder where the temperature $T < -50\,^\circ\text{C}$ and $T > -3\,^\circ\text{C}$). Thus, the nature of dependence (1) is determined by the pore density in the granules. Then, the combustion kinetics and temperature (T_c) also depend on the methane concentration and on the pore density distribution (ρ_p) by the size of r_p (r_p—pore radius) for all granules, i.e., $T_c = f(P(\rho_p))$, where P is the pore size distribution function). Taking into account expression (1) and taking the dependence $q \sim (U_0)^{0.5}$, we obtain expression (2) ($n_2 = n_1 \cdot 0.5 = 0.7 \cdot 0.5 = 0.35$).

$$j = a_2(U_0)^{n_2}, \text{ where } a_2 = 6.8 \cdot 10^{-6}, n_2 = 0.35 \tag{2}$$

3. Experimental Apparatus and Procedure

3.1. Test Section for Organizing Methane Hydrate and Combustion

Most research works on methane hydrate combustion are related to temperature measurements in the powder layer and the flame spread speed determination. The absence of data on the methane flux density and heat flux both in the powder layer and in the gas phase does not allow correctly simulating the processes of dissociation and combustion. To calculate the combustion parameters, it is necessary to know the rate at which methane flows from the powder layer to the diffusion mixing layer. This flow is formed as a result of methane hydrate dissociation and methane formation. In these studies, methane velocity was determined using the weight method (the digital scales Vibra AJH 4200 CE). The dissociation rate j (g/s) was measured as $j = \Delta m / \Delta t$ (where m is the powder mass, and t is the time). The maximum measurement error j was less than 8–9%. Figure 1 shows the experimental setup. Three thermocouples ($y = 1$ mm, $y = 5$ mm, $y = 14$ mm) are located in the center of the working area at different distance y from the upper surface of the powder to measure the temperature profile 1 of methane hydrate in the transverse direction. The vessel walls were made of stainless steel with thickness of 0.7–0.8 mm. Around the tank thermal insulating material was located.

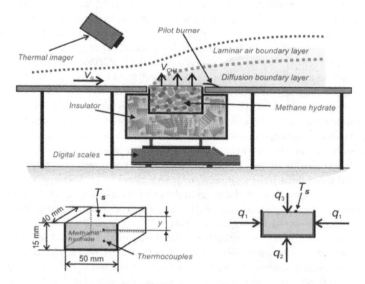

Figure 1. Scheme of experimental setup.

Temperature field on the surface of the powder was measured with the use of a thermal imaging camera (NEC San Instruments). The measurement error of the thermocouple was within 1%. The difference between the temperature values measured by the thermocouple and the thermal imager did not exceed 1.5–2 K.

Methane hydrate combustion was realized at different air flow velocity U_0 from 0 to 0.6 m/s. The mean flow velocity profile coincides with the Blasius velocity profile of a flat laminar boundary layer. The intensity of the turbulent velocity fluctuation e ($e = u_{rms}/U_0$, where u_{rms} is the mean square value of velocity pulsations and U_0 is the average velocity) was less than 2%. The surface temperature range of the samples (T_s) varied from 90 K to 273 K. Before the experiment, the methane powder was taken from liquid nitrogen and placed in a tank. Heating of the powder was due to the heat coming from the ambient medium, i.e., from the air through the top surface of the powder and from the metal walls of the tank. Evaluation of thermal resistance (heat transfer coefficient) shows that heat fluxes both from the upper surface and the side-walls are of the same order of values (without combustion of methane). The influence of the side-walls leads to a significant longitudinal temperature gradient (10–20 K) and the maximum temperature always corresponds to the powder on the surface and near the side-walls. Thus, in addition to measuring the transverse temperature profile, in these experiments the temperature of the entire surface of the powder T_s is measured using the thermal imager. The average temperature of the powder is important to know for the correct simulation of the dissociation rate. In most experimental works, temperatures are measured only in the center of the powder tank. The motion of the leading flame edge starts from the edge of the tank, where the temperature can be 20 K higher and a higher local density of methane flow is realized.

A diffusion layer (methane–air–water vapor) is formed inside the laminar boundary layer of air. The thickness of the diffusion layer is lower than that of the dynamic layer for air (Figure 1).

Methane hydrate was synthesized using the following procedure. The cooled autoclave was loaded with the powder of ice, the autoclave was blown with methane and the methane pressure of 7–10 MPa was set. The autoclave was placed in a thermostat with a temperature of +1 °C. Hydrate was synthesized by slow melting of ice during the day. Further, the autoclave was cooled to the temperature of liquid nitrogen, and the resulting mixture of methane hydrate and ice was ground and reloaded into the autoclave (all at the temperature of liquid nitrogen). After warming up the autoclave to about −100 °C, the methane pressure was set, and the autoclave was again placed in a thermostat with a temperature of +1 °C. This procedure was repeated two times, resulting in an almost pure methane hydrate with an ice content of not more than a few mass %. Before use, the hydrate was stored immersed in liquid nitrogen. The purity of the sample was controlled by powder X-ray diffraction. The average size of methane hydrate particles in all experiments was 0.1–0.2 mm.

3.2. Measurement Using Particle Tracking Velocimetry (PTV) Method

The method of "Particle Tracking Velocimetry" (PTV) was used to determine the gas flow velocity (formed during combustion of methane hydrates). Similar to the Particle Image Velocity (PIV) method, the PTV algorithm is based on determining the most likely particle shift for a very short time delay. The use of the PTV method in these experiments is due to its peculiarities: Unlike the PIV method, the PTV method does not imply the use of a large number of tracing particles. For this reason, the influence of tracers on the process can be neglected. In addition, the PTV method has a higher spatial resolution, as well as a lower sensitivity to uneven seeding of the flow by particles (compared to PIV). TiO_2 powder (particle sizes of 0.1–1 μm) was used as tracing particles for seeding the registration area. The tracers were blown into a container with hydrate in advance (3–5 s before the start of the experiment) and formed a slurry. The registration area was dissected with a sheet of a double pulsed Nd: YAG laser "Qantel EverGreen 70" (wave length—532 nm, pulse energy—74 mJ) with a frequency of 0.25 Hz. To create a flat sheet, an optical nozzle with an opening angle of 45° was used. The thickness of the laser sheet in the registration area was about 200 μm. Images of TiO_2 microparticles in the plane cut by the laser knife were recorded by CCD video camera "ImperX IGV B2020M" (resolution of 2048 × 2048 pix, and bit width of 8 bit). The angle between the optical axis of the camera and the plane of the laser sheet was 90°. A couple of frames were recorded for each moment. The delay between frames in a pair (frame delay) was 50–100 μs. The data processing required the use of "Relaxation method of PTV"—a two-frame PTV with an estimation of the probability of outcomes. Image processing included several successive stages (steps). At the first step, the background was subtracted from the obtained images and extraneous noise was removed (using mathematical algorithms, as well as "Median" and "Average" filters). At the second step, the intensity threshold

"Intensity Threshold" was set to binarize the obtained images. The threshold was set as a percentage of the maximum intensity value (255 for 8-bit images). The areas that underwent binarization marked the position of the particles. At the third step, for each region after binarization, the procedure for finding the center of the particle using the algorithm for calculating the center of mass was carried out. The fourth step was to find a pair for each particle from the first frame in the second frame. The radius for the pair search was set based on the maximum possible displacement of the particles in the image. At the fifth step, the displacement of each particle (l) was determined, after which the absolute value of the velocity of each particle was determined at the known values of the scale coefficient (S) and the frame delay value (dt): $u = S \cdot l/dt$. The last (sixth) step stipulated the elimination of erroneous vectors (using the algorithm "Moving average validation").

The result of each experiment was a set of irregular two-component velocity fields. The systematic error in the determination of particle velocity values by the PTV method depended on the type of optics used and the size of the registration area and did not exceed 1.5% in experiments. An example of an image of tracing particles and an irregular velocity field is shown in Figure 2.

Figure 2. Video image of TiO_2 particles (**left**) and the result of its processing by PTV (**right**).

4. Measurement

4.1. The Dissociation Rate in the Combustion of Methane Hydrate

To determine the dissociation rate of methane hydrate, it is necessary to construct curves of powder mass change over time. Figure 3 shows curves for different external air-flow rates U_0. The powder mass values are divided by the initial mass m_0 (m_0 corresponds to the methane hydrate mass when all methane is in the sample). As the velocity U_0 increases, the slope of the curves grows, i.e., the dissociation rate $j = \Delta m/\Delta t$ increases. The graphs of j change are shown in Figure 4. The rate of methane hydrate dissociation depends on the diffusion, heat transfer, and internal kinetics of the reaction.

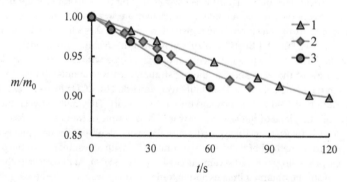

Figure 3. Change of powder mass over time (m_0—initial powder mass): $1 - U_0 = 0$ mm/s; $2 - U_0 = 0.2$ mm/s; and $3 - U_0 = 0.6$ mm/s.

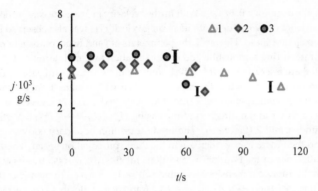

Figure 4. The dissociation rate $j = \Delta m / \Delta t$ over time (m_0—initial powder mass): 1 — $U_0 = 0$ mm/s; 2 — $U_0 = 0.2$ m/s; and 3 — $U_0 = 0.6$ m/s (**I**—maximum error interval).

Figure 5 presents experimental data (points 1) for the dissociation rate j of methane hydrate depending on the velocity of the incoming air flow U_0. The calculated curve 2 is obtained in accordance with expression (2). The difference between the curve 2 and experimental data 1 can be associated with both the experimental error (measurement error j) and the deviation of the dependence $q \sim (U_0)^{0.5}$ from the quadratic form. This deviation is probably due to the presence of both a forced air-flow and a free-convective flow due to high temperature gradient and gas density in the diffusion layer of the mixture. The degree $n = 0.5$ corresponds only to the plane laminar boundary layer when the buoyancy is neglected, i.e., when $St \sim 1/(Re)^{0.5}$ and $q \sim (U_0)^{0.5}$ [41–44], where St and Re are the Stanton number and the Reynolds number, $St = q/(\rho_0 U_0 c_p \Delta T)$, $Re = \rho_0 U_0 x / \mu$, x is the longitudinal coordinate, μ is the dynamic viscosity, and c_p is the heat capacity of the gas mixture. If the characteristic velocities caused by forced flow and buoyancy are comparable (of the same order), their ratio can be estimated using the Richardson number $Ri = Gr/Re^2$ (Gr is the Grashof number).

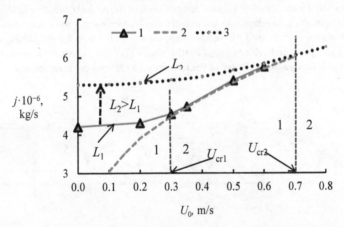

Figure 5. The dissociation rate j of methane hydrate depending on the velocity of the incoming air flow U_0: 1—experiment; 2—calculation by (2); $U_{cr1} = 0.3$ m/s; $U_{cr2} = 0.7$ m/s [29]; 3—Shift curve2 to the top; and L—the characteristic lengths of the working area.

The deviation of experimental data (1) from curve (2) begins at $U_0 < 0.25$–0.3 m/s, since expression (2) does not describe free convection (U_0 is the velocity of the forced convection). Thus, there are at least two dissociation modes of j, depending on U_0, for which the degree n will vary significantly. For 0 m/s $< U_0 < 0.3$ m/s the dissociation rate is quasi-constant ($n = 0$), and for 0.3 m/s $<$

$U_0 < 0.6$–0.7 m/s the degree $n = 0.29$ m/s. With further velocity increase the combustion suppression is observed. The existence of two modes with increasing velocity U_0 was also observed in Reference [29] for the flame propagation speed. However, the parameters of j and flame propagation speed correlate, i.e., the reaction rate during combustion and the dissociation rate are interrelated. At the same time, in Reference [29] there is a certain shift for the boundary of two modes transition. The critical value of U_{cr1} corresponds to velocity, when the experimental curve 1 deviates from the calculated curve 2. U_{cr2} determines the deviation of curve 3 from the calculated curve 2. Curve 3 shows the shift of the experimental curve 1 in the direction of increasing U_{cr2} ($U_{cr2} = 0.7$ m/s). Vertical dotted lines distinguish regimes 1 and 2 (Figure 5). The mode transition boundary corresponded to velocity $U_0 = 0.7$ m/s, and then there was an increase in the flame propagation speed. These differences are related to the difference in the boundary conditions for the two works. It is easy to link possible causes of this shift with two dimensionless criteria $A_1 = A_{1cr} = \delta_d/\delta_m$ (dynamic and diffusion layer thickness ratio) (Figure 6) and $A_2 = A_{2cr} = U_0/U_{conv}$ (ratio of air velocity to convective flow velocity). In Reference [29], there is a longer background (prehistory) and a thicker δ_{d2}. As a result, U_{0m2} will be, on the contrary, less than U_{0m1} in the present work. Let us consider the second reason (A_{2cr}). The convection velocity can be approximately expressed via the characteristic lengths of the working area L in form (3).

$$U_{conv} \sim (2gL\Delta\rho/\rho)^{1/2} \tag{3}$$

Then, the ratio of characteristic convection velocities for two works U_{conv2}/U_{conv1} will be proportional to $(L_2/L_1)^{1/2} = (120\,\text{mm}/25\,\text{mm})^{1/2} = 2.2$. Ignition in [29] begins at the end of the working area, the length of which is 120 mm. In the present work, combustion is realized approximately in the middle of the working section of 50 mm long. Thus, the number A_2 for L_2 will be higher than for L_1. As a result, to realize some critical value of A_{2cr}, it is required to increase the velocity almost 2.2 times in [29]. In this case, external convection U_0 will exert a significant influence on heat and mass transfer and on dissociation rate of methane hydrate. Then, A_{2cr} is implemented for velocity $U_0 = 0.3 \times 2.2 = 0.66$ m/s. The obtained estimated critical velocity (0.66 m/s) corresponds to the velocity in [29], which is 0.7 m/s. The third reason for the differences in the critical velocity (transition modes) is associated with different average temperature of the powder, which will lead to different dissociation rate, methane injection rate, the average combustion temperature, the difference between the combustion temperature, and the surface temperature of the powder ΔT, which will lead to a change in $\Delta\rho$ ($\Delta\rho = \beta\Delta T$) and velocity U_{conv}. Further theoretical and experimental studies are required to determine more precise dependences of the transition mode on these three factors.

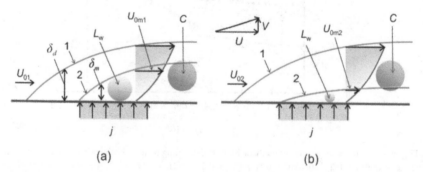

(a)   (b)

Figure 6. Characteristic boundary layers for velocity profile and diffusion profile ($U_{01} < U_{02}$).

As can be seen from the graph, with the growth of U_0 the dissociation rate j increases. However, this growth is an extremely weak function of U_0. In References [28,29] there was a weak increase in the velocity of the flame edge V_c with an increase in the velocity U_0. Obviously, this velocity is a function of the combustion temperature T_c. The reason for this weak influence has been considered above in accordance with expression (2). Another reason for such a weak effect may be due to the geometry of the diffusion and dynamic layer. Figure 6 demonstrates the characteristic profiles for two different velocities of the external incoming air flow U_{01} and U_{02} ($U_{01} < U_{02}$).

The figure shows the dynamic velocity profile 1 for air and the diffusion profile 2 for the gas mixture (methane–air–water vapor). The velocity U_{0m} corresponds to the upper boundary of the mixing layer. Despite the increase in U_{02} compared to U_{01}, the velocity at the boundary of the diffusion layer U_{0m} decreases sharply ($U_{0m2} < U_{0m1}$). In addition, the characteristic scale of vorticity L_w decreases sharply due to the flow instability at combustion. With the fall of U_{0m} (factor 1) and L_w (factor 2), the flow of methane to the combustion region "C" decreases as well. The combustion region "C" for the case (a) covers most of the mixing region (2). At that, there is a high cross-flow of methane j ($V_1 > V_2$). For figure (b), the largest part of the "C" area is outside the diffusion layer (2). Thus, the area of the combustion region, which is inside the diffusion layer, is the third factor that affects combustion. As a result, we obtain the following. The increase in the velocity U_0 for the laminar boundary layer leads to an increase in the heat flux ($q \sim (U_0)^{0.5}$) in the mixing layer.

According to the energy conservation, the heat flux to the powder is equal to the heat flux inside the powder. The increase in heat flux leads to an increase in dissociation rate according to (1) and to an increase in methane concentration within the combustion region. However, the other three factors mentioned above significantly reduce the positive role of U_0 growth. Thus, we can draw two important conclusions from the above: (1) The increase in velocity U_0 leads to a weak increase in the burning rate; and (2) the average temperature of the powder and the uneven temperature distribution in the layer strongly affect the methane concentration in the combustion region. In turn, the second factor (temperature) is extremely sensitive to the pores distribution density in the powder particles due to the phenomenon of self-preservation.

Let us consider the time of combustion start t_{0c} depending on the velocity of the incoming air-flow U_0 (where t_{0c} is the time from the powder placement in the tank to the start of combustion). Methane was ignited periodically every 5 s. However, combustion began only when the methane concentration in the combustion zone reached a certain minimum limiting value. Figure 7 shows that with U_0 increase, the value of t_{0c} decreases. This decrease is due to the fact that with the growth of U_0, q and j increase, which leads to a faster increase in the powder temperature. Thus, the required combustion temperature in the "C" zone and the concentration of methane in the combustion region are achieved faster. Curve 2, generalizing experimental data for t_{0c}, has a quasi-linear form $t_{0c} = -kU_0$ (where parameter k is a function of both properties of boundary layers over the plate (Figure 1) and properties of methane hydrate powder.

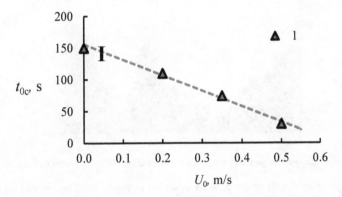

Figure 7. The ignition time t_{0c} depending on velocity of incoming air flow U_0: (1—experimental data; 2 – $t_{0c} = -kU_0$.

4.2. Temperature Measurements in the Powder Layer at Methane Hydrate Dissociation

Since the dissociation rate of methane hydrate and the reaction rate during combustion strongly depend on the temperature, it is important to know the temperature field for the entire volume of the powder for correct modeling.

The photo of the tank with the sample is shown in Figure 8a (top view). Figure 8b shows a thermal image of the powder surface 10 s before combustion. Time $t = 0$ s corresponds to the beginning of combustion of methane hydrate.

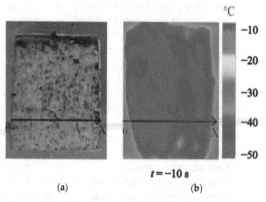

$t = -10$ s

(a)   (b)

Figure 8. (a) Photo of a reservoir with a sample; and (b) thermal imaging photograph of the powder surface (photos were taken 10 s before the start of combustion).

Figure 9 shows a graph of the sample surface temperature change along the longitudinal axis 0X, shown in Figure 8a,b. The maximum temperature of the sample corresponds to the side-walls of the tank ($X = 0$; 40 mm), which indicates that the heat flux from the walls cannot be neglected. Since burning begins near the wall (the location of the pilot burner can be seen in Figure 1), it is important to know not only the average volumetric temperature of the sample, but also the temperature near the pilot burner. In most experimental works only the temperatures on the tank axis are measured.

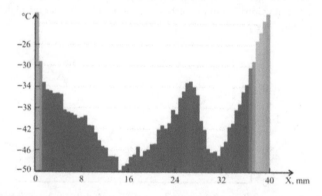

Figure 9. Temperature profile of the powder surface along the 0X line shown in Figure 8a,b.

The temperature change of the powder is shown in Figure 10. Thermocouples were located on the tank axis (Figure 1). For thermocouple 1 $y = 5$ mm (the distance y is measured from the top powder surface, see Figure 1); curve 2 corresponds to $y = 14$ mm. Over time, the temperature increases continuously. It should be noted that the powder temperature should fall due to the methane hydrate dissociation and increase due to the heat flux from the walls of the tank (q_1 and q_2) (Figure 1) and from the ambient medium (q_3 is heat flux to the top surface of the powder). Since temperature T increases for any moment t, it can be concluded that the heat of dissociation is much lower than the heat supplied to the sample from the ambient medium. Time $t = 0$ s corresponds to the beginning of combustion.

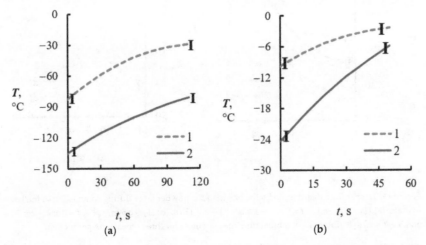

Figure 10. Temperature change in the powder layer over time during combustion of methane hydrate ($1 - y = 5$ mm; $2 - y = 14$ mm, and y—distance from the top surface of the powder layer; thermocouples are located in the center of the tank (Figure 1)): (**a**) the velocity of the incoming flow $U_0 = 0$ m/s (Figure 1); and (**b**) $U_0 = 0.5$ m/s (I—intervals of temperature measurement for three repeated experiments).

As can be seen from the graphs, for $U_0 = 0.5$ m/s, the combustion starts when the temperature of the powder (for $y = 5$ mm) is approximately 70 °C higher than for the option with $U_0 = 0$ m/s. This feature (the temperature of combustion onset) is obvious. With an increase in the external flow velocity, the concentration of methane in the combustion zone C (Figure 6) decreases and combustion occurs when the powder surface temperature and the temperature in the combustion region increases substantially. The occurrence of combustion depends on the concentration of the mixture components, and the temperature of the gas.

Figure 11 shows graphs of temperature distribution (on the tank axis) over the height of the powder layer during combustion of methane hydrate ($y = 0$ mm corresponds to the upper surface of the layer). Time $t = 0$ s corresponds to the beginning of combustion. As can be seen from the graphs, for the entire combustion period there are nonlinear curves for the temperature distribution that characterizes the non-stationary character of the heat equation inside the thick layer. This non-stationary character is already available for the distance $y > 3$–5 mm. For smaller values of y (1 mm $< y <$ 2 mm), the temperature for a few seconds approaches 0 °C and then remains constant due to the ice crust melting on the surface of the granules, i.e., only for the uppermost layer (small height y) a stationary approximation may be considered. However, for dissociation, despite $T = $ const, the flow of methane j in the near-surface region will decrease due to the curvature of the granules [21], i.e., the diffusion problem must be solved in a non-stationary form even under quasi-isothermal conditions.

The above diagrams show the uneven nature of temperature, and accordingly, of the methane hydrate dissociation throughout the powder volume. As a result, most of the burning time is characterized by changing flame height as well as by "effective area" of combustion on the layer surface, which is clearly seen from Figure 12. Features of inhomogeneous combustion are considered in the following paragraph.

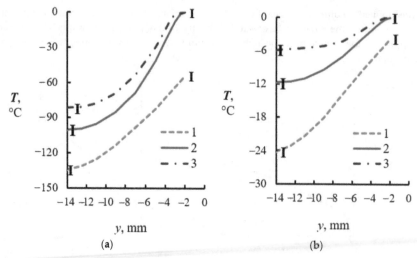

Figure 11. Temperature distribution by the height of the powder layer during combustion of methane hydrate: (a) $U_0 = 0$ m/s; $1 - t = 0$ s; $2 - t = 60$ s; $3 - t = 110$ s; (b) $U_0 = 0.5$ m/s; $1 - t = 0$ s; $2 - t = 30$ s; and $3 - t = 50$ s (I—intervals of temperature measurement for three repeated experiments).

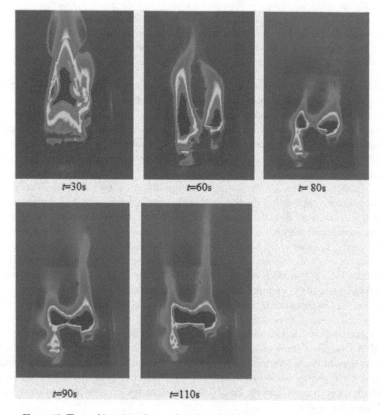

Figure 12. Thermal imaging photos of methane hydrate combustion ($U_0 = 0$ m/s).

4.3. Heterogeneous Character of Methane Hydrate Combustion, Velocity Field at Combustion

When calculating the flow of methane only the full surface of the layer is always taken into account. The average velocity of methane V, directed perpendicular to the layer surface (Figure 13b) is defined as $V = \Delta m / (\rho F \Delta t)$, where m is the mass of methane released, F is layer surface area. The mass can be measured both by weight and volume method and reflects only the average specific flow of methane. Figure 13 shows the "effective area" of combustion based on Figure 12. Only the surface areas of the layer above which methane is burning are highlighted in red. Thus, for $t = 80$ s there are three characteristic areas (F_1, F_2, and F_3) and three characteristic velocities (V_1, V_2, and V_3). "Effective areas (F_{ef})" are determined by imaging using two thermal imagers, located in the direction of two perpendicular axes (Figure 13a). It is extremely difficult to experimentally determine the local velocities of methane V_{ef} over these F_{ef} areas and it is the subject of further research. PIV and PTV methods can be used for these purposes.

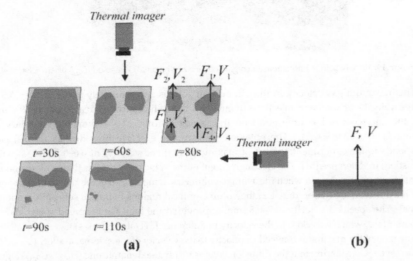

Figure 13. (a) "Effective area" of methane hydrate combustion (red); and (b) average methane velocity V from the entire surface area of layer F.

It should be noted that $F > F_{ef} = F_1 + F_2 + F_3$, and $V > V_{ef} = V_1 + V_2 + V_3$ ($V = V_1 + V_2 + V_3 + V_4$), where V_4 corresponds to the surface F_4, over which there is no combustion. Not all methane (V_4) from the surface ($F4$) enters the combustion region, but only part of it. Therefore, over the powder surface there is always a local uneven flow of methane and uneven local concentration of methane in the field of burning. The changes in local concentrations of methane and buoyancy effect will lead to unstable combustion. This unstable combustion is clearly seen in Figure 12.

Figures 1 and 6 show schematic images of the dynamic (δ_d) and diffusion layer (δ_m) development. The laminar velocity profile is considered in the numerical calculation of combustion. However, this type of profile is valid only for very low rates of injection and with disregard for free convection. The blowing parameter can be estimated as $b = 3\rho_{CH4} V_{CH4} (Re_x)^{0.5} / (\rho_0 U_0) = 0.2$. This value is much lower than the critical $b_{cr} = 1.86$ [41–44], which corresponds to the separation of the wall layer. Thus, the injected methane from the powder can only partially deform the dynamic layer of air. In the presence of combustion, a significant impact on the development of profiles is exerted by buoyancy (density gradient due to temperature difference.) The estimation by the approximate formula shows that the velocity value for free convection in the combustion region is approximately 2–4 m/s. Figure 14 presents the velocity field in the combustion region using the PTV method at an external flow rate $U_0 = 0.5$ m/s.

Figure 14. Velocity field in the combustion region obtained using the PTV method ($U_0 = 0.5$ m/s, $t = 20$ s).

The maximum gas velocity in the combustion zone is approximately equal to $U_c = 3.5$ m/s. Average velocity of methane injection, defined by the entire surface area F, equals 8.4 mm/s. Thus, the injected gas has little effect on the velocity profile of the dynamic boundary layer (3500 mm/s/8.4 mm/s = 417). The obtained velocity value before burning (0.8 mm/s) closely corresponds to the experimental value of maximum methane velocity of 0.6–0.7 mm/s (without combustion) in Reference [31]. The close agreement of the velocities suggests that combustion in the presented experiments began when the average volumetric temperature of the powder corresponded to the maximum dissociation rate, i.e., the volume of local regions of partial self-preservation was minimal. The specified velocity (0.7–0.8 mm/s) provided the high flame spread speed above the powder surface, which was close to the velocity in Reference [31] (about 1 m/s).

The ratio of the maximum convection velocity to the velocity of the external air flow (U_{conv}/U_0) is equal to seven. Considering that the diffusion layer is inside the dynamic one (U_{conv}/U_{0m}) (Figure 6), this ratio will be higher (about 10). In accordance with the directions of the velocity lines and the value of velocity in Figure 14 it can be concluded that buoyancy plays a predominant role in the formation of the velocity field and the type of flow will be fundamentally different from Figures 1 and 6, which are usually taken for modeling. In addition, it is incorrect to take into account only buoyancy, since the lines of the velocity direction significantly deviate in the direction of the velocity U_0. Even low external flow velocities U_0 significantly effect the flow formation at combustion over the methane hydrate layer, and accordingly, the distribution of methane concentrations in the diffusion layer.

5. Conclusions

In this paper, the temperature characteristics inside the powder layer at methane hydrate dissociation and in the presence of the released methane combustion on the powder surface have been experimentally studied. The experiments were carried out at different velocities of the external air-flow from 0 to 0.6 m/s, when there was a laminar flow at the inlet to the working area with the sample. The temperature of the powder during the experiment increased from the temperature of liquid nitrogen to 0 °C due to heat supply from the ambient medium. Measurements were carried out using the weight method, thermal imaging and PTV methods, as well as a high-speed camera.

During combustion, the transverse temperature profile in the powder layer was nonlinear for the most part of the layer height and only in a narrow region (at a depth of about 1 mm from the top surface of the layer) where the powder temperature was quasi-static. Thus, a small granule and a thin layer modeling is possible in the form of a stationary heat equation, and for a thick layer, the non-stationary term cannot be neglected.

Measuring the temperature field of the powder using a thermal imager has shown that there is a highly uneven temperature field on the surface of the layer. The maximum temperature always corresponds to the powder near the side-walls, which is more than 10 °C higher than the average temperature in the layer. Thus, it is incorrect to carry out measurements only on the axial line of the layer, as it is done in most works. The heat flux from the side-walls is higher than for the upper surface of the powder without combustion.

Thermal imaging measurements have revealed the heterogeneous nature of combustion, when only a partial area of the layer is effective over the surface of the powder (combustion is realized above it). Over the rest of the layer surface, the powder temperature is much lower. Thus, simulating combustion on the average rate of methane release is incorrect, since the rate of methane release above the surface layer varies considerably along the surface. As a result, the combustion temperature will be different, i.e., have a non-uniform character, which will lead to non-stationary and non-uniform combustion.

Measurements of the velocity field using the PTV method have shown that the maximum gas velocity in the field of combustion exceeds 3 m/s. To perform stoichiometric ratios, given the methane velocity and oxygen concentration in the air, the velocity of about 0.1–0.2 mm/s is required. In accordance with the foregoing, the excess of oxidizer over the fuel leads to a more than ten-fold violation of the stoichiometric ratio. It is obvious that such a high velocity can be achieved only due to the increased inflow of external air. The methane velocity cannot increase tenfold due to the limited gas diffusion through the pores of methane hydrate granules (diffusion resistance of pores in the formed ice crust). Thus, free convection during combustion leads to a significant excess of the oxidant in the combustion zone, and it is surprising that this air excess results in high velocities of the flame spread speed above the powder surface, which was close to the velocity in Reference [31] (about 1 m/s). For correct modeling, it is necessary to take the real velocity profile rather than the laminar one, but taking into account free convection.

Despite that the velocity profile in the combustion region is formed mainly due to free convection, it is also necessary to take into account the external flow of the forced gas U_0. Even at low velocities U_0, the velocity direction lines are significantly deviated by the forced air-flow.

Author Contributions: M.S.Y., V.I.S., M.V.S., and Y.O.S. performed experiments related to the methane hydrate combustion, PTV and TV measurements. M.A.Y. and I.A.V. performed synthesis and characterization of methane hydrate.

Funding: Experiments related to the methane hydrate combustion, PTV and TV measurements were carried out within the framework of the National Research Tomsk Polytechnic University development program (project VIU-ISHFVP-184/2018) (contributions by M.S.Y., V.I.S., Y.O.S.).

Conflicts of Interest: The authors declare no conflict of interest.

References

1. Rehder, G.; Eckl, R.; Elfgen, M.; Falenty, A.; Hamann, R.; Kahler, N.; Kuhs, W.F.; Osterkamp, H.; Windmeier, C. Methane hydrate pellet transport using the self-preservation effect: A Techno-Economic Analysis. *Energies* **2012**, *5*, 2499–2523. [CrossRef]
2. Cui, Y.; Lu, C.; Wu, M.; Peng, Y.; Yao, Y.; Luo, W. Review of exploration and production technology of natural gas hydrate. *Adv. Geo-Energy Res.* **2018**, *2*, 53–62. [CrossRef]
3. Takahashi, M.; Moriya, H.; Katoh, Y.; Iwasaki, T. Development of natural gas hydrate (NGH) pellet production system by bench scale unit for transportation and storage of NGH pellet. In Proceedings of the 6th International Conference on Gas Hydrates, Vancouver, BC, Canada, 6–10 July 2008.
4. Istomin, V.A.; Yakushev, V.S. *Gas Hydrates in Nature*; Nedra Publisers: Moscow, Russia, 1992.
5. Sum, A.K.; Koh, C.A.; Sloan, E.D. Clathrate hydrates: From laboratory science to engineering practice. *Ind. Eng. Chem. Res.* **2009**, *48*, 7457–7465. [CrossRef]

6. Kuhs, W.F.; Genov, G.; Staykova, D.K.; Hansen, T. Ice perfection and onset of anomalous preservation of gas hydrates. *Phys. Chem. Chem. Phys.* **2004**, *6*, 4917–4920. [CrossRef]

7. Takeya, S.; Yoneyama, A.; Ueda, K.; Mimachi, H.; Takahashi, M.; Sano, K.; Hyodo, K.; Takeda, T.; Gotoh, Y. Anomalously preserved clathrate hydrate of natural gas in pellet form at 253 K. *J. Phys. Chem.* **2012**, *116*, 13842–13848. [CrossRef]

8. Yakushev, V.S.; Istomin, V.A. Gas-hydrates self-preservation effect. In *Physics and Chemistry of Ice*; Maeno, N., Hondoh, T., Eds.; Hokkaido Univ. Press: Sapporo, Japan, 1992; pp. 136–139.

9. Sato, H.; Sakamoto, H.; Ogino, S.; Mimachi, H.; Kinoshita, T.; Iwasaki, T.; Sano, K.; Ohgaki, K. Self-preservation of methane hydrate revealed immediately below the eutectic temperature of the mother electrolyte solution. *Chem. Eng. Sci.* **2013**, *91*, 86–89. [CrossRef]

10. Zhang, G.; Rogers, R.E. Ultra-stability of gas hydrates at 1 atm and 268.2 K. *Chem. Eng. Sci.* **2008**, *63*, 2066–2074. [CrossRef]

11. Takeya, S.; Ripmeester, J.A. Anomalous preservation of CH_4 hydrate and its dependence on the morphology of ice. *ChemPhysChem* **2010**, *11*, 70–73. [CrossRef]

12. Takeya, S.; Yoneyama, A.; Ueda, K.; Hyodo, K.; Takeda, T.; Mimachi, H.; Takahashi, M.; Iwasaki, T.; Sano, K.; Yamawaki, H.; et al. Nondestructive imaging of anomalously preserved methane clathrate hydrate by phase contrast X-ray imaging. *J. Phys. Chem.* **2011**, *115*, 16193–16199. [CrossRef]

13. Stern, L.A.; Circone, S.; Kirby, S.H.; Durham, W.B. Anomalous preservation of pure methane hydrate at 1 atm. *J. Phys. Chem.* **2011**, *105*, 1756–1762. [CrossRef]

14. Stern, L.A.; Cirone, S.; Kirby, S.H.; Durham, W.B. Temperature, pressure and compositional effects on anomalous or "self" preservation of gas hydrates. *Can. J. Phys.* **2003**, *81*, 271–283. [CrossRef]

15. Falenty, A.; Kuhs, W.F. Self-preservation of CO_2 gas hydrates-surface microstructure and ice perfection. *J. Phys. Chem.* **2009**, *113*, 15975–15988. [CrossRef] [PubMed]

16. Shimada, W.; Takeya, S.; Kamata, Y.; Uchida, T.; Nagao, J.; Ebinuma, T.; Narita, H. Texture change of ice on anomalously preserved methane clathrate hydrate. *J. Phys. Chem.* **2005**, *109*, 5802–5807. [CrossRef] [PubMed]

17. Nguyen, A.H.; Koc, M.A.; Shepherd, T.D.; Molinero, V. Structure of the ice-clathrate interface. *J. Phys. Chem.* **2015**, *119*, 4104–4117. [CrossRef]

18. Lv, Y.; Jia, M.; Chen, J.; Sun, C.; Gong, J.; Chen, G.; Liu, B.; Ren, N.; Guo, S.; Li, Q. Self-Preservation effect for hydrate dissociation in water + diesel oil dispersion systems. *Energy Fuels* **2015**, *29*, 5563–5572. [CrossRef]

19. Prasad, P.S.R.; Chari, V.D. Preservation of methane gas in the form of hydrates: Use of mixed hydrates. *J. Nat. Gas Sci. Eng.* **2015**, *25*, 10–14. [CrossRef]

20. Misyura, S.Y.; Donskoy, I.G. Dissociation of natural and artificial gas hydrate. *Chem. Eng. Sci.* **2016**, *148*, 65–77. [CrossRef]

21. Misyura, S.Y. Effect of heat transfer on the kinetics of methane hydrate dissociation. *Chem. Phys. Lett.* **2013**, *583*, 34–37. [CrossRef]

22. Semenov, M.E.; Manakov, A.Y.; Shitz, E.Y.; Stoporev, A.S.; Altunina, L.K.; Strelets, L.A.; Misyura, S.Y.; Nakoryakov, V.E. DSC and thermal imaging studies of methane hydrate formation and dissociation in water emulsions in crude oils. *J. Therm. Anal. Calorim.* **2015**, *119*, 757–767. [CrossRef]

23. Aerov, M.E.; Todes, O.M.; Narinsky, D.A. *Apparatuses with the Steady Grain Layer: Hydraulic and Thermal Fundamentals of Operation*; Khimiya: Leningrad, Russia, 1992.

24. Chen, X.R.; Li, X.S.; Chen, Z.Y.; Zhang, Y.; Yan, K.F.; Lv, Q.-N. Experimental investigation into the combustion characteristics of propane hydrates in porous media. *Energies* **2015**, *8*, 1242–1255. [CrossRef]

25. Nakoryakov, V.E.; Misyura, S.Y. Nonstationary combustion of methane with gas hydrate dissociation. *Energy Fuels* **2013**, *27*, 7089–7097. [CrossRef]

26. Nakoryakov, V.E.; Misyura, S.Y.; Elistratov, S.L.; Manakov, A.Y.; Shubnikov, A.E. Combustion of methane hydrates. *J. Eng. Thermophys.* **2013**, *22*, 87–92. [CrossRef]

27. Nakoryakov, V.E.; Misyura, S.Y.; Elistratov, S.L.; Manakov, A.Y.; Sizikov, A.A. Methane combustion in hydrate systems: Water-methane and water-methane-isopropanol. *J. Eng. Thermophys.* **2013**, *22*, 169–173. [CrossRef]

28. Maruyama, Y.; Yokomori, T.; Ohmura, R.; Ueda, T. Flame spreading over combustible hydrate in a laminar boundary layer. In Proceedings of the 7th International Conference on Gas Hydrate, Edinburgh, UK, 17–21 July 2011.

29. Nakamura, Y.; Katsuki, R.; Yokomori, T.; Ohmura, R.; Takahashi, M.; Iwasaki, T.; Uchida, K.; Ueda, T. Combustion characteristics of methane hydrate in a laminar boundary layer. *Energy Fuels* **2009**, *23*, 1445–1449. [CrossRef]

30. Kitamura, K.; Nakajo, K.; Ueda, T. Numerical calculation of a diffusion flame formed in the laminar boundary layer over methane-hydrate. In Proceedings of the Fourth International Conference on Gas Hydrates, Yokohama, Japan, 19–23 May 2002; pp. 1055–1058.

31. Maruyama, Y.; Fuse, M.J.; Yokomori, T.; Ohmura, R.; Watanabe, S.; Iwasaki, T.; Iwabuchi, W.; Ueda, T. Experimental investigation of flame spreading over pure methane hydrate in a laminar boundary layer. *Proc. Combust. Inst.* **2013**, *34*, 2131–2138. [CrossRef]

32. Wu, F.H.; Padilla, R.E.; Dunn-Rankin, D.; Chen, G.B.; Chao, Y.C. Thermal structure of methane hydrate fueled flames. *Proc. Combust. Inst.* **2017**, *36*, 4391–4398. [CrossRef]

33. Yoshioka, T.; Yamamoto, Y.; Yokomori, T.; Ohmura, R.; Ueda, T. Experimental study on combustion of methane hydrate sphere. *Exp. Fluids* **2015**, *56*, 192. [CrossRef]

34. Bar-Kohany, T.; Sirignano, W.A. Transient combustion of methane-hydrate sphere. *Combust. Flame* **2016**, *163*, 284–300. [CrossRef]

35. Misyura, S.Y. Efficiency of methane hydrate combustion for different types of oxidizer flow. *Energy* **2016**, *103*, 430–439. [CrossRef]

36. Cuenot, B.; Poinsot, T. Effect of curvature and unsteadiness in diffusion flames implications for turbulent diffusion combustion. *Proc. Combust. Inst.* **1994**, *25*, 1383–1390. [CrossRef]

37. Dehaech, S.; Van Parys, H.; Hubin, A.; Van Beeck, J.P.A.J. Laser marked shadowgraphy: A novel optical planar technique for the study of microbubbles and droplets. *Exp. Fluids* **2009**, *47*, 333–341. [CrossRef]

38. Kuznetsov, G.V.; Piskunov, M.V.; Volkov, R.S.; Strizhak, P.A. Unsteady temperature fields of evaporating water droplets exposed to conductive, convective and radiative heating. *Appl. Therm. Eng.* **2018**, *131*, 340–355. [CrossRef]

39. Hagiwara, Y.; Sakamoto, S.; Tanaka, M.; Yoshimura, K. PTV measurement on interaction between two immiscible droplets and turbulent uniform shear flow of carrier fluid. *Exp. Therm. Fluid Sci.* **2002**, *26*, 245–252. [CrossRef]

40. Westerweel, J. Fundamentals of digital particle image velocimetry. *Meas. Sci. Technol.* **1997**, *8*, 1379–1392. [CrossRef]

41. Kutateladze, S.S. *Fundamentals of Heat Transfer*; Arnold, E., Ed.; Wiley: London, UK, 1963.

42. Volchkov, E.P.; Makarov, M.S.; Makarova, S.N. Heat and mass diffusion fluxes on a permeable wall with foreign-gasblowing. *Int. J. Heat Mass Transf.* **2012**, *55*, 1881–1887. [CrossRef]

43. Kutateladze, S.S.; Leont'ev, A.I. *Heat Transfer, Mass Transfer, and Friction in Turbulent Boundary Layers*; Hemisphere Publishing Corporation: New York, NY, USA, 1989.

44. Volchkov, E.P. Concerning the heat and mass transfer features on permeable surfaces. *Int. J. Heat Mass Transf.* **2006**, *49*, 755–762. [CrossRef]

MDPI
St. Alban-Anlage 66
4052 Basel
Switzerland
Tel. +41 61 683 77 34
Fax +41 61 302 89 18
www.mdpi.com

Energies Editorial Office
E-mail: energies@mdpi.com
www.mdpi.com/journal/energies

Printed in July 2019
by Rotomail Italia S.p.A., Vignate (MI) - Italy